Prestressed Concrete

Problems and Solutions

THIRD EDITION

conciliate 和解.
entrench 巩固
precipitate 沉淀

Prestressed Concrete

Problems and Solutions

THIRD EDITION

N Krishna Raju
BE MSc (Engg) PhD MI Struct.E C Engg MIE

Emeritus Professor of Civil Engineering
MS Ramaiah Institute of Technology
Bengaluru, Karnataka

CBS Publishers & Distributors Pvt Ltd

New Delhi • Bengaluru • Chennai • Kochi • Kolkata • Mumbai
Hyderabad • Nagpur • Patna • Pune • Vijayawada

Disclaimer

Science and technology are constantly changing fields. New research and experience broaden the scope of information and knowledge. The author has tried his best in giving information available to him while preparing the material for this book. Although, all efforts have been made to ensure optimum accuracy of the material, yet it is quite possible some errors might have been left uncorrected. The publisher, the printer and the author will not be held responsible for any inadvertent errors, omissions or inaccuracies.

Prestressed Concrete
Problems and Solutions
Third Edition

ISBN: 978-81-239-2542-4

Third Edition: 2015
Reprint: 2017
First Edition: 1995
Second Edition: 2000

Published by Satish Kumar Jain and Produced by Varun Jain for
CBS Publishers & Distributors Pvt Ltd
4819/XI Prahlad Street, 24 Ansari Road, Daryaganj, New Delhi 110 002, India.
Ph: 23289259, 23266861, 23266867 Website: www.cbspd.com
Fax: 011-23243014 e-mail: delhi@cbspd.com; cbspubs@airtelmail.in.

Corporate Office: 204 FIE, Industrial Area, Patparganj, Delhi 110 092
Ph: 4934 4934 Fax: 4934 4935 e-mail: publishing@cbspd.com; publicity@cbspd.com

Branches

- **Bengaluru:** Seema House 2975, 17th Cross, K.R. Road, Banasankari 2nd Stage, Bengaluru 560 070, Karnataka
  Ph: +91-80-26771678/79 Fax: +91-80-26771680 e-mail: bangalore@cbspd.com
- **Chennai:** No. 7, Subbaraya Street, Shenoy Nagar, Chennai 600 030, Tamil Nadu
  Ph: +91-44-26680620/26681266 Fax: +91-44-42032115 e-mail: chennai@cbspd.com
- **Kochi:** Ashana House, 39/1904, AM Thomas Road, Valanjambalam, Ernakulam 682 016, Kochi, Kerala
  Ph: +91-484-4059061-62-64-65 Fax: +91-484-4059065 e-mail: kochi@cbspd.com
- **Kolkata:** No. 6/B, Ground Floor, Rameswar Shaw Road, Kolkata-700014 (West Bengal), India
  Ph: +91-33-2289-1126, 2289-1127, 2289-1128 e-mail: kolkata@cbspd.com
- **Mumbai:** 83-C, Dr E Moses Road, Worli, Mumbai-400018, Maharashtra
  Ph: +91-22-24902340/41 Fax: +91-22-24902342 e-mail: mumbai@cbspd.com

Representatives

- **Hyderabad** 0-9885175004
- **Nagpur** 0-9021734563
- **Patna** 0-9334159340
- **Pune** 0-9623451994
- **Vijayawada** 0-9000660880

Printed at: Magic International Pvt. Ltd., Greater Noida, UP

Preface to the Third Edition

The subject of prestressed concrete is gaining more importance over the last decade mainly due to its increased use in the construction of bridges. The second edition of the book was well received by the students of civil and structural engineering mainly to prepare for their semester examinations and also competitive examinations like the combined engineering services examinations conducted by the Union Public Service Commission.

At present, India has embarked on a gigantic highway project involving the golden quadrilateral connecting the north–south and east–west corridors with access to the capital cities of various states. Naturally, this massive highway project necessarily involves the design and construction of innumerable number of prestressed concrete bridges to cross the east and west flowing rivers.

The present revised third edition includes additions to most of the chapters in the form of new problems culled from different sources. The material presented conforms to the revised specifications of the Indian Standard Codes such as IS: 1343-2012, IRC: 18-2000, IRC: 6-2000, IRC: 21-2000 and IS: 456-2000.

Many practical problems generally encountered in several types of prestressed concrete structures like continuous girder bridges and cable stayed bridges have been included for better understanding of the subject by the students. The revised edition will also be very useful for engineering teachers to set model question papers for various types of competitive examinations conducted by UPSC like IRSE, CES and other central services. It will also help as an invaluable source material for the engineering teachers to set question papers for the BE/ BTech and master's degree examinations of the civil and structural engineering streams. A glimpse of suggested reading and reference list is also included at the end of the book.

I gratefully acknowledge the help rendered by my colleagues, students and teachers of various engineering institutions who communicated valuable and informative material for the revised edition.

I am deeply indebted to the publishers (CBSPD) for their constant encouragement and effective cooperation in bringing out the third edition.

N Krishna Raju

Preface to the First Edition

Prestressed concrete has established itself as an unique material in the construction of building, hydraulic, environmental, highway, nuclear and marine structures. Hence, graduates of every civil engineering programme must have a minimum requirement, a basic understanding of the variety of problems encountered in prestressed concrete.

The book covers the various type of problems along with concise solutions presented in a logical sequence. The problems have been selected from the examination papers of various universities in India. The book presents in 13 chapters, a concise exposition of the integrated design principles, illustrated by worked out examples relevant to design practice, conforming to the Indian standard code of practice for prestressed concrete IS: 1343-1980.

The various problems and solutions presented in the text have been the outcome of authors experience in teaching the subject for undergraduate and postgraduate students in the civil and structural engineering streams for a number of years. The international system (SI) of units have been used throughout and the notationals generally conform to the prescribed norms set by the European Concrete Committee and since adopted by the British and Indian Standard Codes of practice.

The book is intended for civil and structural engineering students, preparing for the degree and competitive examinations and it should also serve as a reference book for teachers and structural engineers.

The author is grateful to Ms Amrutha for the preparation of typescript and Mr Prasad for the production of figures at a short notice. Sincere thanks are due to M/s CBS Publishers for their effective cooperation. Finally the author welcomes constructive criticism and suggestions which will be helpful in updating the text.

N Krishna Raju

Notations

A	Cross sectional area of concrete member
A_p	Area of prestressing tendons
D_c	Density of concrete
E_c	Modulus of elasticity of concrete
E_s	Modulus of elasticity of steel
F_{bst}	Bursting tension
I	Second moment of area of gross section
I_r	Second moment of area of cracked transformed section
K	Friction coefficient for wave effect
L	Effective span
M	Bending moment, generally
M_{cr}	Cracking moment
M_u	Ultimate moment
M_g	Bending moment due to dead loads
M_q	Bending moment due to live loads
N_d	Design tensile force
P	Prestressing force
P_k	Characteristic load in tendon near anchorage
T	Torque
V	Shear force
V_u	Ultimate shear force
Z	Section modulus
Z_t	Section modulus of top fibre
Z_b	Section modulus of bottom fibre
a	Deflection
b	Breadth of section or compression force
b_w	Breadth of web
d	Effective depth of tension reinforcement
e	Eccentricity of prestressing force
f_c	Compressive stress
f_{ci}	Compressive strength of concrete at transfer

f_{ck} Characteristic compressive strength of concrete
f_{ct} Allowable compressive stress in concrete at transfer
f_{cw} Allowable compressive stress in concrete at working loads
f_v Transverse tensile stress
f_{pu} Characteristic tensile strength of tendons
f_{tt} Allowable tensile stress in concrete at transfer
f_{tw} Allowable tensile stress in concrete under working loads
g Distributed dead load or acceleration due to gravity
h Overall depth of the section
n Number of turns of wire winding
q Distributed live load
S Spacing of stirrup or links
t Thickness of flange
$w_{\min}$ Minimum uniformly distributed load
w_{ud} Ultimate design load
y_o Half depth of an anchorage block
y_{po} Half depth of loaded area of an anchorage block
y_t Distance of highest point above centroid of concrete section
α Angle
α_e Modular ratio of steel to concrete
β Dimensionless coefficient
γ_f Partial safety factor for loads
γ_m Partial safety factor for material strength
ε Strain
ε_{cs} Limiting shrinkage strain in concrete
η Loss ratio
θ Rotation of beam at supports
μ Coefficient of friction
τ_c Ultimate shear stress in concrete
τ_v Shear stress due to transverse shear
f Creep coefficient

Contents

1

Analysis of Prestress and Bending Stresses

Problem 1.1

A rectangular concrete beam, 100 mm wide by 250 mm deep spanning over 8 m is prestressed by a straight cable carrying an effective prestressing force of 250 kN located at an eccentricity of 40 mm. The beam supports a live load of 1.2 kN/m.

a. Calculate the resultant stress distribution for the central cross-section of the beam. The density of concrete is 24 kN/m^3.

b. Find the magnitude of prestressing force with an eccentricity of 40 mm which can balance the stresses due to dead and live loads at the bottom fibre of the central section of the beam.

Solution

The concrete beam prestressed by an eccentric tendon is shown in Fig. 1.1.

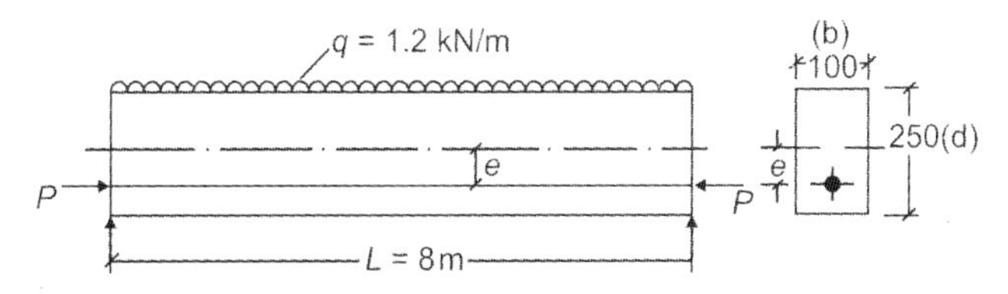

Fig. 1.1

Prestressing force $P = 250$ kN

Cross-sectional area $A = (100 \times 250) = 25 \times 10^3$ mm^2

Section modulus $$Z = \left(\frac{bd^2}{6}\right) = \left(\frac{100 \times 250^2}{6}\right)$$

$$= 1.04 \times 10^6 \text{ mm}^3$$

Eccentricity $e = 40$ mm

Self weight of beam $= g = (0.1 \times 0.25 \times 24) = 0.6$ kN/m

$\therefore$ Total load on beam $= (g + q)$

$= (1.2 + 0.6) = 1.8$ kN/m

B.M. at centre of span $M = \left(\dfrac{1.8 \times 8^2}{8}\right) = 14.4$ kN.m

$\therefore$ Stress due to loads $= \left(\dfrac{M}{Z}\right) = \left(\dfrac{14.4 \times 10^6}{1.04 \times 10^6}\right)$

$= \pm 13.8$ N/mm^2

Prestress $= \left(\dfrac{P}{A}\right) = \left(\dfrac{250 \times 10^3}{25 \times 10^3}\right) = 10$ N/mm^2

$$\left(\frac{Pe}{Z}\right) = \left(\frac{250 \times 10^3 \times 40}{1.04 \times 10^6}\right) = 96 \text{ N/mm}^2$$

a. *Resultant stresses*

At top $= (10 - 9.6 + 13.8) = 14.2$ N/mm^2 (compression)

At bottom $= (10 + 9.6 - 13.8) = 5.8$ N/mm^2 (compression)

b. If P = Prestressing force, then

$$\left[\frac{P}{A} + \frac{Pe}{Z}\right] = \left(\frac{M}{Z}\right)$$

$$P\left[\frac{1}{25 \times 10^3} + \frac{40}{1.04 \times 10^6}\right] = \left(\frac{14.4 \times 10^6}{1.04 \times 10^6}\right)$$

Solving $P = 176 \times 10^3$ N $= 176$ kN.

Problem 1.2

A prestressed concrete beam supports a live load of 4 kN/m over a simply supported span of 8 m. The beam has an I-section with an overall depth of 400 mm. The thickness of the flanges and web are 60 and 80 mm, respectively. The width of the flange is 200 mm. The beam is to be prestressed by an effective prestressing force of 235 kN at a suitable eccentricity such that the resultant stresses at the soffit of the beam at the centre of span is zero.

a. Find the eccentricity required for the force.
b. If the tendon is concentric, what should be the magnitude of the prestressing force for the resultant stress to be zero at the bottom fiber of the central span section.

Solution

The prestressed concrete beam of symmetrical I-section is shown in Fig. 1.2.

$$P = 235 \text{ kN}$$
$$e = ?$$

For the I-section

$$I = \left[\frac{200 \times 400^3}{12} - \frac{120 \times 280^3}{12}\right]$$
$$= 8.475 \times 10^8 \text{ mm}^4$$

$$Z = \left(\frac{8.475 \times 10^8}{200}\right) = 4.23 \times 10^6 \text{ mm}^3$$

$$A = 2(200 \times 60) + (280 \times 80) = 46400 \text{ mm}^2$$

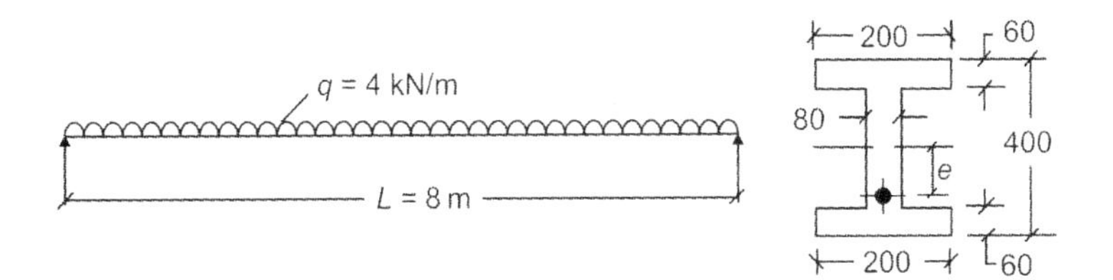

Fig. 1.2

Self weight of beam $= (0.0464 \times 1 \times 24) = 1.1136$ kN/m

$\therefore$ Total load $= (4 + 1.1136) = 5.1136$ kN/m

$$\text{Moment at centre } M = \left(\frac{5.1136 \times 8^2}{8}\right) = 40.9 \text{ kN.m}$$

$$\text{Stress at bottom} = \left(\frac{40.9 \times 10^6}{4.23 \times 10^6}\right) = 9.66 \text{ N/mm}^2 \text{ (tension)}$$

a. For zero stress at bottom

$$\left(\frac{P}{A}\right) + \left(\frac{Pe}{Z}\right) = \left(\frac{M}{Z}\right)$$

$$\left(\frac{235\times10^3}{46400}\right)+\left(\frac{235\times10^3\times e}{4.23\times10^6}\right)=9.66$$

Solving $e = 84$ mm

b. If tendon is concentric

$$\frac{P}{A}=\frac{M}{Z}$$

$$\therefore \quad P=\left(\frac{MA}{Z}\right)=\left(\frac{40.9\times10^6\times46400}{4.23\times10^6}\right)$$

Solving $P = 450 \times 10^3$ N

or $P = 450$ kN

Problem 1.3

A prestressed concrete beam, 200 mm wide and 300 mm deep, is used over an effective span of 6 m to support an imposed load of 4 kN/m. The density of concrete is 24 kN/m^3. At the quarter span section of the beam, find the magnitude of:

a. The concentric prestressing force necessary for zero fiber stress at the soffit when the beam is fully loaded; and

b. The eccentric prestressing force located 100 mm from the bottom of the beam which would nullify the bottom fibre stress due to loading.

Solution

The prestressed concrete beam of rectangular section and eccentrically prestressed is shown in Fig. 1.3.

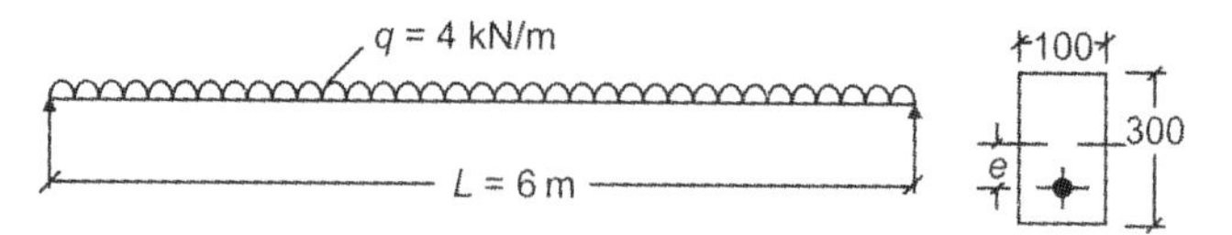

Fig. 1.3

$$A = (200 \times 300) = 6 \times 10^4 \text{ mm}^2$$

$$Z=\left(\frac{bd^2}{6}\right)=\left(\frac{200\times300^2}{6}\right)=3\times10^6 \text{ mm}^3$$

Self weight of beam $g = (0.2 \times 0.3 \times 24) = 1.44$ kN/m

$\therefore$ Total load $w = (g + q) = (4 + 1.44) = 5.44$ kN/m

At $\frac{1}{4}$ span, moment

$$M = \left(\frac{3}{32}\right)wL^2$$

$$= \left(\frac{3}{32} \times 5.44 \times 6^2\right) = 18.36 \text{ kN.m}$$

a. $$\text{Stress} = \left(\frac{M}{Z}\right)$$

$$= \left(\frac{18.36 \times 10^6}{3 \times 10^6}\right) = 6.12 \text{ N/mm}^2$$

$\therefore$ $$\frac{P}{A} = \frac{M}{Z}$$

$\therefore$ $$P = \left(\frac{6.12 \times 6 \times 10^4}{1000}\right) = 367.2 \text{ kN}$$

b. If $$e = 200 \text{ mm}$$

$$\left(\frac{P}{A} + \frac{Pe}{Z}\right) = \left(\frac{M}{Z}\right)$$

$$P\left[\frac{1}{6 \times 10^4} + \frac{200}{3 \times 10^6}\right] = 6.12$$

Solving $$P = 183.6 \times 10^3 \text{ N}$$

$$P = 183.6 \text{ kN}$$

Problem 1.4

A concrete beam of symmetrical I-section spanning 8 m has the width and thickness of flanges equal to 200 and 60 mm, respectively. The overall depth of beam is 400 mm. The thickness of web is 80 mm. The beam is prestressed by a parabolic cable with an eccentricity of 150 mm at the centre and zero at the supports with an effective force of 100 kN. The live load on the beam is 2 kN/m. Draw the stress distribution diagram at the central section for;

a. Prestress + self weight (density of concrete = 24 kN/m^3)

b. Prestress + self weight + live load

Solution

The prestressed beam of symmetrical I-section is shown in Fig. 1.4.

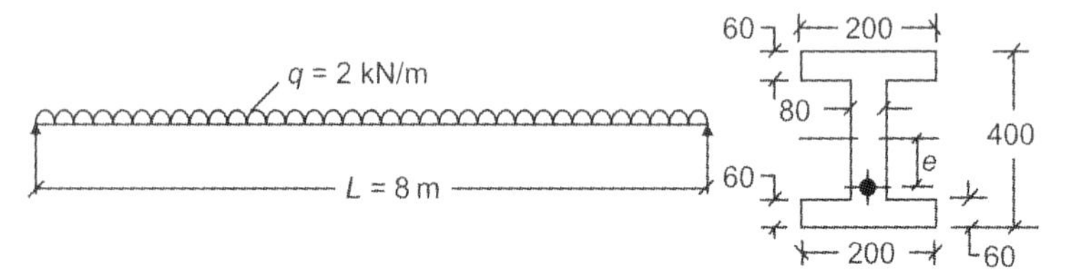

Fig. 1.4

$$P = 100 \text{ kN}$$
$$e = 150 \text{ mm}$$
$$q = 2 \text{ kN/m}$$
$$A = 46400 \text{ mm}^2 = 0.0464 \text{ m}^2$$
$$I = 8.475 \times 10^8 \text{ mm}^4$$
$$Z = 4.23 \times 10^6 \text{ mm}^3$$

Self weight of beam = $(0.046 \times 24) = 1.1136$ kN/m

$$\therefore \quad M_g = \left(\frac{1.1136 \times 8^2}{8}\right) = 8.90 \text{ kN.m}$$

$$M_q = \left(\frac{2 \times 8^2}{8}\right) = 16 \text{ kN.m}$$

$$\text{Prestress: } \left(\frac{P}{A}\right) = \left(\frac{100 \times 10^3}{46400}\right) = 2.155 \text{ N/mm}^2$$

$$\left(\frac{Pe}{Z}\right) = \left(\frac{100 \times 10^3 \times 150}{4.23 \times 10^6}\right) = 3.54 \text{ N/mm}^2$$

Stress due to self weight

$$= \left(\frac{M_g}{Z}\right) = \pm\left(\frac{8.9 \times 10^6}{4.23 \times 10^6}\right)$$
$$= 2.10 \text{ N/mm}^2$$

$$\text{Live load stress} = \left(\frac{M_q}{Z}\right) = \pm\left(\frac{16 \times 10^6}{4.23 \times 10^6}\right)$$
$$= 3.78 \text{ N/mm}^2$$

Resultant stresses

a. Prestress + Self weight

At top $= (2.155 - 3.54 + 2.10)$

$= 0.7\ \text{N/mm}^2$ (compression)

At bottom $= (2.155 + 3.54 - 2.10)$

$= 3.6\ \text{N/mm}^2$ (compression)

b. Prestress + Self weight + Live load

At top $= (2.155 - 3.54 + 2.10 + 3.78)$

$= 4.5\ \text{N/mm}^2$ (compression)

At bottom $= (2.155 + 3.54 - 2.10 - 3.78)$

$= -0.2\ \text{N/mm}^2$ (tension)

Problem 1.5

A concrete beam with a double over hang, has the middle span equal to 10 m and the equal overhang on either side is 2.5 m. Determine the profile of the prestressing cable with an effective force of 250 kN which can balance a uniformly distributed load of 8 kN/m on the beam, which includes the self weight of the beam.

Sketch the cable profile marking the eccentricity of cable at support and mid span.

Solution

The double over hang beam prestressed by a parabolic cable is shown in Fig. 1.5.

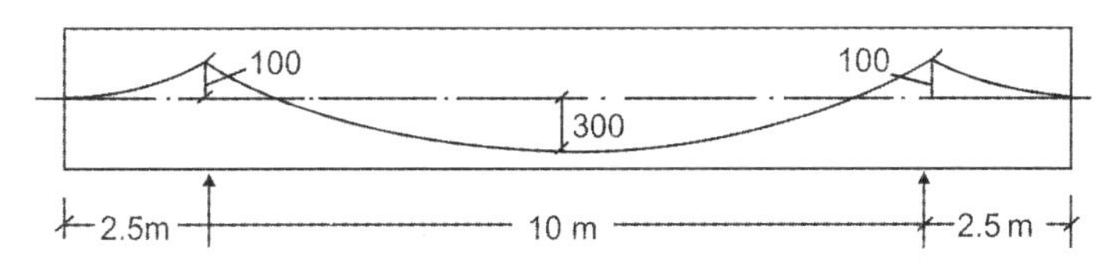

Fig. 1.5

$P = 250$ kN $\qquad w = 8$ kN/m

$L = 10$ m $\qquad a = 2.5$ m

Maximum B.M. at centre of span

$$= \left(\frac{wL^2}{8} - \frac{wa^2}{2}\right) = \left(\frac{8\times 10^2}{8} - \frac{8\times 2.5^2}{2}\right)$$

$= 75$ kN.m

B.M. at support section

$$= \left(\frac{wa^2}{2}\right) = \left(\frac{8 \times 2.5^2}{2}\right) = 25 \text{ kN.m}$$

At the centre of span:

If e = eccentricity

$$P.e = M = 75 \times 10^6$$

$$\therefore \quad e = \left(\frac{75 \times 10^6}{250 \times 10^3}\right) = 300 \text{ mm}$$

At support section

$$P.e = 25 \times 10^6$$

$$\therefore \quad e = \left(\frac{25 \times 10^6}{250 \times 10^3}\right) = 100 \text{ mm}$$

Problem 1.6

A prestressed concrete beam, 120 mm wide by 300 mm deep, is prestressed by a cable which has an eccentricity of 100 mm at the centre of span section. The span of the beam is 6 m. If the beam supports two concentrated loads of 10 kN each at one-third span points, determine the magnitude of the prestressing force in the cable for load balancing for the following cases;

a. Considering live loads but neglecting self weight of beam; and
b. Considering both self weight of beam and live load.
   ($D_c = 24 \text{ kN/m}^3$)

Solution

The concrete beam of rectangular section, prestressed by a trapezoidal cable is shown in Fig. 1.6.

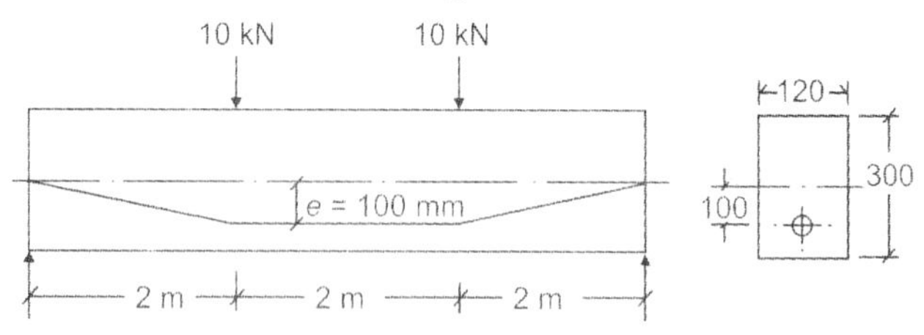

Fig. 1.6

Bending moment at centre of span

$$= M = (10 \times 2) = 20 \text{ kN.m}$$

Given e = eccentricity = 100 mm

$P.e = M$

$$\therefore \quad P = \left(\frac{M}{e}\right) = \left(\frac{20}{0.1}\right) = 200 \text{ kN}$$

Self weight of beam = $(0.12 \times 0.3 \times 24) = 0.864$ kN/m

$$\text{Moment} = M_g = \left(\frac{0.864 \times 6^2}{8}\right) = 3.888 \text{ kN.m}$$

Total B.M. at centre = $(20 + 3.888) = 23.888$ kN.m

$\therefore \quad P.e = 23.888$

$$\therefore \quad P = \left(\frac{23.888}{0.1}\right) = 238.88 \text{ kN}$$

Problem 1.7

A post tensioned slab spanning in one direction over 9 m is 300 mm deep with straight bars at a depth of 250 mm. The slab is subjected to two line loads of 15 kN spread over a width of 300 mm applied along the third points of the span parallel to the supports. Neglecting the reduction in concrete area due to ducts, calculate the increase in steel stress due to applied loads.

a. When the bars are efficiently grouted so that the strain in steel and adjacent concrete are equal; and
b. When the bars are ungrouted and can move in ducts without friction. Modulus of elasticity of steel and concrete are 210 and 35 kN/mm^2 respectively.

Solution

The prestressed beam of rectangular section with an eccentric tendon and supporting concentrated live loads is shown in Fig. 1.7.

$$\text{Modular ratio, } \alpha_e = \left(\frac{E_s}{E_c}\right) = \left(\frac{210}{35}\right) = 6$$

$$e = 100 \text{ mm} = y$$

$$I = \left(\frac{300 \times 300^3}{12}\right) = 56.25 \times 10^6 \text{ mm}^4$$

BM at centre of span $M = (15 \times 3) = 45$ kN.m

a. *Bonded Beam*

$$\alpha_c = \text{Stress in concrete at level of cable}$$

$$\alpha_c = \left(\frac{My}{I}\right) = \left(\frac{45\times10^6\times100}{56.25\times10^6}\right) = 8\ \text{N/mm}^2$$

$\because$ Stress in steel $= (\alpha_e \cdot \alpha_c)$

$= (6 \times 8) = 48\ \text{N/mm}^2$

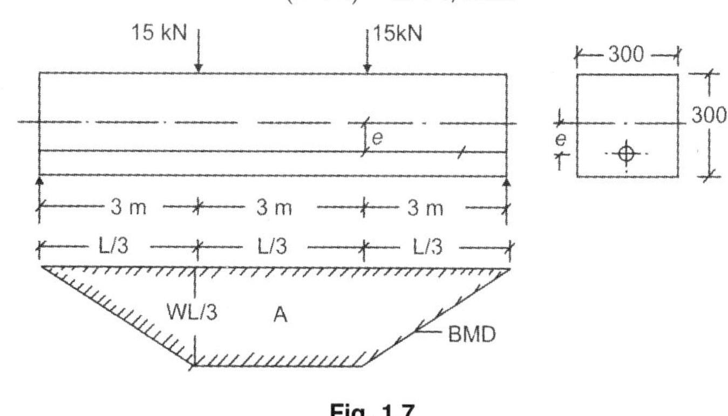

Fig. 1.7

b. *Unbonded beam*

$$\text{Stress in steel} = \left(\frac{\alpha_e \cdot y \cdot A}{I \cdot L}\right)$$

A = Area of B.M.D. due to loads

$$A = 2\left(\frac{1}{2}\cdot\frac{L}{3}\cdot\frac{WL}{3}\right)+\left(\frac{WL}{3}\cdot\frac{L}{3}\right) = \frac{2WL^2}{9}$$

$$\therefore \text{Stress in steel} = \left(\frac{\alpha_e \cdot y \cdot A}{I \cdot L}\right)$$

$$= \left(\frac{6\times100\times2\times15\times10^3\times9000^2}{9\times56.25\times10^6\times9000}\right)$$

$$= 320\ \text{N/mm}^2$$

Problem 1.8

A concrete beam, 120 mm wide and 300 mm deep, is prestressed by a straight cable carrying an effective force of 180 kN at an eccentricity of 50 mm. The beam spanning over 6 m supports a total uniformly distributed load of 4 kN/m which includes the self weight of the beam. The initial stress in the tendons is

1000 N/mm². Determine the percentage increase of stress in the tendons due to loading on the beam.

$$E_s = 210 \text{ kN/mm}^2$$
$$E_c = 35 \text{ kN/mm}^2$$

Solution

The concrete beam of rectangular section, prestressed by an eccentric tendon is shown in Fig. 1.8.

$$P = 180 \text{ kN} \qquad e = 50 \text{ mm} = y$$
$$E_s = 210 \text{ kN/mm}^2 \qquad E_c = 35 \text{ kN/mm}^2$$

$$\therefore \qquad \alpha_e = \left(\frac{E_s}{E_c}\right) = \left(\frac{210}{35}\right) = 6$$

$$I = \left(\frac{210 \times 300^3}{12}\right) = 27 \times 10^7 \text{ mm}^4$$

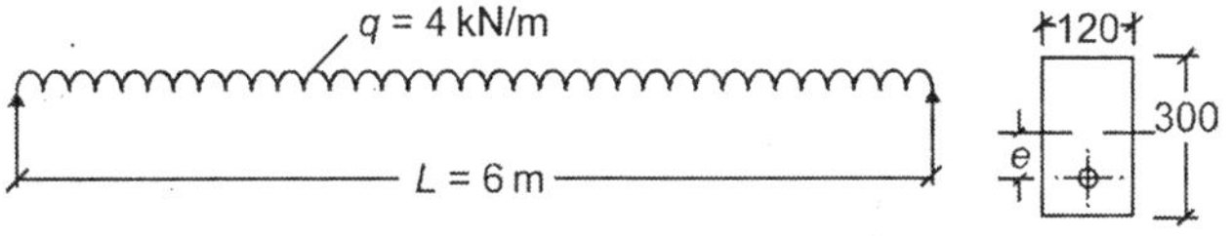

Fig. 1.8

Increase of stress in steel

$$= \left(\frac{\alpha_e y \cdot wL^2}{12\, I}\right)$$

$$= \left(\frac{6 \times 50 \times 4 \times 6000^2}{12 \times 27 \times 10^7}\right) = 13.3 \text{ N/mm}^2$$

Percentage increase of stress of steel

$$= \left(\frac{1000 - 13.3}{1000}\right) 100 = 0.98\%$$

Problem 1.9

A rectangular concrete beam of cross-section 300 mm deep and 200 mm wide is prestressed by fifteen, 5 mm diameter wires located 65 mm from the bottom of the beam and three, 5 mm wires, 25 mm from the top. Assuming the effective stress in the steel as 840 N/mm².

a. Calculate the stresses at the extreme fibres of the mid span section when the beam is carrying its own weight over a span of 6 m and

b. If a uniformly distributed working load of 6 kN/m is imposed and the modulus of rupture of concrete is 6.5 N/mm^2, obtain the maximum working stress in concrete and estimate the load factor against cracking. The density of concrete is 24 kN/m^3.

Solution

The concrete beam of rectangular section and prestressed by high tensile wires at top and bottom is shown in Fig. 1.9.

$$\text{Centroid of wires} = y = \left[\frac{(3\times 275)+(15\times 65)}{18}\right] = 100 \text{ mm}$$

$$Z = \left(\frac{200\times 300^2}{6}\right) = 3\times 10^6 \text{ mm}^3$$

$\therefore$

$$e = [150 - 100] = 50 \text{ mm}$$

$$A = 6\times 10^4 \text{ mm}^2$$

$$P = \frac{\left[(\pi\times 5^2)/4\times 18\right](840)}{1000} = 297 \text{ kN}$$

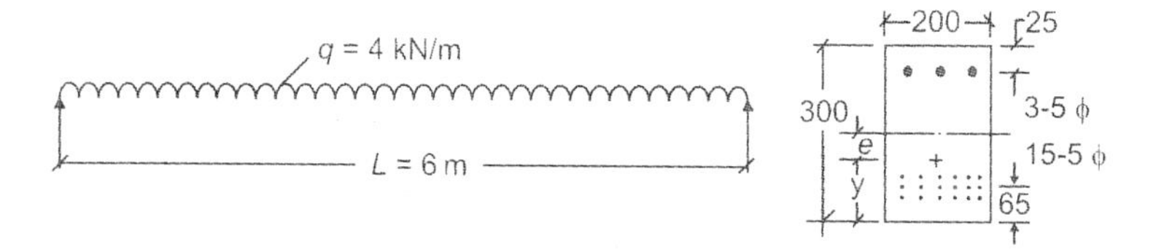

Fig. 1.9

Self weight of beam

$$= (0.2\times 0.3\times 24) = 1.44 \text{ kN/m}$$

$$M_g = (0.125\times 1.44\times 6^2) = 6.48 \text{ kN.m}$$

$$\text{Prestress: } \left(\frac{P}{A}\right) = 4.95 \text{ N/mm}^2$$

$$\left(\frac{Pe}{Z}\right) = \pm 4.95 \text{ N/mm}^2$$

$$\left(\frac{M_g}{Z}\right) = \left(\frac{6.48 \times 10^6}{3 \times 10^6}\right) = 2.16 \text{ N/mm}^2$$

$$M_q = (0.125 \times 6 \times 6^2) = 27 \text{ kN.m}$$

$$\therefore \quad \left(\frac{M_q}{Z}\right) = \left(\frac{72 \times 10^6}{3 \times 10^6}\right) = 9 \text{ N/mm}^2$$

a. Stress at top = (4.95 – 4.95 + 2.16)
= 2.16 N/mm^2 (compression)

Stress at bottom = (4.95 + 4.95 – 2.16)
= 7.74 N/mm^2 (compression)

b. Stress at top = (4.95 – 4.95 + 2.16 + 9)
= 11.16 N/mm^2 (compression)

Stress at bottom = (4.95 + 4.95 – 2.16 – 9)
= 1.26 N/mm^2 (tension)

Cracking stress = (6.5 – 1.26) = 5.24 N/mm^2

$$\text{Extra moment} = M_{cr} = (\alpha.Z) = \left(\frac{5.24 \times 3 \times 10^6}{10^6}\right)$$

= 15.72 kN.m

Working moment = (27 + 6.48) = 33.48 kN.m

Cracking moment = (15.72 + 6.48 + 72) = 49.2 kN.m

∴ Factor of safety against cracking

$$= \left(\frac{\text{Cracking moment}}{\text{Working moment}}\right)$$

$$= \left(\frac{48.2}{33.48}\right) = 1.469$$

Problem 1.10

A concrete beam of symmetrical I-section shown in Fig. 1.10 is prestressed by a cable carrying a force of 120 kN at an eccentricity of 150 mm at centre of span section. The beam supports a live load of 2.5 kN/m over the entire span of 8 m. Determine the resultant stresses at mid span section for the following cases of loading:

Case a. Prestress + self weight

Case b. Prestress + self weight + live load

(Bangalore University, 1992)

Solution

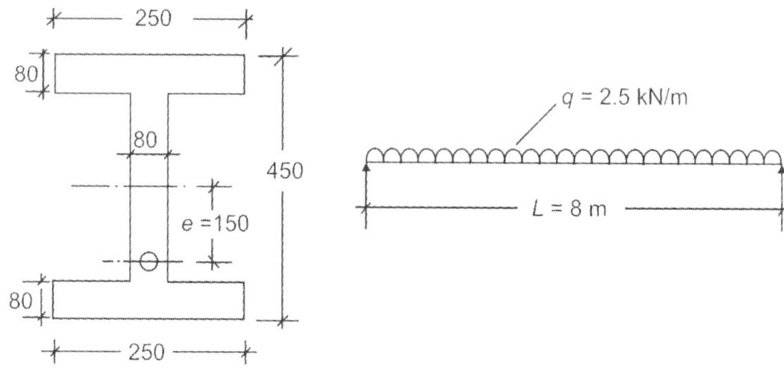

Fig. 1.10

$P = 120$ kN $\quad A = (2 \times 250 \times 80) + (290 \times 80) = 63200$ mm^2

$e = 150$ mm $\quad g = (0.0632 \times 24) = 1.516$ kN/m

$q = 2.5$ kN/m $\quad I = \dfrac{(250 \times 450^3)}{12} - \dfrac{(170 \times 290^3)}{12}$

$$= 1.553 \times 10^9 \text{ mm}^4$$

$$Z = Z_b = Z_t = \frac{(1.553 \times 10^9)}{(225)} = 6.9 \times 10^6 \text{ mm}^3$$

$$M_g = (0.125 \times 1.516 \times 8^2) = 12.56 \text{ kN.m}$$

$$M_q = (0.125 \times 2.5 \times 8^2) = 20 \text{ kN.m}$$

$$\left(\frac{P}{A}\right) = \frac{(120 \times 10^3)}{63200} = 1.90 \text{ kN.m}$$

$$\left(\frac{Pe}{Z}\right) = \frac{(120 \times 10^3 \times 150)}{(6.9 \times 10^6)} = 2.6 \text{ N/mm}^2$$

$$\left(\frac{M_g}{Z}\right) = \frac{(12.56 \times 10^6)}{(6.9 \times 10^6)} = 1.82 \text{ N/mm}^2$$

$$\left(\frac{M_g}{Z}\right) = \frac{(12.56 \times 10^6)}{(6.9 \times 10^6)} = 2.89 \text{ N/mm}^2$$

Resultant stresses

Case a. Top fibre $= f_t = (1.90 - 2.6 + 1.82)$

$= 1.12\ \text{N/mm}^2$ (compression)

Bottom fibre $= F_b = (1.90 + 2.6 - 1.82)$

$= 2.68\ \text{N/mm}^2$ (compression)

Case b. Top fibre $= f_t = (1.90 - 2.6 + 1.82 + 2.89)$

$= 4.01\ \text{N/mm}^2$ (compression)

Bottom fibre $= f_b = (1.90 + 2.6 - 1.82 - 2.89)$

$= -0.21\ \text{N/mm}^2$ (tension)

Problem 1.11

A post tensioned concrete beam having an unsymmetrical I-section as shown in Fig. 1.11 is used to support live loads of 11 kN/m over a span of 30 m. The prestressing force of 3200 kN is located at an eccentricity of 580 mm at the centre of span section. Determine the extreme fibre stresses at mid span section when the beam supports dead and live loads assuming the loss of prestress as 15 percent.

Solution

$P = 3200$ kN $\quad I = (72490 \times 10^5)\ \text{mm}^4$

$e = 580$ mm $\quad Z_t = (127 \times 10^6)\ \text{mm}^3$

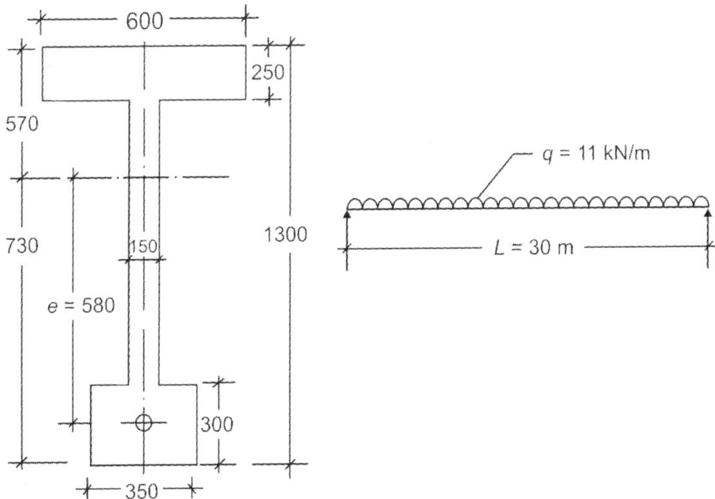

Fig. 1.11

$A = 367500 \text{ mm}^2 \qquad Z_b = (99 \times 10^6) \text{ mm}^3$

$y = 570 \text{ mm} \qquad \text{Loss ratio} = \eta = 0.85$

$g = (0.3675 \times 24) = 8.8 \text{ kN/m}$

$M_g = (0.125 \times 8.8 \times 30^2) = 990 \text{ kN.m}$

$q = 11 \text{ kN/m}$

$M_q = (0.125 \times 11 \times 30^2) = 1237.5 \text{ kN.m}$

$$\left(\frac{P}{A}\right) = \frac{(3200 \times 10^3)}{(367500)} = 8.7 \text{ N/mm}^2$$

$$\left(\frac{Pe}{Z_t}\right) = \frac{(3200 \times 10^3 \times 580)}{(127 \times 10^6)} = 14.6 \text{ N/mm}^2$$

$$\left(\frac{Pe}{Z_b}\right) = \frac{(3200 \times 10^3 \times 580)}{(99 \times 10^6)} = 18.74 \text{ N/mm}^2$$

$$\left(\frac{M_g}{Z_t}\right) = \frac{(990 \times 10^6)}{(127 \times 10^6)} = 7.8 \text{ N/mm}^2$$

$$\left(\frac{M_g}{Z_b}\right) = \frac{(990 \times 10^6)}{(99 \times 10^6)} = -10 \text{ N/mm}^2$$

$$\left(\frac{M_q}{Z_t}\right) = \frac{(1237.5 \times 10^6)}{(127 \times 10^6)} = 9.74 \text{ N/mm}^2$$

$$\left(\frac{M_q}{Z_b}\right) = \frac{(1237.5 \times 10^6)}{(99 \times 10^6)} = -12.5 \text{ N/mm}^2$$

Resultant stresses at mid span section

Top fibre $= f_t = \left(\frac{P}{A}\right) - \left(\frac{Pe}{Z_t}\right) + \left(\frac{M_g}{Z_t}\right) + \left(\frac{M_g}{Z_t}\right)$

$= (8.7 - 14.6 + 7.8 + 9.74)$

$= 11.76 \text{ N/mm}^2$ (compression)

Bottom fibre $= f_b = \left(\frac{P}{A}\right) + \left(\frac{Pe}{Z_t}\right) - \left(\frac{M_g}{Z_b}\right) - \left(\frac{M_q}{Z_b}\right)$

$= (8.7 + 18.74 - 10 - 12.5)$

$= 4.94 \text{ N/mm}^2$ (compression)

Problem 1.12

A simply supported beam of prestressed concrete spanning over 10 m is of rectangular section 500 mm wide by 750 mm deep. The beam is prestressed by a parabolic cable having an eccentricity of 200 mm at the centre of span and zero at the end supports. The effective force in the cable is 1600 kN. If the beam supports a total uniformly distributed load of 40 kN/m, which includes the self weight of the beam,

a. Evaluate the extreme fibre stresses at the mid span section.

b. Calculate the force required in the cable having the same eccentricity to balance a total load of 50 kN/m on the beam.

Solution

$$P = 1600 \text{ N} \quad A = (300 \times 750) = 375 \times 10^3 \text{ mm}^2$$

$$e = 200 \text{ mm}$$

$$Z_b = Z_t = Z = \left(\frac{bD^2}{6}\right) = \frac{(500 \times 750^2)}{6}$$

$$= 46.87 \times 10^6 \text{ mm}^3$$

$$L = 10 \text{ m}$$

$$\text{Total load} = w = 40 \text{ kN/m}$$

$$\text{Moment at mid span} = M = (0.125 \times w \times L^2)$$

$$= (0.125 \times 40 \times 10^2) = 500 \text{ kN.m}$$

Case a. $$\left(\frac{P}{A}\right) = \frac{(1600 \times 10^3)}{(375 \times 10^3)} = 4.26 \text{ N/mm}^2$$

$$\left(\frac{Pe}{Z}\right) = \frac{(1600 \times 10^3 \times 200)}{(46.87 \times 10^6)} = 6.82 \text{ N/mm}^2$$

$$\left(\frac{M}{Z}\right) = \frac{(500 \times 10^6)}{(46.87 \times 10^6)} = 10.66 \text{ N/mm}^2$$

$$\text{Stress at top} = f_t = \left[\left(\frac{P}{A}\right) - \left(\frac{Pe}{Z}\right) + \left(\frac{M}{Z}\right)\right]$$

$$= [4.26 - 6.82 + 10.66]$$

$$= 8.10 \text{ N/mm}^2 \text{ (compression)}$$

$$\text{Stress at bottom} = f_b = \left[\left(\frac{P}{A}\right)+\left(\frac{Pe}{Z}\right)-\left(\frac{M}{Z}\right)\right]$$

$$= [4.26 + 6.82 - 10.66]$$

$$= 0.42 \text{ N/mm}^2 \text{ (compression)}$$

Case b. If P = Force in cable to balance a total load of 50 kN/m on the beam

$$\left[\left(\frac{P}{A}\right)+\left(\frac{Pe}{Z}\right)\right] = \left(\frac{M}{Z}\right)$$

$$P\left[\frac{1}{375\times10^3}+\frac{200}{46.87\times10^6}\right] = \left[\frac{\left(0.125\times50\times10^2\times10^6\right)}{46.87\times10^6}\right]$$

Solving, P = 3125000 N = 3125 kN.

Problem 1.13

A post tensioned prestressed concrete beam is designed to support uniformly distributed dead and live loads of 6000 N/m and 9000 N/m, respectively. The beam is simply supported over a span of 12 m. The beam is prestressed by 48 wires of 5 mm diameter at an average eccentricity of 250 mm. The cross-section of the beam is a symmetrical I-section 300 mm by 750 mm overall with flange and web thickness of 120 and 100 mm respectively. Determine the maximum stresses developed at various stages of loading if the beam is prestressed by steel stressed to an initial stress of 1000 N/mm^2. Loss factor = 0.85. **(Bangalore University, 1983)**

Solution

$$P = \frac{(48\times20\times1000)}{10^3} = 960 \text{ kN}$$

$$e = 250 \text{ mm}$$

$$A = (2\times300\times120)+(510\times100)$$

$$= 123\times10^3 \text{ mm}^3$$

$$g = (0.123\times24) = 2.952 \text{ kN/m}$$

$$M_g = (0.125\times2.952\times12^2) = 54 \text{ kN.m}$$

$$I = \left[\frac{\left(300\times750^3\right)}{12}\right]-\left[\frac{\left(200\times510^3\right)}{12}\right]$$

$$Z = Z_t = Z_b = \frac{(8300 \times 10^6)}{(375)} = 22.1 \times 10^6 \text{ mm}^3$$

$$\text{Dead load} = w = 6 \text{ kN/m}$$

$$M_w = (0.125 \times 6 \times 12^2) = 108 \text{ kN.m}$$

$$\text{Live load} = q = 9 \text{ kN/m}$$

$$M_q = (0.125 \times 9 \times 12^2) = 162 \text{ kN.m}$$

Prestress

$$\left(\frac{P}{A}\right) = \frac{(960 \times 10^3)}{(123 \times 10^3)} = 7.8 \text{ N/mm}^2$$

$$\left(\frac{Pe}{Z}\right) = \frac{(960 \times 10^3 \times 250)}{(22.1 \times 10^6)} = 10.85 \text{ N/mm}^2$$

$$\left(\frac{M_g}{Z}\right) = \frac{(54 \times 10^6)}{(22.1 \times 10^6)} = 2.44 \text{ N/mm}^2$$

$$\left(\frac{M_w}{Z}\right) = \frac{(108 \times 10^6)}{(22.1 \times 10^6)} = 4.88 \text{ N/mm}^2$$

$$\left(\frac{M_q}{Z}\right) = \frac{(162 \times 10^6)}{(22.1 \times 10^6)} = 7.33 \text{ N/mm}^2$$

Stresses due to (prestress + self weight)

$$f_t = \left(\frac{P}{A}\right) - \left(\frac{Pe}{Z}\right) + \left(\frac{M_g}{Z}\right) = 7.8 - 10.85 + 2.44$$

$$= -0.61 \text{ N/mm}^2 \text{ (tension)}$$

$$f_b = \left(\frac{P}{A}\right) + \left(\frac{Pe}{Z}\right) - \left(\frac{M_g}{Z}\right) = 7.8 + 10.85 - 2.44$$

$$= 16.21 \text{ N/mm}^2 \text{ (compression)}$$

Stresses due to (prestress + self weight + dead load + live load)

$$f_t = 0.85\left[\left(\frac{P}{A}\right)-\left(\frac{Pe}{Z}\right)\right]+\left(\frac{M_g}{Z}\right)+\left(\frac{M_w}{Z}\right)+\left(\frac{M_q}{Z}\right)$$

$$= -2.59 + 2.44 + 4.88 + 7.33$$

$$= 12.06 \text{ N/mm}^2 \text{ (compression)}$$

$$f_b = 0.85\left[\left(\frac{P}{A}\right)+\left(\frac{Pe}{Z}\right)\right]-\left(\frac{M_g}{Z}\right)-\left(\frac{M_w}{Z}\right)-\left(\frac{M_q}{Z}\right)$$

$$= 15.85 - 2.44 - 4.88 - 7.33$$

$$= 1.2 \text{ N/mm}^2 \text{ (compression).}$$

Problem 1.14

A concrete beam with a single overhang is simply supported at *A* and *B* over a span of 8 m and the overhang *BC* is 2 m. The beam is of rectangular section 300 mm wide by 900 mm deep and supports a uniformly distributed live load of 3.52 kN/m over the entire length in addition to its self weight. Determine the profile of the prestressing cable with an effective force of 500 kN which can balance the dead and live loads on the beam. Sketch the profile of the cable along the length of the beam.

Solution

The given single overhang beam with dead and live loads is shown in Fig. 1.12(a)

$$P = 500 \text{ kN}$$

$$\text{Self weight} = (0.3 \times 0.9 \times 24) = 6.48 \text{ kN/m}$$

$$q = 3.52 \text{ kN/m}$$

$$\text{Live load} = 3.52 \text{ kN/m}$$

$$\text{Total load} = w = 10.00 \text{ kN/m}$$

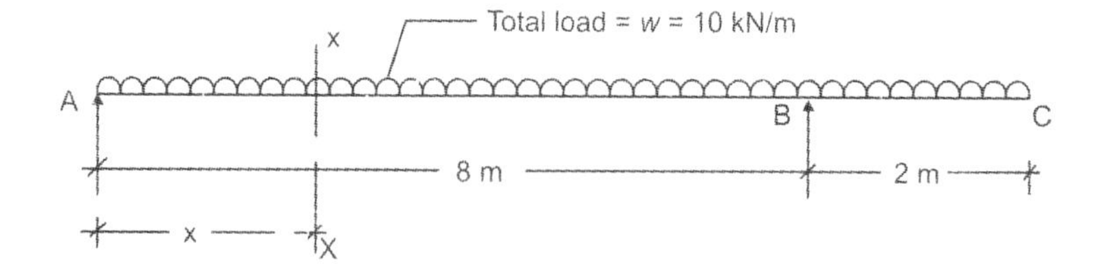

(a) Loaded beam

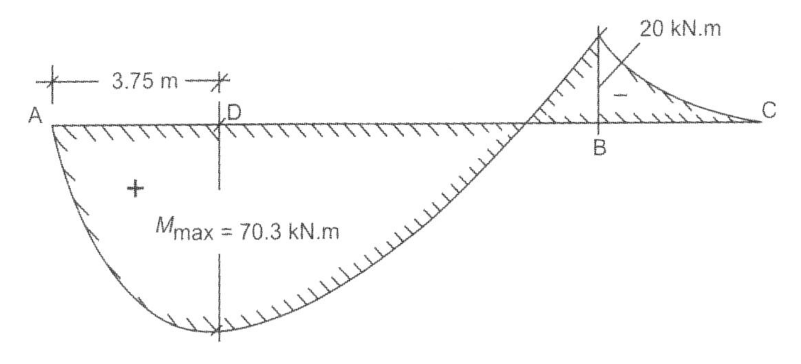

(b) Bending moment diagram

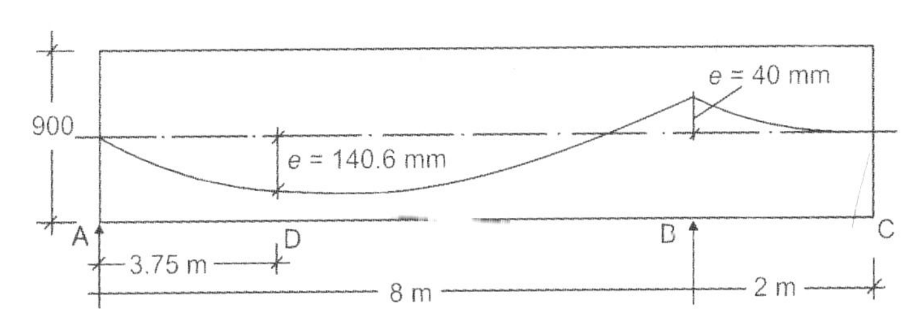

(c) Cable profile

Fig. 1.12

Let R_A and R_B are the reactions at supports A and B.

$$(R_B \times 8) = (10 \times 10 \times 5) = 500$$

$\therefore$ $R_B = 62.5$ kN and $R_A = 37.5$ kN

Moment at $B = M_B = (0.5 \times 10 \times 2^2) = 20$ kN.m

In span AB, $M_x = (37.5x - 0.5 \times 10 \times x^2) = (37.5x - 5x^2)$

For maximum bending moment, $\left(\dfrac{dM_x}{dx}\right) = 0$

$\therefore$ $37.5 - 10x = 0$ $\qquad \therefore$ $x = 3.75$ m

Maximum bending moment

$$= M_{max} = M_D = (37.5 \times 3.75 - 0.5 \times 10 \times 3.75^2)$$

$$= 70.3 \text{ kN.m [Refer Fig. 1.12 (b)]}$$

For load balancing, $M = Pe$, where e = eccentricity

At D, $\qquad e = \left(\dfrac{M_D}{P}\right) = \dfrac{(70.3 \times 10^6)}{(500 \times 10^3)} = 140.6$ mm

At B, $$e = \left(\frac{M_B}{P}\right) = \frac{(20 \times 10^6)}{(500 \times 10^3)} = 40 \text{ mm}$$

The cable profile is parabolic and is shown in Fig. 1.12(c).

Problem 1.15

A concrete beam with a double overhang has the middle span of 8 m and the equal overhang on either side is 2 m. The beam is of rectangular section 400 mm wide by 800 mm deep. The beam supports a total uniformly distributed load of 10 kN/m which includes the self weight of the beam. The beam is also prestressed by a parabolic cable with an effective force of 200 kN and having an eccentricity of 300 mm towards the soffit at centre of span and an eccentricity of 100 mm towards the top at supports. Compute the resultant stress distribution at centre of span section.

Solution

Total load $= w = 10$ kN/m

Span length $= L = 8$ m

Overhang length $= a = 2$ m

Cross-sectional area $= A = (400 \times 800) = 32 \times 10^4 \text{ mm}^2$

Section modulus $$= Z = \left(\frac{bD^2}{6}\right) = \frac{(400 \times 800^2)}{6}$$

$$= 42.66 \times 10^6 \text{ mm}^3$$

Prestressing force $= P = 200$ kN, Eccentricity $= e = 300$ mm

Maximum bending moment at centre of span

$$= M = \left[\left(\frac{wL^2}{8}\right) - \left(\frac{wa^2}{2}\right)\right]$$

$$M = \left[\frac{(10 \times 8^2)}{8} - \frac{(10 \times 2^2)}{2}\right] = 60 \text{ kN.m}$$

Resultant stresses at mid span section

Stress at top $$= f_t = \left[\left(\frac{P}{A}\right) - \left(\frac{Pe}{Z}\right) + \left(\frac{M}{Z}\right)\right]$$

$$= \left[\left(\frac{200 \times 10^3}{32 \times 10^4}\right) - \left(\frac{200 \times 10^3 \times 300}{42.66 \times 10^6}\right) + \left(\frac{60 \times 10^6}{42.66 \times 10^6}\right)\right]$$

$$= 0.625 \text{ N/mm}^2 \text{ (compression)}$$

Stress at bottom $= f_b\left[\left(\frac{P}{A}\right)+\left(\frac{Pe}{Z}\right)-\left(\frac{M}{Z}\right)\right]$

$= 0.625\ \text{N/mm}^2$ (compression).

Problem 1.16

A concrete beam of rectangular section 200 mm wide by 300 mm deep is prestressed by means of high tensile wires of area 300 mm², located 65 mm from the bottom of the beam and wires of area 60 mm², located 25 mm from the top of the beam. The effective stress in the wires is 840 N/mm². Compute the resultant stresses at mid span section when the beam is supporting its own weight together with a live load of 6 kN/m, assuming the density of concrete as 24 kN/m³ and the span of the beam as 6 m.

Solution

Distance of the centroid of the prestressing force from base.

$$y = \left[\frac{(300\times 65)+(60\times 275)}{360}\right] = 100\ \text{mm}$$

Eccentricity $= e(150-100) = 50$ mm

Prestressing force $= P = (360 \times 840) = 3 \times 10^5$ N

Area of cross-section $= A = (300 \times 200) = 6 \times 10^4\ \text{mm}^2$

Section modulus $= Z = \frac{(200\times 300^2)}{6} = 3 \times 10^6\ \text{mm}^3$

Self weight of beam $= (0.2 \times 0.3 \times 24) = 1.44$ kN/m

Live load on the beam $= 6$ kN/m

Total load on the beam $= (1.44 + 6) = w = 7.44$ kN/m

Maximum moment at centre of span $= (0.125 \times 7.44 \times 6^2)$

$= 33.48$ kN.m

Resultant stresses at mid span section

Stress at top $= f_t = \left(\frac{P}{A}\right)-\left(\frac{Pe}{Z}\right)+\left(\frac{M}{Z}\right)$

$$= \left[\left(\frac{3\times 10^3}{6\times 10^4}\right)-\left(\frac{3\times 10^5\times 50}{3\times 10^6}\right)+\left(\frac{33.48\times 10^6}{3\times 10^6}\right)\right]$$

$= 11.16\ \text{N/mm}^2$ (compression)

$$\text{Stress at bottom} = f_b = \left[\left(\frac{P}{A}\right)+\left(\frac{Pe}{Z}\right)-\left(\frac{M}{Z}\right)\right]$$

$$= \left[\left(\frac{3\times10^5}{6\times10^4}\right)+\left(\frac{3\times10^5\times50}{6\times10^6}\right)-\left(\frac{33.48\times10^6}{3\times10^6}\right)\right]$$

$$= 1.16\ \text{N/mm}^2\ \text{(compression)}$$

Problem 1.17

The I-section used as a post tensioned beam shown in Fig. 1.13 is simply supported over a span of 25 m. Determine the permissible uniformly distributed live load the beam can carry from the following data

Initial prestress in steel = 1125 N/mm^2

Loss of prestress = 225 N/mm^2

Permissible stress in concrete under working loads is limited to 17 N/mm^2 (compression) and 1.7 N/mm^2 (tension).

Moment of inertia of the I-section about the centroidal axis = 15×10^9 mm^4

Area of cross-section of I-section = 12×10^4 mm^2

(Bangalore University; 1996)

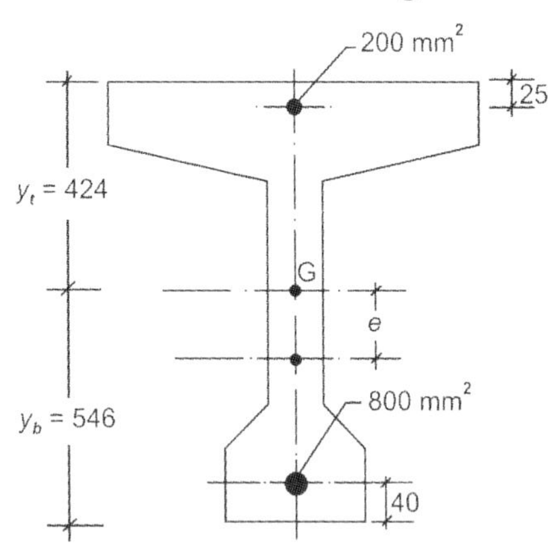

Fig. 1.13

Solution

Given data: $A = 12 \times 10^4$ mm^2 $\quad y_t = 424$ mm

$I = 15 \times 10^9$ mm^4 $\quad y_b = 546$ mm

$Z_t = (I/y_t) = 35.37 \times 10^6$ mm^3

Span = L = 25 m

$Z_b = (I/y_b)$ = 27.47×10^6 mm^3

Distance of centroid from bas $= \dfrac{[(200 \times 945) + (800 \times 40)]}{1000}$

= 221 mm

$\therefore$ Eccentricity $= e = (546 - 221) = 325$ mm

Effective stress in steel $= (1125 - 225) = 900$ N/mm^2

Total area of steel $= A_p = (800 + 200) = 1000$ mm^2

Pretressing force $= P = (1000 \times 900) = 9 \times 10^5$ N

Self weight $= g = (0.12 \times 24) = 2.88$ kN/m

Self weight moment $= M_g = (0.125 \times 2.88 \times 25^2) = 225$ kN.m

Let q = Live load on beam expressed in kN/m

Then, Moment due to live load $= M_q = (0.125 \times q \times 25^2)$

$= 78.125\, q$ kN.m

$$\left(\frac{P}{A}\right) = \frac{(9 \times 10^5)}{12 \times 10^4} = 7.5 \text{ N/mm}^2$$

$$\left(\frac{Pe}{Z_t}\right) = \frac{(9 \times 10^5 \times 325)}{(35.37 \times 10^6)} = 8.26 \text{ N/mm}^2$$

$$\left(\frac{Pe}{Z_b}\right) = \frac{(9 \times 10^5 \times 325)}{(27.47 \times 10^6)} = 10.64 \text{ N/mm}^2$$

$$\left(\frac{M_g}{Z_t}\right) = \frac{(225 \times 10^6)}{(35.37 \times 10^6)} = 6.36 \text{ N/mm}^2$$

$$\left(\frac{M_g}{Z_b}\right) = \frac{(225 \times 10^6)}{(27.47 \times 10^6)} = 8.19 \text{ N/mm}^2$$

$$\left(\frac{M_q}{Z_t}\right) = \frac{(78.125 \times 10^6)}{(35.37 \times 10^6)} = 2.208 \text{ qN/mm}^2$$

$$\left(\frac{M_q}{Z_b}\right) = \frac{(78.125 \times 10^6)}{(27.47 \times 10^6)} = 2.84 \text{ qN/mm}^2$$

The permissible stress conditions yield the following equations:

$$\text{Stress at top} = f_t = 17 = \left[\left(\frac{P}{A}\right) - \left(\frac{Pe}{Z_t}\right) + \left(\frac{M_g}{Z_t}\right) + \left(\frac{M_q}{Z_t}\right)\right]$$

Stress at bottom $= f_b = -1.7 = \left[\left(\frac{P}{A}\right)+\left(\frac{Pe}{Z_b}\right)-\left(\frac{M_g}{Z_b}\right)-\left(\frac{M_q}{Z_b}\right)\right]$

Substituting the values of stresses, we have the relations,

$$f_t = 17 = (7.5 - 8.26 + 6.36 + 2.28\,q)$$

Solving, $q = 5$ kN/m

$$f_b = -1.7 = (7.5 + 10.64 - 8.19 - 2.84\,q)$$

Solving, $q = 4.10$ kN/m

Hence the maximum permissible live load on the beam is 4.10 kN/m.

Problem 1.18

An unsymmetrical I-section concrete beam shown in Fig. 1.14 is used to support a live load of 2 kN/m over a span of 8 m. The effective prestressing force is 100 kN located 50 mm from the soffit of the beam.

a. Compute the stress in concrete at the centre of span soffit of the beam under working loads.
b. If the modulus of rupture of concrete is 5 N/mm^2, determine the load factor against cracking.

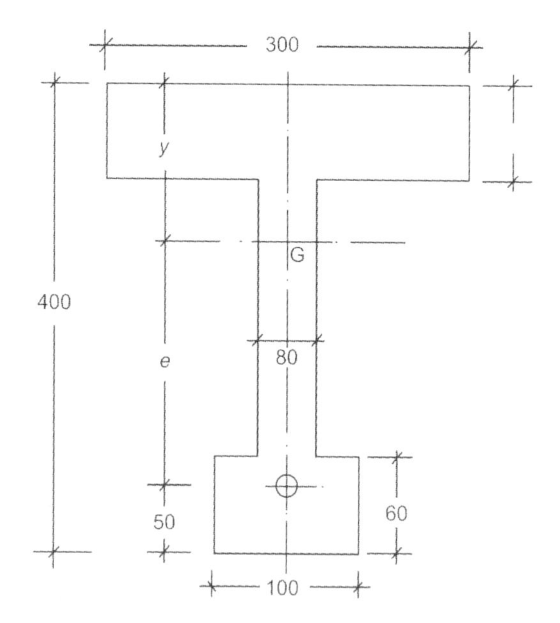

Fig. 1.14

Solution

Prestressing force $= P = 100$ kN

Area of cross-section $= A = 46400\ \text{mm}^2$

Distance of centroid from top $= y = 156$ mm

Eccentricity $= e = 194$ mm

Second moment of area $= I = 75.8 \times 10^7\ \text{mm}^4$

Section modulus of top fibre $= Z_t = \dfrac{(75.8 \times 10^7)}{(156)}$

$= 485 \times 10^4\ \text{mm}^3$

Section modulus of bottom fibre $= Z_b = \dfrac{(75.8 \times 10^7)}{(244)}$

$= 310 \times 10^4\ \text{mm}^3$

Self weight $= g = (0.0464 \times 24) = 1.12$ kN/m

Self weight moment $= M_g = (0.125 \times 1.12 \times 8^2)$

$= 8.96$ kN.m

Live load moment $= M_q = (0.125 \times 2 \times 8^2) = 16$ kN.m

Stress in concrete at soffit of beam is given by

$$f_b = \left[\left(\frac{P}{A}\right) + \left(\frac{Pe}{Z_b}\right) - \left(\frac{M_g}{Z_b}\right) - \left(\frac{M_q}{Z_b}\right)\right]$$

$$= \left[\frac{(100 \times 10^3)}{(46400)} + \frac{(100 \times 10^3 \times 194)}{310 \times 10^4} - \frac{(8.96 \times 10^6)}{(310 \times 10^4)} - \frac{(16 \times 10^6)}{(310 \times 10^4)}\right]$$

$= 0.35\ \text{N/mm}^2$ (compression)

Cracking stress at soffit $= f_{cr} = (0.35 + 5) = 5.35\ \text{N/mm}^2$

Additional moment required to develop cracks at soffit is

$$M = (f_{cr} \cdot Z_b) = (5.35 \times 310 \times 10^4)$$

$$= 16.585 \times 10^6 \text{ N.mm} = 16.585 \text{ kN.m}$$

Working moment $= M_w = (M_g + M_q) = (8.96 + 16)$

$= 24.96$ kN.m

Cracking moment $= M_{cr} = (24.96 + 16.585) = 41.545$ kN.m

Load factor against cracking $= \left(\dfrac{M_{cr}}{M_w}\right) = \left(\dfrac{41.545}{24.96}\right) = 1.66.$

Problem 1.19

A double overhang beam of concrete with supports 8 m apart has equal overhangs of 2 m. The beam is of rectangular section 200 mm wide by 500 mm deep. The beam is prestressed by a parabolic cable having an eccentricity of 200 mm towards the soffit at centre of span and an eccentricity of 66.66 mm towards the top at support sections. The beam supports a uniformly distributed live load of 7.6 kN/m in addition to its self weight. Determine the pressure line and locate its position along the length of the beam. The effective prestressing force in the cable is 300 kN. If the modulus of rupture of concrete is 4 N/mm^2, estimate the load factor against cracking.

Solution

Fig. 1.15 shows the beam with the cable profile and loads.

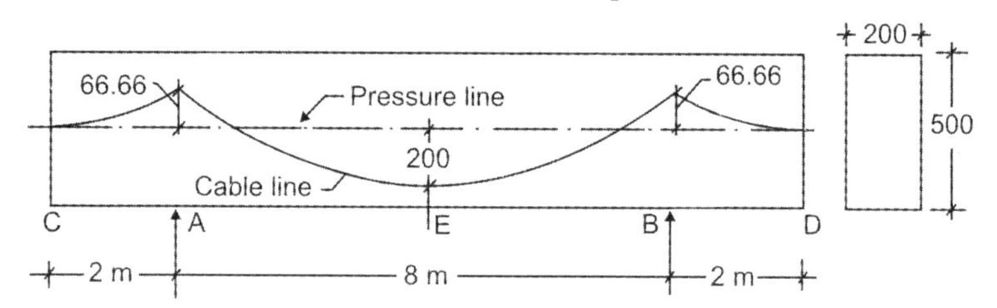

Fig. 1.15

$$P = 300 \text{ kN}$$

$$\text{Self weight} = g = (0.2 \times 0.5 \times 24) = 2.4 \text{ kN/m}$$

$$e_E = 200 \text{ mm} \qquad e_A = e_B = 66.66 \text{ mm}$$

$$\text{Live load} = q = 7.6 \text{ kN/m}$$

$$\text{Total load} = w = 10.0 \text{ kN/m}$$

Section modulus

$$= Z_t = Z_b = \frac{(bD^2)}{6} = \frac{(200 \times 500^2)}{6}$$

$$= 8.33 \times 10^6 \text{ mm}^3$$

Area of cross-section

$$= A = (200 \times 500) = 10^5 \text{ mm}^2$$

Moment at centre of span

$$= M_E = \left[\frac{(10 \times 8^2)}{8} - \frac{(10 \times 2^2)}{2}\right] = 60 \text{ kN.m}$$

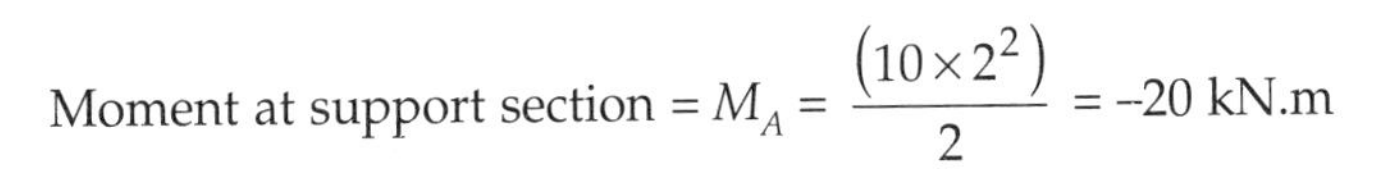

Moment at support section $= M_A = \dfrac{(10 \times 2^2)}{2} = -20$ kN.m

Shift of pressure line at $E = \left(\dfrac{M_E}{P}\right) = \dfrac{(60 \times 10^6)}{(300 \times 10^3)} = 200$ mm

Shift of pressure line at $A = \left(\dfrac{M_A}{P}\right) = \dfrac{(20 \times 10^6)}{(300 \times 10^3)} = 66.66$ mm

Pressure line is shown in Fig. 1.15

$$\text{Stress at soffit at } E = f_b = \left[\left(\frac{P}{A}\right) + \left(\frac{Pe}{Z_b}\right) - \left(\frac{M_E}{Z_b}\right)\right]$$

$$= (3 + 7.2 - 7.2)$$

$$= 3 \text{ N/mm}^2 \text{ (compression)}$$

Cracking stress at soffit at $E = (3 + 4) = 7$ N/mm^2

$$\text{Additional moment required} = M = (f.z.) \frac{(7 \times 8.33 \times 10^6)}{10^6}$$

$$= 58.31 \text{ kN.m}$$

$$\text{Cracking moment} = M_{cr} = (M_E + M)$$

$$= (60 + 58.31) = 118.31 \text{ kN.m}$$

$$\text{Factor of safety against cracking} = \left(\frac{M_{cr}}{M_E}\right) = \left(\frac{118.31}{60}\right) = 1.97$$

Problem 1.20

A prestressed concrete girder of 30 m span is of unsymmetrical section as shown in Fig. 1.16. The girder supports a uniformly distributed dead load of 34.87 kN/m over the entire span and concentrated loads of 15 kN each at 5 m intervals. The maximum live load bending moment at centre of span is estimated to be 2974 kN.m. The girder is prestressed by a parabolic cable with an effective force of 5145 kN located at 200 mm from the soffit of the beam at centre of span section. Determine the stress distribution at centre of span section under service loads.

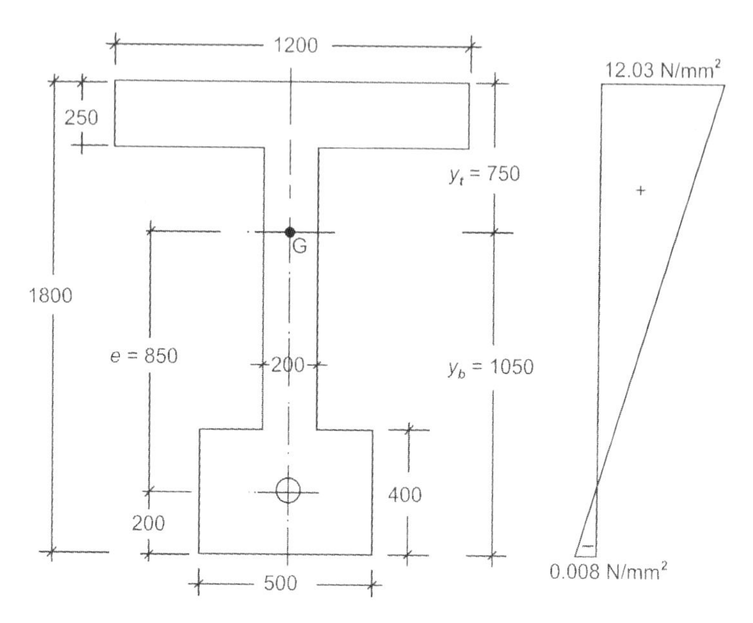

Fig. 1.16

Solution

$$P = 5145 \text{ kN} \qquad I = 2924 \times 10^8 \text{ mm}^4$$

$$e = 850 \text{ mm} \qquad Z_t = \left(\frac{I}{Y_t}\right) = 3.89 \times 10^8 \text{ mm}^3$$

$$A = 0.73 \times 10^4 \text{ mm}^2 \quad Z_b = \left(\frac{I}{Y_b}\right) = 2.78 \times 10^8 \text{ mm}^3$$

$$M_g = [(0.125 \times 34.87 \times 30^2) + (0.25 \times 15 \times 30) + (15 \times 10) + (15 \times 15)]$$

$$= 4261 \text{ kN.m}$$

$$M_q = 2074 \text{ kN.m}$$

$\therefore$ Total service load moment $= M_w = (4261 + 2074) = 6335$ kN.m

$$\left(\frac{P}{A}\right) = \frac{(5145 \times 10^3)}{(0.73 \times 10^6)} = 7.05 \text{ N/mm}^2$$

$$\left(\frac{Pe}{Z_t}\right) = \frac{(5145 \times 10^3 \times 850)}{(3.89 \times 10^8)} = 11.24 \text{ kN/mm}^2$$

$$\left(\frac{Pe}{Z_b}\right) = \frac{\left(5145\times10^3\times850\right)}{\left(2.78\times10^8\right)} = 15.72\ \text{N/mm}^2$$

$$\left(\frac{M_w}{Z_b}\right) = \frac{\left(6335\times10^6\right)}{\left(2.78\times10^8\right)} = 22.78\ \text{N/mm}^2$$

$$\left(\frac{M_w}{Z_t}\right) = \frac{\left(6335\times10^6\right)}{\left(3.89\times10^8\right)} = 16.28\ \text{N/mm}^2$$

Stresses at centre of span section are computed as

$$f_t = \left[\left(\frac{P}{A}\right)-\left(\frac{Pe}{Z_t}\right)+\left(\frac{M_w}{Z_t}\right)\right]$$

$$= [7.05 + 11.24 - 16.28]$$

$$= 12.09\ \text{N/mm}^2\ \text{(compression)}$$

$$f_b = \left[\left(\frac{P}{A}\right)+\left(\frac{Pe}{Z_b}\right)-\left(\frac{M_w}{Z_b}\right)\right]$$

$$= [7.05 + 15.72 - 22.78]$$

$$= 0.008\ \text{N/mm}^2\ \text{(tension)}$$

The stress distribution at centre of span section is shown in Fig. 1.16.

Problem 1.21

A beam of symmetrical I-section spanning 8 m has a flange width and depth of 250 mm and 80 mm, respectively. The overall depth of the beam is 450 mm and thickness of the web is 80 mm. The beam is prestressed by a parabolic cable with an eccentricity of 150 mm at the centre of the span and concentric at supports. The superimposed live load on the beam is 2.5 kN/m.

a. Determine the effective force in the cable for balancing the dead and live loads on the beam.

b. Calculate the shift of pressure line from the tendon centre line at the centre of span section.

c. Find the pressure line distribution along the length of the beam for the condition of (prestress + dead load + live load).

Solution

The properties of the symmetrical I-section are as follows:

$$A = 0.063 \text{ m}^2 \qquad g = 1.57 \text{ kN/m}$$
$$I = 1.553 \times 10^9 \text{ mm}^4 \qquad q = 2.50 \text{ kN/m}$$
$$Z = 6.9 \times 10^6 \text{ mm}^3 \qquad L = 8 \text{ m}$$
$$e = 150 \text{ mm}$$

The bending moment at the centre of span section is computed as,

$$M_g = (0.125 \times 1.57 \times 8^2) = 12.56 \text{ kN.m}$$
$$M_q = (0.125 \times 2.50 \times 8^2) = 20.00 \text{ kN.m}$$

Total bending moment

$$= M = (M_g + M_q) = (12.56 + 20.00) = 32.56 \text{ kN.m}$$

If P = tendon force, then for load balancing condition, we have the relation,

$$P = (M/e) = [(32.56 \times 10^3)/150] = 217 \text{ kN}$$

Shift of pressure line at centre of span section

$$= (M/P) = [(32.56 \times 10^3)/(217 \times 10^3)]$$
$$= 150 \text{ mm}$$

The pressure line coincides with centroidal axis.

Problem 1.22

A prestressed concrete beam of 200 mm wide by 300 mm deep supports a live load of 2.56 kN/m over an effective span of 10 m. The high tensile wires housed in ducts are located at a constant eccentricity of 100 mm. Calculate the increase in steel stress due to the loading for the following two cases:

a. The ducts are grouted so that the strains in the steel and adjacent concrete are equal.
b. The ducts are ungrouted so that the tendons can freely move in ducts without friction.

Assume the modulus of elasticity of steel and concrete as 210 and 35 kN/mm^2 respectively.

Solution

Self weight of the beam = $g = (0.2 \times 0.3 \times 24) = 1.44$ kN/m

Live load on the beam = $q = 2.56$ kN/m

Total load on the beam = $w_d = (g + q) = 4.00$ kN/m

Second moment of area of cross-section

$$= I = [(200 \times 300^3)/12] = (45 \times 10^7) \text{ mm}^4$$

Bending moment at the centre of the span section

$$= M = (0.125 \times 4 \times 10^2) = 50 \text{ kN.m}$$

a. Bonded beam

$$\text{Stress in concrete} = \left(\frac{My}{I}\right) = \left(\frac{50 \times 10^6 \times 100}{45 \times 10^7}\right)$$

$$= 11.1 \text{ N/mm}^2$$

$$\text{Stress in steel} = \left(\frac{\alpha_e My}{I}\right) (6 \times 11.1) = 66.6 \text{ N/mm}^2$$

b. Unbonded beam

$$\text{Stress in steel} = \left(\frac{\alpha_e y w_d L^2}{12\, I}\right) = \left(\frac{6 \times 100 \times 4 \times (10 \times 1000)^2}{12 \times 45 \times 10^7}\right)$$

$$= 44.4 \text{ N/mm}^2.$$

Problem 1.23

A rectangular beam of cross-section 120 mm wide by 300 mm deep is prestressed by a straight cable carrying an effective force of 180 kN at a constant eccentricity of 50 mm. The beam supports an imposed load of 3.14 kN/m over a span of 6 m. If the modulus of rupture of concrete is 5 N/mm^2, estimate the load factor against cracking assuming the self weight of concrete as 24 kN/m^3.

Solution

$P = 180$ kN $\qquad I = (27 \times 10^7)$ mm^4

$e = 50$ mm $\qquad Z = (18 \times 10^5)$ mm^3

$A = (36 \times 10^3)$ mm^3 $\qquad g = (0.12 \times 0.3 \times 24) = 0.86$ kN/m

$q = 3.14$ kN/m

Total load $= w = (g + q) = (0.86 + 3.14) = 4$ kN/m

Stress due to prestress:

$$(P/A) = [(180 \times 10^3)/(36 \times 10^3)] = 5 \text{ N/mm}^2$$

$$(Pe/Z) = [(180 \times 10^3 \times 50)/(18 \times 10^5)] = 5 \text{ N/mm}^2$$

Stress at soffit due to moment

$$= \left(\frac{M}{Z}\right) = \left[\frac{(18 \times 10^6)}{(18 \times 10^5)}\right] = 10 \text{ N/mm}^2$$

Stress at the bottom fiber at working load

$$= (5 + 5 - 10) = 0 \text{ N/mm}^2$$

Stress corresponding to cracking at the soffit = 5 N/mm^2

Extra moment required to create this stress

$$= (5 \times 18 \times 10^5) = (9 \times 10^6) \text{ N.mm} = 9 \text{ kN.m}$$

Hence the cracking moment = (18 + 9) = 27 kN.m
Load factor against cracking = $(M_{cr}/M_w) = (27/18) = 1.5$

Problem 1.24

A concrete beam of rectangular section 150 mm wide by 300 mm deep spans over 10 m. The prestressing cable carrying an effective force of 500 kN is concentric at supports and linearly varies to an eccentricity of 50 mm at the centre of span. Find the magnitude of a concentrated live load Q located at the centre of span for the following conditions at the central span section:

a. If the load counteracts the bending effect of the prestressing force only neglecting the self weight of the beam.
b. If the pressure line passes through the upper kern of the section under the action of the external load, self weight and prestress.

Solution

Area of cross-section $= A = (150 \times 300) = (45 \times 10^3)\ \text{mm}^2$

Section modulus $= Z = \left(\frac{150 \times 300^2}{6}\right) = (225 \times 10^4)\ \text{mm}^3$

Self weight of the beam = $g = (0.15 \times 0.3 \times 24) = 1.08$ kN/m

For the centre of span section, we have

$$P = 500\ \text{kN and } e = 50\ \text{mm}$$

For load balancing,

a. $Q = 2P \sin\theta = 2P \tan\theta = \left(\frac{2 \times 500 \times 50}{5 \times 1000}\right)$

b. Moment due to self weight = $(0.125 \times 1.08 \times 10^2) = 13.5$ kN.m

Stress due to self weight $= \left(\frac{13.5 \times 10^6}{225 \times 10^4}\right) = 6\ \text{N/mm}^2$

Stress due to prestressing $= \left[\frac{P}{A} + \frac{Pe}{Z}\right]$

$$= \left[\frac{500 \times 10^3}{45 \times 10^3} + \frac{500 \times 10^3 \times 50}{225 \times 10^4}\right]$$

$$= 22.22\ \text{N/mm}^2$$

If Q is the concentrated load at the centre of span, then moment at centre of span = $[(Q \times L)/4] = 2.5\,Q$

Bending stress $= \left[\frac{(2.5Q) \times 10^6}{225 \times 10^6}\right]$

If the pressure line passes through the upper kern at the section, stress at bottom fibre = 0.

Hence, we have the relation, $\left[\dfrac{(2.5Q)\times 10^6}{225\times 10^6}\right] + 6 = 22.22$

Solving, $Q = 14.60$ kN.

Problem 1.25

A concrete beam with a double overhang is made up a main span of 8 m and an equal overhang on either side of 4 m. The beam supports a uniformly distributed load of 10 kN/m which includes the self weight of the beam. Determine the profile of the cable carrying an effective force of 200 kN for balancing the load on the beam. Also determine the eccentricity of this cable at the centre of span and supports.

Solution

Total uniformly distributed load on the beam = w = 10 kN/m
Central span = L = 8 m and cantilever span on either side = a = 4 m
Effective prestressing force in the cable = P = 200 kN
Maximum bending moment at the centre of span

$$= M = \left[\frac{wL^2}{8} - \frac{wa^2}{2}\right] = \left[\frac{10\times 8^2}{8} - \frac{10\times 4^2}{2}\right] = 0$$

If e = eccentricity of the prestressing force at the centre of span section (for load balancing),

$$M = Pe \quad \text{or} \quad 0 = Pe \qquad \therefore e = 0$$

Hence, the cable coincides with the centroidal axis at the centre of span section.

Maximum bending moment at supports

$$= M = [w\, a^2/2] = [(10 \times 4^2)/2] = 80 \text{ kN.m}$$

If e = eccentricity of cable at supports, then we have the relation,

$$M = Pe$$
$$80 = [200 \times e]$$

Solving, the eccentricity at support section

$$= e = [80/200] = 0.4 \text{ m} = 400 \text{ mm.}$$

Problem 1.26

A rectangular concrete beam of section 200 mm wide by 600 mm deep is spanning over 10 m. The beam is prestressed by a parabolic cable concentric at supports and with an eccentricity of 100 mm towards the soffit of the beam at the centre of span. The force in the cable is 500 kN. Estimate the stresses developed due to prestress only at the centre of span and quarter span section.

Solution

Width of section $= b = 200$ mm

Depth of section $= d = 600$ mm

Area of section $= A = (200 \times 600) = (12 \times 10^4)$ mm^4

Section modulus of top and bottom fibres

$$= Z = [b\,d^2/6] = [(200 \times 600^2)/6] = (12 \times 10^6)\ \text{mm}^3$$

Eccentricity of the cable at the centre of span $= e = 100$ mm

Eccentricity of parabolic cable at quarter span

$$= (0.75\,e) = (0.75 \times 100) = 75\ \text{mm}$$

Prestressing force in the cable $= P = 500$ kN

a. Prestress at centre of span section

Stress at top fibre

$$f_t = \left[\frac{P}{A} - \frac{Pe}{Z}\right] = \left[\frac{500 \times 10^3}{12 \times 10^4} - \frac{500 \times 10^3 \times 100}{12 \times 10^6}\right] = 0$$

Stress at bottom fibre

$$f_b = \left[\frac{P}{A} - \frac{Pe}{Z}\right] = \left[\frac{500 \times 10^3}{12 \times 10^4} + \frac{500 \times 10^3 \times 100}{12 \times 10^6}\right]$$

$$= 8.32\ \text{N/mm}^2\ \text{(compression)}$$

b. Prestress at quarter span section

Stress at top fibre

$$f_t = \left[\frac{P}{A} - \frac{Pe}{Z}\right] = \left[\frac{500 \times 10^3}{12 \times 10^4} - \frac{500 \times 10^3 \times 75}{12 \times 10^6}\right]$$

$$= 1.04\ \text{N/mm}^2\ \text{(compression)}$$

Stress at bottom fibre

$$f_b = \left[\frac{P}{A} + \frac{Pe}{Z}\right] = \left[\frac{500 \times 10^3}{12 \times 10^4} + \frac{500 \times 10^3 \times 75}{12 \times 10^6}\right]$$

$$= 7.28\ \text{N/mm}^2\ \text{(compression)}$$

Problem 1.27

A prestressed concrete beam is of tee section having a flange width and depth of 1500 and 40 mm respectively. The rib is 200 mm wide by 360 mm deep. The beam spans over a length of 15 m and is prestressed by a Freyssinet cable comprising 18 wires of 7 mm diameter, stressed to 1100 N/mm^2. The cable is located at a distance of 101 mm from the soffit of the rib. Assuming the density of concrete as 24 kN/m^3, estimate the maximum stresses developed at the top and bottom fibres of the tee section at the centre of span under service load conditions, if the beam supports a uniformly distributed service live load of 3 kN/m.

Solution

Section properties:

Width of flange = 1500 mm Width of rib = 200 mm

Depth of flange = 40 mm Depth of rib = 360 mm

The centroid of the tee section is located at a distance of 129 mm from top and 271 mm from the soffit of the tee section.

Hence, $y_t = 129$ mm and $y_b = 271$ mm

Second moment of area

$= I = (2183 \times 10^6)\ \text{mm}^4$

Section modulus of top fibre

$= Z_t = [I/y_t] = [(2183 \times 10^6)/129]$

$= (16.92 \times 10^6)\ \text{mm}^3$

Section modulus of the bottom fibre

$= Z_b = [I/y_b] = [(2183 \times 10^6)/271] = (8.05 \times 10^6)\ \text{mm}^3$

Area of cross-section = $A = 0.132\ \text{m}^2$

Self weight of beam = $g = (0.132 \times 24) = 3.168$ kN/m

$$\text{Prestressing force in cable} = P = \left[\frac{18 \times 38.4 \times 1100}{1000}\right] = 760\ \text{kN}$$

Eccentricity of the prestressing force = $e = (271 - 101) = 170$ mm

Dead load moment at the centre of span

$= M_g = [0.125 \times 3.168 \times 15^2] = 89.100$ kN.m

Live load moment at centre of span

$= M_q = [0.125 \times 3 \times 15^2] = 84.37$ kN.m

Service load moment = $M_w = [89.10 + 84.37] = 173.47$ kN.m

Prestress at top fibre

$$= f_{tp} = \left[\frac{P}{A} - \frac{Pe}{Z_t}\right] = \left[\frac{760 \times 10^3}{132 \times 10^3} - \frac{760 \times 10^3 \times 170}{16.92 \times 10^6}\right]$$

$= -1.85\ \text{N/mm}^2$ (tension)

Prestress at bottom fibre

$$= f_{bp} = \left[\frac{P}{A} - \frac{Pe}{Z_b}\right] = \left[\frac{760 \times 10^3}{132 \times 10^3} + \frac{760 \times 10^3 \times 170}{8.05 \times 10^6}\right]$$

$= 21.74$ N/mm² (compression)

Service load stresses at top and bottom fibres of the central span section:

$$\text{At top} = f_{tw} = \left[\frac{M_w}{Z_t}\right] = \left[\frac{173.47 \times 10^6}{16.92 \times 10^6}\right]$$

$= 10.25$ N/mm² (compression)

Final stresses at top and bottom fibres due to prestress and service loads:

At top $= [f_{tp} + f_{tw}] = [-1.85 + 10.25]$

$= 18.40$ N/mm² (compression)

At bottom $= [f_{bp} + f_{bw}] = [21.74 - 21.54]$

$= 0.20$ N/mm² (compression)

Problem 1.28

A tee beam spanning over 5 m has a flange 400 mm wide and 40 mm thick and a rib 100 mm wide by 200 mm deep. The beam supports a uniformly distributed live load of 4.4 kN/m. If the beam is prestressed by a cable carrying an effective force of 180 kN located at 75 mm from the soffit of the beam, calculate the stresses developed at the centre of span section under the service loads.

Solution

Section properties:

Cross-section area $= A = [(400 \times 40) + (100 \times 200)] = 36000$ mm²

Centroid of the section is located at 87 mm from the top of section

$y_t = 87$ mm and $y_b = 153$ mm

Second moment of area of section

$= I = [19575 \times 10^4]$ mm⁴

$Z_t = (I/y_t) = [(19575 \times 10^4)/87] = (225 \times 10^4)$ mm³

$Z_b = (I/y_b) = [(19575 \times 10^4)/153] = (128 \times 10^4)$ mm³

Self weight of the beam $= g = (0.036 \times 24) = 0.86$ kN/m

Live load on the beam $= q = 4.4$ kN/m

Effective span of the beam $= L = 5$ m

Dead load bending moment at the centre of span

$$= M_g = \left[\frac{gL^2}{8}\right] = \left[\frac{0.86 \times 5^2}{8}\right] = 2.68 \text{ kN.m}$$

Live load bending moment at the centre of span

$$= M_q = \left[\frac{qL^2}{8}\right] = \left[\frac{4.4 \times 5^2}{8}\right] = 13.75 \text{ kN.m}$$

Prestressing force = P = 180 kN and eccentricity

$$= e = (y_b - 75) = (153 - 75) = 78 \text{ mm}$$

Prestress at top fibre

$$= f_{tp} = \left[\frac{P}{A} - \frac{Pe}{Z_t}\right] = \left[\frac{180 \times 10^3}{36000} - \frac{180 \times 10^3 \times 78}{225 \times 10^4}\right]$$

$$= -1.24 \text{ N/mm}^2 \text{ (tension)}$$

Prestress at bottom fibre

$$= f_{bp} = \left[\frac{P}{A} + \frac{Pe}{Z_b}\right] = \left[\frac{180 \times 10^3}{36000} + \frac{180 \times 10^3 \times 78}{128 \times 10^4}\right]$$

$$= 15.96 \text{ N/mm}^2 \text{ (compression)}$$

Dead load stress at top

$$= \left[\frac{M_g}{Z_t}\right] = \left[\frac{2.68 \times 10^6}{225 \times 10^4}\right]$$

$$= 1.19 \text{ N/mm}^2 \text{ (compression)}$$

Dead load stress at soffit

$$= \left[\frac{M_g}{Z_b}\right] = \left[\frac{2.68 \times 10^6}{128 \times 10^4}\right] = 2.09 \text{ N/mm}^2 \text{ (tension)}$$

Live load stress at top

$$= \left[\frac{M_q}{Z_t}\right] = \left[\frac{13.75 \times 10^6}{225 \times 10^4}\right]$$

$$= 6.11 \text{ N/mm}^2 \text{ (compression)}$$

Live load stress at soffit

$$= \left[\frac{M_q}{Z_b}\right] = \left[\frac{13.75 \times 10^6}{128 \times 10^4}\right] = 10.74 \text{ N/mm}^2 \text{ (tension)}$$

Resultant stresses due to prestress, dead loads and live loads are computed as,

Stress at top fibre

$$= [-1.24 + 1.19 + 6.11]$$

$$= 6.06 \text{ N/mm}^2 \text{ (compression)}$$

Stress at bottom fibre

$$= [15.96 - 2.09 - 10.74]$$

$$= 3.13 \text{ N/mm}^2 \text{ (compression)}$$

2

Losses of Prestress

Problem 2.1

A pre-tensioned beam of rectangular cross-section, 150 mm wide and 300 mm deep, is prestressed by 8 wires of 7 mm wires located 100 mm from the soffit of the beam. If the wires are initially tensioned to a stress of 1100 N/mm^2, calculate their stress at transfer and the effective stress after all losses, given the following data:

	Upto time of transfer	Total
Relaxation of steel	35 N/mm^2	70 N/mm^2
Shrinkage of concrete	100×10^{-6}	300×10^{-6}
Creep coefficient	—	1.6
$E_s = 210$ kN/mm^2	$E_c = 31.5$ kN/mm^2	

Solution

The pre-tensioned beam of rectangular section is shown in Fig. 2.1.

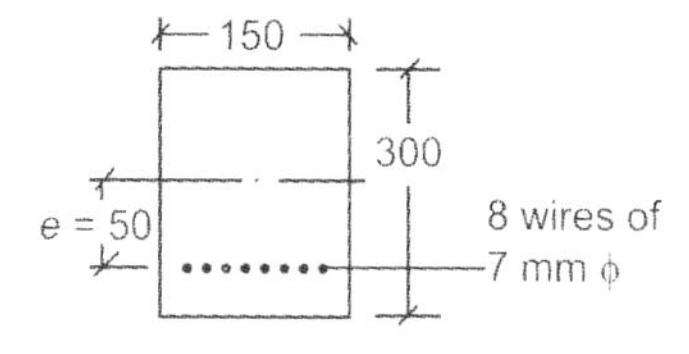

Fig. 2.1

$$\text{Initial stress} = 1100 \text{ N/mm}^2$$
$$E_s = 210 \text{ kN/mm}^2$$
$$E_c = 31.5 \text{ kN/mm}^2$$
$$\phi = \text{Creep coefficient} = 1.6$$
$$a_s = (E_s/E_c) = 6.66$$
$$A_c = 45000 \text{ mm}^2$$
$$I = 33.75 \times 10^7 \text{ mm}^4$$

Area of wires $= (8 \times 38.48) = 308 \text{ mm}^2$

$$P = \left(\frac{308 \times 1100}{1000}\right) = 339 \text{ kN}$$

Stress in concrete at level of steel

$$\alpha_c = \left[\frac{339 \times 10^3}{45000} + \frac{339 \times 10^3 \times 50 \times 50}{33.75 \times 10^7}\right]$$

$$= 10.04 \text{ N/mm}^2$$

a. *At transfer*

Loss due to relaxation $= 35 \text{ N/mm}^2$

Loss due to elastic deformation $= (\alpha_c \cdot \sigma_c) = (6.66 \times 10.04)$
$= 66.5 \text{ N/mm}^2$

Loss due to shrinkage $= (100 \times 10^{-6} \times 210 \times 10^2)$
$= 21 \text{ N/mm}^2$

Total loss $= 122.8 \text{ N/mm}^2$

$\therefore$ Stress in steel at transter $= (1100 - 122.8)$
$= 977.5 \text{ N/mm}^2$

b. *Effective stress after all losses*

Relaxation loss $= 70.0 \text{ N/mm}^2$

Shrinkage loss $= (300 \times 10^{-6} \times 210 \times 10^3) = 63.0$

Elastic deformation loss $= 66.5$

Creep of concrete loss $(\phi \times \sigma_c \cdot E_s)$
$= (1.6 \times 10.04 \times 210 \times 10^3) = 106.9$

Total loss $= 306.4 \text{ N/mm}^2$

Stress in steel after all loss $= (1100 - 306.4) = 793.6 \text{ N/mm}^2$

Problem 2.2

A prestressed concrete pile of cross-section, 250 mm by 250 mm contains 60 pre-tensioned wires each of 2 mm diameter, uniformly distributed over the section. The wires are initially tensioned on the prestressing bed with a total force of 300 kN. If $E_s = 210 \text{ kN/mm}^2$ and $E_c = 32 \text{ kN/mm}^2$, calculate the respective stresses in steel and concrete immediately after the transfer of prestress assuming that upto this point, the only loss of stress is that due to elastic shortening.

If the concrete undergoes a further shortening due to shrinkage of 200×10^{-6} per unit length, while there is a relaxation of 5% of steel stress due to creep of steel, find the greatest tensile stress which can occur in a pile, 20 m long when lifted at two points 4 m from each end. Assume creep coefficient as 1.6.

Solution

The concrete pile prestressed by 60 high tensile wires is shown in Fig. 2.2.

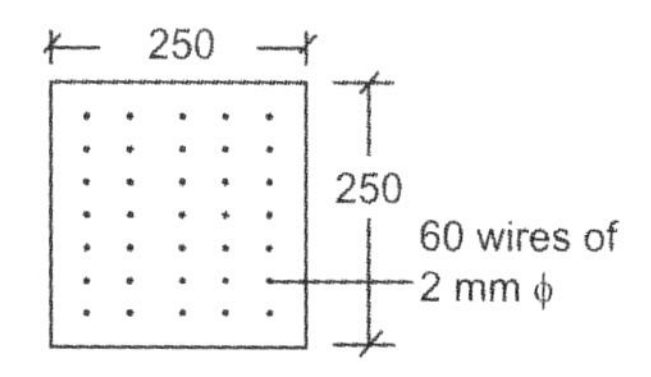

Fig. 1.2

$$A_s = \left(\frac{60 \times \pi \times 2^2}{4}\right) = 188.49 \text{ mm}^2$$

Total force $P = 300$ kN

$$\text{Initial stress in wires} = \left(\frac{300 \times 10^3}{188.49}\right) = 1590 \text{ N/mm}^2$$

$$\alpha_e = \left(\frac{E_s}{E_c}\right) = \frac{210}{32} = 6.56$$

$$\text{Stress in concrete} = \left(\frac{P}{A_c}\right) = \left(\frac{300 \times 10^3}{250 \times 250}\right) = 4.8 \text{ N/mm}^2$$

Loss due to elastic deformation = (6.56 × 4.8) = 31.5 N/mm^2

Stress in steel = (1590 – 31.5) = 1558.5 N/mm^2

Force in steel after loss = (1558.5 × 188.49)/1000 = 293.76 kN

$$\therefore \text{Stress in concrete} = \left(\frac{293.76 \times 10^3}{250 \times 250}\right) = 4.7 \text{ N/mm}^2$$

Various other losses

Loss due to creep = (1.6 × 4.7 × 6.56) = 49.3 N/mm^2

Loss due to shrinkage = $(200 \times 10^{-6} \times 210 \times 10^3)$ = 42.0 N/mm^2

Relaxation loss = [(5/100) × 1558.5] = 77.9 N/mm^2

∴ Total loss = 169.2 N/mm^2

∴ Total loss of stress in steel = (31.5 + 169.2) = 200.7 N/mm^2

Final stress in steel $= (1590 - 200.7) = 1389.3 \text{ N/mm}^2$

$\therefore \quad P = (1389.3 \times 188.49) = 261.89 \times 10^3 \text{ N}$

$$\text{Stress in concrete} = \left(\frac{261.86 \times 10^3}{250 \times 250}\right) = 4.18 \text{ N/mm}^2$$

The pile of 20 m length with the lifting points 4 m from each end shown in Fig. 2.3.

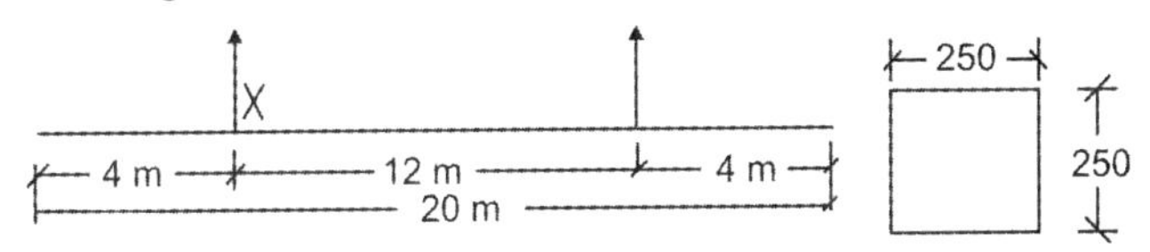

Fig. 2.3

Self weight of pile $= (0.25 \times 0.25 \times 24) = 1.5 \text{ kN/m}$

$$\text{Bending moment at } X = \left(\frac{1.5 \times 4^2}{2}\right) = 12 \text{ kN.m}$$

$$I = \left(\frac{250^4}{12}\right) \text{mm}^4$$

$$\therefore \quad \text{Extreme fibre stress} = \left(\frac{M \cdot y}{I}\right) = \left(\frac{12 \times 10^6 \times 125}{250^4/12}\right)$$

$$= 4.6 \text{ N/mm}^2$$

$$\therefore \quad \text{Tensile stress in concrete} = [4.18 - 4.6]$$

$$= -0.42 \text{ N/mm}^2 \text{ (tension)}$$

Problem 2.3

A post-tensioned cable of a beam 10 m long is initially tensioned to a stress of 1000 N/mm^2 at one end. If the tendons are curved so that the slope is 1 in 24 at each end with an area of 600 mm^2. Calculate the loss of prestress due to friction given, the following data:

Coefficient of friction between duct and cable = 0.55

Friction coefficient for wave effect = 0.0015/m

During anchoring, if there is a slip of 3 mm at the jacking end, calculate the final force in the cable and the percentage loss of prestress due to friction and slip.

Solution

The concrete beam post-tensioned by a curved cable is shown in Fig. 2.4.

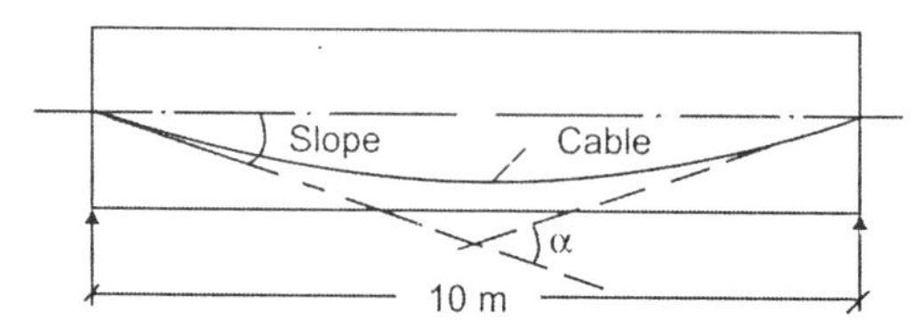

Fig. 2.4

Slope of cable 1 in 24

$$\text{Total change of slope} = \alpha = \left(\frac{2\times 1}{24}\right) = \left(\frac{1}{12}\right)$$

$$P = \left(\frac{600\times 1000}{1000}\right) = 600 \text{ kN}$$

$$K = 0.0015/\text{m}$$

$$\mu = 0.55$$

a. *Loss due to friction*

$$\text{Loss of stress} = P_o(\mu\alpha + K\,.\,x)$$

$$= 1000\left(0.55\times\frac{1}{12} + 0.0015\times 10\right)$$

$$= 60 \text{ N/mm}^2$$

b. *Loss due to slip*

$$\text{Slip} = 3 \text{ mm} = \left(\frac{PL}{AE}\right)$$

$$\therefore \quad P = \left(\frac{3\times 600\times 210\times 10^3}{10\times 1000}\right)$$

$$= 37800 \text{ N} = 37.8 \text{ kN}$$

$$\therefore \quad \text{Loss of stress} = \left(\frac{37800}{600}\right) = 63 \text{ N/mm}^2$$

Loss of force due to friction $= (600 \times 60) = 36{,}000 \text{ N} = 36 \text{ kN}$

$\therefore$ Total loss of force $= (36 + 37.8) = 73.8 \text{ kN}$

$\therefore$ Final force $= (600 - 73.8) = 526.2 \text{ kN}$

$$\text{Percentage loss of stress} = \left(\frac{60+63}{1000}\right)\times 100 = 12.3\%$$

Problem 2.4

A post-tensioned concrete beam with a cable of 24 parallel wires (total area = 800 mm^2) is tensioned with 2 wires at a time. The cable with zero eccentricity at the ends and 150 mm at the centre is on a circular curve. The span of the beam is 10 m and the cross-section is 200 mm wide and 450 mm deep. The wires are to be stressed from one end to a value of f_1 to overcome frictional loss and then released to a value of f_2 so that immediately after anchoring, an initial prestress of 840 N/mm^2 would be obtained. Compute f_1, f_2 and the design stress in steel after all losses, given the following data:

Coefficient of friction for 'curvature' effect = 0.6
Friction coefficient for 'wave' effect = 0.003/m
Deformation and slip of anchorage = 1.25 mm
E_s = 210 kN/mm^2 E_c = 28 kN/mm^2
Shrinkage of concrete = 0.0002
Relaxation of steel stress = 3 per cent of the initial stress

Solution

The concrete beam with the curved post-tensioned cable is shown in Fig. 2.5.

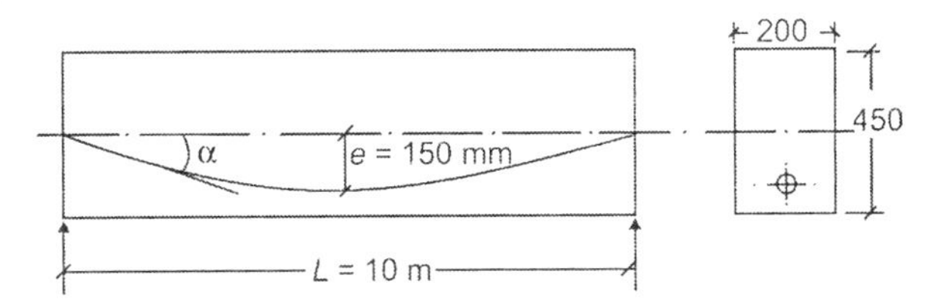

Fig. 2.5

$$\text{Total change of slope} = \left(\frac{2\times 4e}{L}\right) = \left(\frac{2\times 4\times 0.15}{10}\right) = 0.12$$

$$\text{Loss due to friction} = P_o(\mu\alpha + K.x)$$
$$= P_o(0.6 \times 0.12 + 0.003 \times 10) = 0.102\, P_o$$

$$\text{Loss due to slip} = 1.25 \text{ mm}$$

$$\therefore \quad \left(\frac{PL}{AE}\right) = 1.25$$

$$\therefore \quad P = \left(\frac{1.25\times 800\times 210\times 10^2}{10\times 1000}\right) = 21000 \text{ N}$$

$\therefore$ Loss due to slip $= \left(\dfrac{21000}{800}\right) = 26.2\ \text{N/mm}^2$

If the initial stress after anchoring $= 840\ \text{N/mm}^2$

Stress at far end $= (840 + 26.2)$

$$f_2 = 866.2\ \text{N/mm}^2$$

Stress at jacking end $= f_1 = (866.2 + 0.102 \times 866.2)$

$$= 954\ \text{N/mm}^2$$

Loss due to shrinkage $= (0.0002 \times 210 \times 10^3) = 42\ \text{N/mm}^2$

Loss due to relaxation $= \left(\dfrac{3}{100} \times 840\right) = 25.2\ \text{N/mm}^2$

Initial total force $= (840 \times 800) = 672 \times 10^3$ N

Stress in concrete at level of steel at centre of span

$$\sigma_c = \left[\frac{672 \times 10^3}{200 \times 450} + \frac{(672 \times 10^3 \times 150)150}{\dfrac{200 \times 450^3}{12}}\right]$$

$$\sigma_c = 17.41\ \text{N/mm}^2$$

$\therefore$ Loss of stress $= \alpha_e \cdot \sigma_c = (6 \times 17.41) = 104.4\ \text{N/mm}^2$

$\therefore$ Total loss $= (42 + 25.2 + 104.4) = 171.6\ \text{N/mm}^2$

$\therefore$ Final stress $= (840 - 171.6) = 668.4\ \text{N/mm}^2$

Problem 2.5

A pre-tensioned beam 250 mm wide and 300 mm deep is prestressed by 12 wires each 7 mm diameter initially stressed to 1200 N/mm^2 with their centroids located 100 mm from the soffit. Estimate the final percentage loss of stress due to elastic deformation, creep shrinkage and relaxation using IS: 1343-80 code with the following data:

Relaxation of steel stress $= 90\ \text{N/mm}^2$

$E_s = 210\ \text{kN/mm}^2$

$E_c = 35\ \text{kN/mm}^2$

Creep coefficient $(\phi) = 1.6$

Residual shrinkage strain $= 3 \times 10^{-4}$

Solution

The concrete beam eccentrically prestressed is shown in Fig. 2.6.

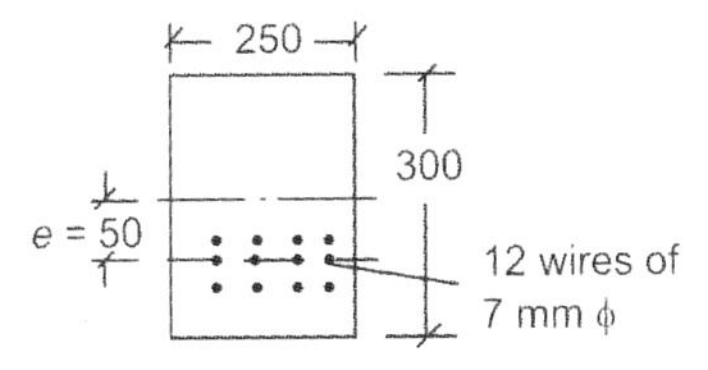

Fig. 2.6

Initial stress in steel $= 1200 \text{ N/mm}^2$

$$E_s = 210 \text{ kN/mm}^2$$

$$E_c = 5700\sqrt{45} = 38.2 \text{ N/mm}^2$$

$$\alpha_e = (E_s/E_c) = 5.49$$

$$P = \left(\frac{12 \times \pi \times 7^2}{4}\right) \times 1200 = 554176 \text{ N}$$

Stress in concrete at level of steel

$$\sigma_c = \left[\frac{554176}{250 \times 300} + \frac{554176 \times 50 \times 50}{\frac{350 \times 200^3}{12}}\right]$$

$$= 9.84 \text{ N/mm}^2$$

Loss due to elastic deformation

$$= \alpha_e \,.\, \sigma_c = (5.49 \times 9.84) = 54 \text{ N/mm}^2$$

Loss due to relaxation $= 5\% = \left(\frac{5}{100} \times 1200\right) = 60 \text{ N/mm}^2$

Loss due to shrinkage

$$= (300 \times 10^{-6})\,(210 \times 10^3) = 63 \text{ N/mm}^2$$

Loss of stress in steel due to creep of concrete

$$= (\phi \,.\, \sigma_c \,.\, \alpha_e) = (1.6)\,(9.84)\,(5.49)$$

$$= 86 \text{ N/mm}^2$$

$\therefore$ Total losses $= (60 + 63 + 54 + 86) = 263 \text{ N/mm}^2$

Percentage loss $= \left(\frac{263}{1200}\right) \times 100 = 22\%$

Problem 2.6

In a post-tensioned beam of length 12 m, a cable is laid symmetrically with its central 6 m length horizontal and the two straight end portions sloping up at an angle with the horizontal whose tangent is equal to 0.075. The cable is tensioned by jacking at one end and is anchored at the remote end of the beam. At the jacking end the measured stress is 1040 N/mm^2. The 'wobble' coefficient K may be assumed as 0.004/m. Calculate the stress in the cable at the remote end and the two points where the alignment of the cable changes. Assume the coefficient of friction between cable and duct of 0.40. What is the percentage loss of prestress between the jacking end and the anchored end.

Solution

The concrete beam post tensioned by the trapezoidal cable is shown in Fig. 2.7.

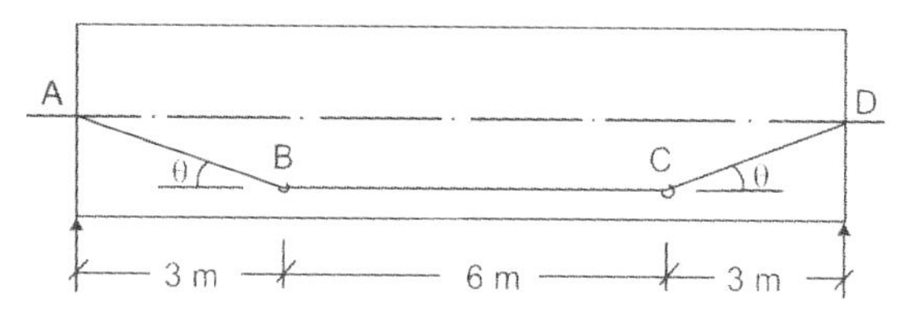

Fig. 2.7

Slope $= \theta = 0.075$

$K = 0.004$

Stress at jacking end $A = 1040$ N/mm^2

Loss of stress between A and $B = P_o(\mu\alpha + K\,.\,x)$

$= 1040(0.4 \times 0.075 + 0.004 \times 3)$

$= 44$ N/mm^2

$\therefore$ Stress at first kink $= (1040 - 44) = 996$ N/mm^2

Loss of stress between B and C $= 996\,(K\,.\,x)$

$= 996\,(0.004 \times 6) = 24$ N/mm^2

Stress at second kink $= (996 - 24) = 972$ N/mm^2

Loss of stress between C and $D = 44$ N/mm^2

Stress at anchored end D $= 972 - 44 = 928$ N/mm^2

Total loss $= (1040 - 928) = 112$ N/mm^2

Percentage loss $= \left(\dfrac{112}{1040}\right) \times 100 = 10.8\%$

Problem 2.7

A prestressed concrete beam 300 mm wide and 600 mm deep is prestressed with tendons of area 250 mm^2 located at a constant eccentricity of 100 mm and carrying an initial stress of 1050 N/mm^2. The span of the beam is 10.5 m.

Calculate the percentage loss of stress in tendons if

i. the beam is pre-tensioned and

ii. the beam is post-tensioned, using the following data

Modular ratio	= 6
Anchorage slip	= 1.5 mm
Friction coefficient for wave effect	= 0.0015/m

Ultimate creep strain = 40×10^{-6} and 20×10^{-6} mm/mm per N/mm^2 for pre-tensioned and post tensioned member.

Shrinkage of concrete = 300×10^{-6} for pre-tensioned and 200×10^{-6} for post tensioned member.

Relaxation of steel stress = 2.5 percent

Solution

$$P = (1050 \times 250) = 262500 \text{ N},$$

$$A = (300 \times 600) = 180000 \text{ mm}^2$$

$$A_s = 250 \text{ mm}^2; \quad f_s = 1050 \text{ N/mm}^2;$$

$$m = 6 \qquad \Delta = 1.5 \text{ mm};$$

$E_{cc} = 40 \times 10^{-6}$ and 20×10^{-6} for pre-tensioned and post-tensioned members respectively.

Relaxation loss = 2.5 percent of initial stress

$$\varepsilon_{sh} = 300 \times 10^{-6} \text{ for pre-tensioned and}$$

$$= 200 \times 10^{-6} \text{ for post-tensioned member}$$

$$E_s = 210 \text{ kN/mm}^2$$

$$I = \left(\frac{300 \times 600^3}{12}\right) = 54 \times 10^8 \text{ mm}^4$$

Stress in concrete at the level of steel is computed as

$$f_c = \left(\frac{262500}{180000}\right) + \left(\frac{262500 \times 100 \times 100}{(54 \times 10^4)}\right)$$

$$= 1.944 \text{ N/mm}^2$$

$$= 1.944 \text{ N/mm}^2 \text{ [Table 2.1 shows the losses]}$$

Table 2.1: Computation of losses of prestress

Sl. No.	Type of loss	Equation	Loss of stress in	
			Pre-tensioned beam (N/mm^2)	Post-tensioned beam (N/mm^2)
1.	Elastic deformation	$(m \cdot f_c)$	$(6 \times 1.944) = 11.664$	No loss of stress
2.	Relaxation of steel stress	2.5%	$\left(\frac{2.5}{100}\right)(1050) = 26.250$	$\left(\frac{2.5}{100}\right)(1050) = 26.250$
3.	Creep of concrete	$(E_{cc} \cdot f_c \cdot E_s)$	$(40 \times 10^{-6}) \times (1.944)\,(210 \times 10^3)$ $= 16.329$	$(20 \times 10^{-6}) \times (1.944)\,(210 \times 10^3)$ $= 8.164$
4.	Shrinkage of concrete	$(\varepsilon_{sc} \cdot E_s)$	$(300 \times 10^6) \times (210 \times 10^3) = 63.00$	$(200 \times 10^6) \times (210 \times 10^3) = 42.00$
5.	Friction loss	$(f_s \cdot K \cdot x)$	No loss of stress	$(1050 \times 0.0015 \times 10.5) = 16.53$
6.	Anchorage slip	$\left(\frac{E_s \Delta}{L}\right)$	No loss of stress	$\left\{\frac{(210 \times 10^3)(1.5)}{(10.5 \times 10^3)}\right\} = 30.00$
		Total loss of stress	$= 117.243$	$= 122.944$
		Percentage loss of stress	$= \left\{\frac{117.243 \times 100}{1050}\right\}$ $= 11.1$ percent	$\left\{\frac{122.944 \times 100}{1050}\right\}$ $= 11.7$ percent

Problem 2.8

A post-tensioned prestressed concrete beam of span length 10 m has a rectangular section 300 mm wide by 800 mm deep. The beam is prestressed by a parabolic cable concentric at the supports and with an eccentricity of 250 mm at the centre of span. The cross-sectional area of high tensile wires in the cable is 500 mm^2. The wires are stressed by using a jack at the left end so that the initial force in the cable at the right end is 250 kN. Using the following data

a. Calculate the total loss of stress in the wires

b. The jacking force required at the left end.

Coefficient of friction for 'curvature' effect	= 0.55
Friction coefficient for 'wave' effect	= 0.003/m
Anchorage slip at the jacking end	= 3 mm
Relaxation of steel stress	= 4 percent
Shrinkage of concrete	= 0.0002
Creep coefficient	= 2.2
Modulus of elasticity of steel	= 210 kN/m^2
Modulus of elasticity of concrete	= 35 kN/m^2
Modular ratio	$= \left(\frac{E_S}{E_C}\right) = 6$

Solution

$$\text{Cumulative angle between tangents} = 2\left(\frac{4e}{L}\right)$$

$$= 2\left(\frac{4 \times 0.250}{10}\right) = 0.2$$

$$P_x = P_o[1 - (\mu\alpha + K \,.\, x)]$$

$$250 = P_o[1 - (0.55 \times 0.2 + 0.003 \times 10)]$$

a. Solving $P_o = 290$ kN $\quad f_s = \dfrac{0.5(250 + 290)}{500} = 540 \text{ N/mm}^2$

b. Loss of stress

$$\text{Relaxation loss} \quad 4\% = \left(\frac{4}{100} \times 540\right) = 21.6 \text{ N/mm}^2$$

Stress in concrete at level of steel

$$= \left[\frac{(270 \times 10^3)}{(300 \times 800)}\right] + \left[\frac{270 \times 10^3 \times 250 \times 250}{1.28 \times 10^{10}}\right]$$

$$= 2.54 \text{ N/mm}^2$$

Loss of stress due to creep $= [2.2 \times 2.54 \times 6] = 33.5\ \text{N/mm}^2$

Loss due to shrinkage $= (0.0002 \times 210 \times 10^3) = 42.0\ \text{N/mm}^2$

$$\text{Loss due to slip of anchorage} = \frac{(210 \times 10^3)3}{(10 \times 10^3)} = 63\ \text{N/mm}^2$$

Loss due to friction $= (540)\ (0.14) = 75.6\ \text{N/mm}^2$

Total loss of stress in steel $= 235\ \text{N/mm}^2$

Problem 2.9

A rectangular beam 180 mm wide by 400 mm deep is simply supported over a span of 8 mm and is reinforced with 3 wires of 8 mm diameter. The wires are located at a constant eccentricity of 80 mm and are subjected to an initial stress of 1200 N/mm^2. Calculate the percentage loss of stress in the wires if the beam is (a) Pre-tensioned and (b) Post-tensioned. $E_s = 210$ kN/mm^2, modular ratio = 6, slip at anchorage = 0.8 mm, friction coefficients = 0.002/m, relaxation of steel stress = 6%. Adopt creep and shrinkage coefficients as per IS: 1343 code specifications.

(Bangalore University, 1993)

Solution

$$A = (180 \times 400) = 72000\ \text{mm}^2,$$
$$A_s = (3 \times 50) = 150\ \text{mm}^2$$
$$A_s = (150 \times 1200) = 180000\ \text{mm}^2$$
$$I = \frac{(180 \times 400^3)}{12} = 96 \times 10^7\ \text{mm}^4$$
$$E_s = 210\ \text{kN/mm}^2,\ m = 6 \quad \text{Silp} = \Delta = 0.8\ \text{mm}$$
$$\phi = 1.6$$

$$\text{Stress at level of steel} = \left[\left(\frac{180000}{72000}\right) + \frac{(180000 \times 80 \times 80)}{96 \times 10^7}\right]$$
$$= 3.7\ \text{N/mm}^2$$

Computation of losses of stress

Type of loss	Pre-tensioned beam	Post-tensioned beam
Elastic deformation	$(6 \times 3.7) = 22.2\ \text{N/mm}^2$	No loss
Relaxation loss	$(0.06 \times 1200) = 72.00$	'do' = 72.00
Creep of concrete	$(1.6 \times 3.7 \times 6) = 35.50$	'do' = 35.50
Shrinkage loss	$(210 \times 10^3 \times 300 \times 10^{-6})$ = 63.0	$(210 \times 10^3 \times 200 \times 10^{-6})$ = 42.0

Slip at anchorage	$\frac{(0.8 \times 210 \times 10^3)}{8000} = 21.0$	
Friction loss	$(1200 \times 0.002 \times 8) = 19.2$	
Total loss of stress	$= 192.7\ \text{N/mm}^2$	$= 189.70$
Percentage loss	$= \frac{(19.27 \times 100)}{1200}$	$= \frac{(189.7 \times 100)}{1200}$
of stress	$= 16.05\%$	$= 15.8\%$

Problem 2.10

A concrete beam of span 10 m is post-tensioned by a cable with an initial stress of 1200 N/mm^2 from one end. The cable is concentric at supports and has an eccentricity of 100 mm at the centre of span. Coefficient of friction due to curvature effect is 0.5 and friction coefficient for Wobble effect = 0.003/m. Compute the loss of stress in the tendons assuming the cable to have

a. Circular profile
b. Parabolic profile

Solution

a. *Circular profile cable*

Let R = Radius of curvature of cable

$$R^2 = (R - 0.1)^2 + 5^2$$

Solving $R = 125.05$ m

$$\text{Slope of cable at support} = \Theta = \left(\frac{5}{125.05}\right) = 0.03998 \simeq 0.04$$

Total change of slope from left to right end = $(2 \times 0.04) = 0.08$

Hence $\alpha = 0.08$ and $K = 0.003/\text{m}$

Loss of stress due to friction

$$= f_s[\mu\alpha + K \, . \, x]$$
$$= 1200[(0.50 \times 0.08) + (0.003 \times 10)]$$
$$= 84\ \text{N/mm}^2$$

b. *Parabolic cable profile*

Eccentricity of cable at centre of span = e = 100 mm

$$\text{Total change of slope} = \alpha = 2\,\Theta = \frac{(2 \times 4 \times 100)}{10} \times 10^3 = 0.08$$

Loss of stress due to friction

$$= f_s[\mu\alpha + K \,.\, x]$$
$$= 1200[(0.5 \times 0.08) + (0.003 \times 10)]$$
$$= 84 \text{ N/mm}^2$$

Hence, we can conclude that the loss of stress due to friction is the same for both circular and parabolic cable profile.

Problem 2.11

A cylindrical water tank, 50 m external diameter, is to be prestressed circumstantially by means of high tensile wires (E_s = 210 kN/mm^2) jacked at 4 points, 90° apart. If the minimum stress in the wires immediately after tensioning is to be 500 N/mm^2 and the coefficient of friction is 0.5, calculate

a. the maximum stress to be applied to the wires at the jack, and
b. the expected extension at the jack.

Solution

Let P_o = stress in wires at the jacking end

P_x = stress in wires at the farther end (90 degree apart)

$P_x = P_o e^{-\mu\alpha}$, where e = 2.7183 and

$$\alpha = 90° = \left(\frac{\pi}{2}\right)$$

$$500 = P_o(2.7183)^{-(0.5 \times 3.14/2)}$$

Solving P_o = 1102 N/mm^2

Average stress in wires $= 0.5(1102 + 500)$

$= 801 \text{ N/mm}^2$

Length of wires $= 0.25[3.14 \times 50 \times 1000]$

$= 39250$ mm

Extension at jack $= \dfrac{(801 \times 39250)}{(210 \times 10000)} = 149.7$ mm.

Problem 2.12

A prestressed concrete girder of 30 m span is prestressed by a parabolic cable concentric at supports and having an eccentricity of 850 mm at centre of span. The effective stress in tendons is 1200 N/mm^2. The coefficient of friction between tendons and cable duct is 0.5 and the friction coefficient for wave effect is 0.0015/m.

If the anchorage slip is 5 mm, compute the loss of stress in tendons due to friction and anchorage slip. Assump $E_s = 210$ kN/mm^2.

Solution

Span of girder = L = 30 m

Eccentricity at centre of span = e = 850 mm

$$\text{Total change of slope} = 2\left(\frac{4e}{L}\right) = \frac{2(4 \times 850)}{(30 \times 10^3)} = 0.226$$

Loss of stress due to friction is computed as follows:

$$\text{Friction loss} = f_s[\mu\alpha + K \,.\, x]$$
$$= 1200[(0.5 \times 0.226) + (0.0015 \times 30)]$$
$$= 190 \text{ N/mm}^2$$

$$\text{Anchorage slip loss} = \frac{(E_s \Delta)}{L}$$

$$= \frac{(210 \times 10^3 \times 5)}{(30 \times 10^3)} = 35 \text{ N/mm}^2$$

Total loss of stress in tendons due to friction and anchorage slip

$$= (190 + 35) = 225 \text{ N/mm}^2$$

$$\text{Percentage loss of stress} = \left(\frac{225 \times 100}{1200}\right) = 18.75 \text{ percent}$$

Problem 2.13

A prestressed concrete girder of 40 m span is prestressed by a parabolic cable concentric at supports and having an eccentricity of 1000 mm at centre of span. The stress in the wires is 1200 N/mm^2. Ultimate shrinkage strain = 0.0002, E_s = 210 kN/mm^2, coefficient of friction = 0.5, anchorage slip = 8 mm. Estimate the percentage loss of stress in the wires.

Solution

Given data: $E_s = 210$ kN/mm^2 $\quad \varepsilon_{sh} = 0.0002$

$\mu = 0.5 \quad f_s = 1200$ N/mm^2

$$\text{Change of slope of cable} = \left(\frac{8e}{L}\right) = \frac{(8 \times 1000)}{(40 \times 10^3)}$$

$$\alpha = 0.2$$

Loss of stress due to shrinkage $= \varepsilon_{sh} \,.\, E_s$

$= (0.0002 \times 210 \times 10^3)$

$= 42 \text{ N/mm}^2$

Loss of stress due to friction $= f_s[\mu\alpha]$

$= 1200[0.5 \times 0.2]$

$= 120 \text{ N/mm}^2$

Loss of stress due to anchorage slip $= \dfrac{(E_s\Delta)}{L} = \dfrac{(210 \times 10^3 \times 8)}{(40 \times 10^3)}$

$= 42 \text{ N/mm}^2$

Total loss of stress $= (42 + 120 + 42) = 204 \text{ N/mm}^2$

Percentage loss of stress $= \dfrac{(204 \times 100)}{1200} = 17$ percent

Problem 2.14

A prestressed concrete beam of rectangular section 300 mm wide and 600 mm deep is prestressed by 5 nos of 8 mm high tensile wires at 500 mm from top and 15 nos of 8 mm high tensile wires at 83 mm from the soffit. The initial stress in the wires is 1200 N/mm^2. Estimate the loss of stress in wires due to elastic shortening of concrete. Assume modular ratio = m = 6.

Solution

Let y = Distance of the centroid of the wires from the soffit. Taking moments about the base of the section, we have

$$y = \left[\frac{(5 \times 550) + (15 \times 83)}{(20)}\right] = 200 \text{ mm}$$

$$I = \left(\frac{bd^3}{12}\right) = \frac{(300 \times 600^3)}{12} = 54 \times 10^8 \text{ mm}^4$$

Eccentricity $= e = (300 - 200) = 100$ mm

Prestressing force $= P = (20 \times 50 \times 1200) = 12 \times 10^5$ N

Cross-sectional area $= A = (300 \times 600) = 18 \times 10^4 \text{ mm}^2$

Stress in concrete at the level of top wires

$$= f_c = \left[\left(\frac{12 \times 10^5}{18 \times 10^4}\right) - \left(\frac{12 \times 10^5 \times 100 \times 250}{54 \times 10^8}\right)\right]$$

$= 1.11 \text{ N/mm}^2$

Stress in concrete at the level of bottom wires

$$= f_c = \left[\left(\frac{12\times10^5}{18\times10^4}\right)-\left(\frac{12\times10^5\times100\times217}{54\times10^8}\right)\right]$$

$$= 11.48 \text{ N/mm}^2$$

Loss of stress in top wires $m \cdot f_c = (6 \times 1.11) = 6.66 \text{ N/mm}^2$

Loss of stress in top wires $m \cdot f_c = (6 \times 11.48) = 68.88 \text{ N/mm}^2$

Problem 2.15

A concrete beam is post-tensioned by a cable carrying an initial stress of 1200 N/mm^2. The slip at the jacking end is 8 mm. Modulus of elasticity of steel is 210 kN/mm^2. Estimate the percentage loss of stress in steel due to anchorage slip only if the length of the beam is (a) 10 m and (b) 50 m.

Solution

Given data:

$$E_s = 210 \text{ kN/mm}^2$$
$$f_s = 1200 \text{ N/mm}^2$$
$$\Delta = 8 \text{ m}$$
$$L = 10 \text{ m and } 50 \text{ m}$$

Loss of stress due to anchorage slip $= \dfrac{(E_S\Delta)}{L}$

a. For 10 m span beam, loss of stress due to anchorage slip is

$$= \frac{(210\times10^3)8}{(10\times10^3)} = 168 \text{ N/mm}^2$$

Percentage loss of stress $= \dfrac{(168\times100)}{1200} = 14$ percent

b. For 50 m span beam, loss of stress due to anchorage slip is

$$= \frac{(210\times10^3)8}{(50\times10^3)} = 33.6 \text{ N/mm}^2$$

Percentage loss of stress $= \dfrac{(33.6\times100)}{1200} = 2.8$ percent

In short span beams, the loss of stress due to anchorage slip is significantly higher than in log span beams.

Problem 2.16

A concrete beam having a rectangular cross-section, 100 mm wide by 300 mm deep is prestressed by 5 high tensile wires of 7 mm diameter located at an eccentricity of 50 mm. The initial stress in the wires is 1200 N/mm^2. Calculate the loss of stress in steel due to:

a. Elastic shortening of concrete
b. Creep of concrete using the ultimate creep strain method and the creep coefficient method (IS: 1343–2012)

   Adopt the following data:

$$E_c = 35 \text{ kN/mm}^2 \qquad E_S = 210 \text{ kN/mm}^2$$

$$A = (3 \times 10^4) \text{ mm}^2 \qquad \alpha_e = (E_S/E_C) = 6$$

   Ultimate creep strain = ε_{cc} = (41×10^{-6}) mm/mm per N/mm^2

   Prestressing force = $P = (5 \times 38.5 \times 1200) = (23 \times 10^4)$ N

   Second moment of area of section = $I = (225 \times 10^6)$ mm^4

   Creep coefficient = $\Phi = 1.6$

Solution

Stress in concrete at the level of steel is computed as,

$$f_c = \left[\frac{23\times10^4}{3\times10^4} + \frac{(23\times10^4\times50)\,50}{225\times10^6}\right] = 10.2 \text{ N/mm}^2$$

Modular ratio $= \alpha_e = (E_S/E_c) = 6$

a. Loss of stress in steel due to elastic shortening of concrete

$$= (\alpha_e f_c) = (6 \times 10.2) = 60.2 \text{ N/mm}^2$$

b. Loss of stress in steel due to creep of concrete

   1. Ultimate creep strain method

      Loss of stress in steel = $(\varepsilon_{cc} f_c E_S)$

      $= [(41 \times 10^{-6})(10.2)(210 \times 10^3) = 88$ N/mm^2

   2. Creep coefficient method

      Loss of stress in steel = $(\Phi f_c \alpha_e) = [1.6 \times 10.2 \times 6]$

      $= 97.92$ N/mm^2

Problem 2.17

A post-tensioned pre-stressed concrete beam of rectangular section 100 mm wide and 300 mm deep is stressed by a parabolic cable concentric at the supports and an eccentricity of 50 mm at the centre of span. The area of high tensile steel in cable is 200 mm^2 and initial stress in the cable is 1200 N/mm^2. If the ultimate creep strain is

30×10^{-6} mm/mm per N/mm^2 of stress and modulus of elasticity of steel is 210 kN/mm^2, compute the average loss of stress in steel due to creep of concrete.

Solution

Section properties:

$$A = 30000 \text{ mm}^2$$
$$P = (200 \times 1200) = 240000 \text{ N}$$
$$I = (225 \times 10^6) \text{ mm}^4$$
$$e = 50 \text{ mm at the centre of span and zero at supports}$$

Stress in concrete at the level of steel:

$$\text{At support section} = \left[\frac{240000}{30000}\right] = 8 \text{ N/mm}^2$$

At the centre of span section

$$= \left[\frac{240000}{30000} + \frac{(24 \times 10^4 \times 50)50}{225 \times 10^6}\right] = 10.7 \text{ N/mm}^2$$

Average stress at the level of steel

$$= f_c = [8 + (2/3) \times 2.7] = 9.8 \text{ N/mm}^2$$

∴ Loss of stress in the cable due to creep of concrete is

$$= [\varepsilon_{cc} f_c E_s] = [(30 \times 10^{-6})(9.8)(210 \times 10^3)]$$
$$= 62 \text{ N/mm}^2$$

Problem 2.18

A post-tensioned concrete beam of rectangular section, 100 mm wide by 300 mm deep and spanning 10 m is stressed by successive tensioning and anchoring of two cables numbered 1 and 2 respectively. The cross-sectional area of each cable is 200 mm^2 and the initial stress in the cable is 1200 N/mm^2. Modular ratio is 6. The first cable is parabolic with an eccentricity of 50 mm below the centroidal axis the centre of the span and 50 mm above the centroidal axis at support sections. The second cable is also parabolic with an eccentricity of 50 mm towards the soffit at the centre of span and concentric at support sections. Estimate the percentage loss of stress in each of these cables due to elastic deformation of concrete only if they are successively tensioned and anchored.

Solution

Area of cross-section

$$= A = (100 \times 300) = (3 \times 10^4) \text{ mm}^2$$

Second moment of area

$$= I = (225 \times 10^6)\ \text{mm}^4$$

Modular ratio $= \alpha_e = 6$

Prestressing force in each cable

$$= P = (200 \times 1200) = (240 \times 10^3)\ \text{N}$$

When cable 1 is tensioned and anchored, there will be no loss of stress due to elastic deformation of concrete. When cable 2 is tensioned and anchored, stress at the level of cable 1 is computed as,

$$\text{Stress at support section} = \left[\frac{240 \times 10^3}{3 \times 10^4}\right] = 8\ \text{N/mm}^2$$

Stress at the centre of span

$$= \left[\frac{240 \times 10^3}{3 \times 10^4}\right] + \left[\frac{(240 \times 10^3) \times 50 \times 50}{225 \times 10^4}\right]$$

$$= 10.7\ \text{N/mm}^2$$

The stress distribution varies parabolically from support to the centre of span.

Average stress in concrete $= f_c = [8 + (2/3)\ 2.7] = 9.8\ \text{N/mm}^2$

Hence, loss of stress in cable 1 when cable 2 is stressed and anchored

$$= (\alpha_e \times f_c) = (6 \times 9.8) = 58.8\ \text{N/mm}^2$$

$$\text{Percentage loss of stress} = \left[\frac{58.8 \times 100}{1200}\right] = 4.9\%.$$

Problem 2.19

A concrete beam is post-tensioned by a cable carrying an initial stress of 1200 N/mm^2. The slip at the jacking end was observed to be 6 mm. The modulus of elasticity of steel wires in the cable is 210 kN/mm. Compute the percentage loss of stress in steel due to the slip of anchorage if the length of the beam is (a) 30 m and (b) 3 m.

Solution

$$\text{Loss due to anchorage slip} = \left[\frac{E_s \Delta}{L}\right]$$

where E_s = Modulus of elasticity of steel = 210 kN/mm^2

L = Length of the beam = 30 m and 3 m

Δ = Slip at anchorage = 6 mm

f_s = Initial stress in the cable

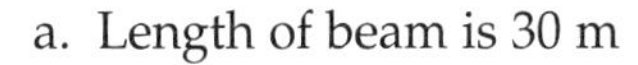

a. Length of beam is 30 m

Loss of stress due to anchorage slip $= \left[\dfrac{(210\times10^3)(6)}{30\times1000}\right]$

$= 42\ \text{N/mm}^2$

Percentage loss of stress $= \left[\dfrac{42\times100}{1200}\right] = 3.5\%$

b. Length of beam is 3 m

Loss of stress due to anchorage slip $= \left[\dfrac{(210\times10^3)(6)}{30\times1000}\right]$

$= 420\ \text{N/mm}^2$

Percentage loss of stress $= \left[\dfrac{420\times100}{1200}\right] = 35\%$

Short length beams suffer comparatively more losses due to anchorage slip.

Problem 2.20

A concrete beam *AB* of 20 m span is post-tensioned by a cable carrying a stress of 1000 N/mm^2 at the jacking end A. The cable is parabolic between the supports *A* and *B* and is concentric at the supports with an eccentricity of 400 mm at the centre of span. The coefficient of friction between duct and cable is 0.35 and friction coefficient for wave effect is 0.15 for 100 m. Calculate the stress in steel allowing for losses due to friction and wave effect at the following points:

a. Assuming the jacking end as *A*, compute the effective stress at *B*.

b. If the cable is tensioned from both ends *A* and *B*, calculate the minimum stress after losses in the cable and its location.

Solution

Span of beam $AB = L = 20$ m

Eccentricity at the centre of span $= e = 400$ mm

Coefficient of friction $= \mu = 0.35$

Friction coefficient for wave effect $= K = 0.15$ for 100 m or 0.0015/m

The cable is parabolic between the supports *A* and *B* with an eccentricity of 400 mm at the centre of span *C*.

Slope of cable at the end support A

$$= \left[\frac{4e}{L}\right] = \left[\frac{4 \times 400}{20 \times 1000}\right] = 0.08$$

Cumulative angle between the tangents at A and B

$= \alpha = (2 \times 0.08) = 0.16$ radians

Loss of stress due to friction $= P_o\,[\mu\alpha + Kx]$

where, x = distance from the jacking end to the point under consideration.

a. Loss of stress between A and B ($x = L = 20$ m)

$= 1000\,[(0.35 \times 0.16) + (0.0015 \times 20] = 59$ N/mm^2

Effective stress at $B = [1000 - 59] = 941$ N/mm^2

b. If the cable is tensioned simultaneously from both ends A and B, the minimum stress will occur at the centre of span C.

Cumulative angle between A and $C = \alpha = (2 \times 0.04) = 0.08$ radians.

Loss of stress between A and C ($x = 0.5\,L = 10$ m)

$= 1000\,[(0.35 \times 0.08) + (0.0015 \times 10] = 43$ N/mm^2

Effective stress at the centre of span $C = [1000 - 43]$

$= 953$ N/mm^2.

3 Deflections

Problem 3.1

A simply supported concrete beam of span 8 m and rectangular cross-section, 125 mm wide and 250 mm deep, is prestressed by a single cable in which the total tensile force is 220 kN. The centre line of the cable is parallel to the axis of the beam and 75 mm above the soffit over the middle third of the span and is curved upward in a parabola over the outer third of the span to a distance of 175 mm above the soffit at the supports. If the modulus of elasticity of concrete is 35 kN/mm^2 and the density of concrete is 24 kN/m^3, calculate:

a. The upward deflection at mid-span due to prestress only;
b. The deflection when the beam is supporting its own weight;
c. The magnitude of concentrated loads Q placed at the third point of the span, which would result in a limiting short term deflection of span/500.

Solution

The pre-tensioned beam of rectangular section is shown in Fig. 3.1.

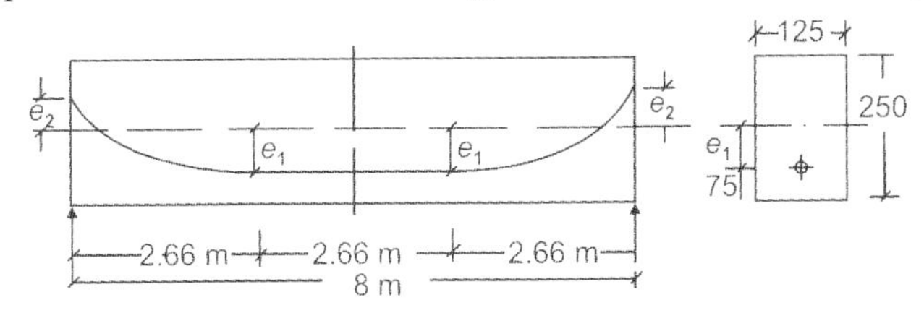

Fig. 3.1

$$e_1 = (125 - 75) = 50 \text{ mm}$$
$$e_2 = 50 \text{ mm}$$
$$E_c = 35 \text{ kN/mm}^2$$

$$P = 220 \text{ kN}$$

$$I = \left(\frac{125 \times 250^3}{12}\right) = 1.62 \times 10^8 \text{ mm}^4$$

a. *Maximum central deflection due to prestress*

$$a_p = \frac{-P(e_1 + e_2)}{12\,EI}\,[5L_1^{\,2} + 12L_1L_2 + 6L_2^{\,2}] + \frac{Pe_2L^2}{8\,EI}$$

Here $L_1 = 2.66$ m

$L_2 = 1.33$ m

$L = 8$ m

$$\therefore \quad a_p = \left[\frac{-220(50+50)}{12 \times 35 \times 1.62 \times 10^8}\right](5) \times (2.66 \times 10^3)^2$$

$$+ (12 \times 2.66 \times 10^3) \times (1.33 \times 10^3)$$

$$+ 6(1.33 \times 10^3)^2 + \left(\frac{220 \times 50 \times 8000^2}{8 \times 35 \times 1.62 \times 10^8}\right)$$

$$= -1.65 \text{ mm (upwards)}$$

b. *Deflection due to (self weight + prestress)*

Self weight of beam $= g = (0.125 \times 0.25 \times 24)$

$= 0.75$ kN/m

$= 0.00075$ kN/mm

$$\therefore \quad a_g = \left(\frac{5gL^4}{384\,EI}\right) = \left(\frac{5 \times 0.00075 \times 8000^4}{384 \times 35 \times 1.62 \times 10^8}\right)$$

$= 7.05$ mm (downwards)

$\therefore$ Net deflection $= (-1.65 + 7.05)$

$= 5.4$ mm (downwards)

The beam supporting concentrated loads of magnitude Q at one third span points is shown in Fig. 3.2.

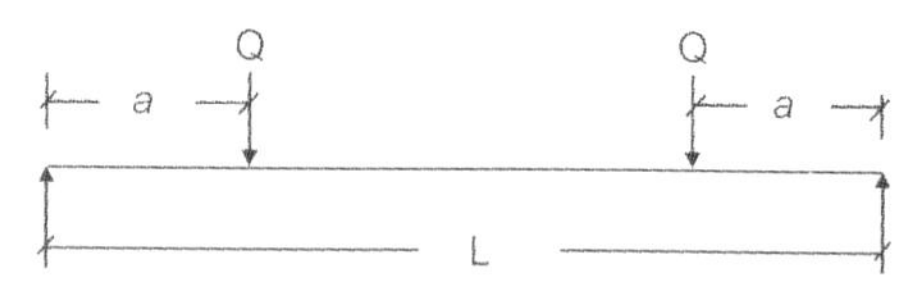

Fig. 3.2

c. *Deflection at centre of span*

$$= \frac{Qa}{24\,EI}\,[3L^2 - 4a^2]$$

Limiting deflection

$$= \left(\frac{\text{span}}{500}\right) = \left(\frac{8000}{500}\right) = 16 \text{ mm}$$

$$\therefore (16 - 5.4) = \frac{Q(2.66 \times 10^3)}{24 \times 35 \times 1.62 \times 10^8}[3(8000)^2 - 4(2.66 \times 1000)^2]$$

Solving $Q = 3.31$ kN

Problem 3.2

A concrete beam with a rectangular section, 100 mm wide and 300 mm deep, is stressed by 3 cables, each carrying an effective force of 240 kN. The span of the beam is 10 m. The first cable is parabolic with an eccentricity of 50 mm below the centroidal axis at the centre of span and 50 mm above the centroidal axis at the supports and an eccentricity of 50 mm at the centre of span. The third cable is straight with a uniform eccentricity of 50 mm below the centroidal axis.

If the beam supports a uniformly distributed live load of 5 kN/m and $E_c = 38$ kN/mm^2, estimate the instantaneous deflection at the following stages:

a. Prestress + self weight of beam, and
b. Prestress + self weight + live load

Solution

The concrete beam of rectangular section prestressed by three cables is shown in Fig. 3.3.

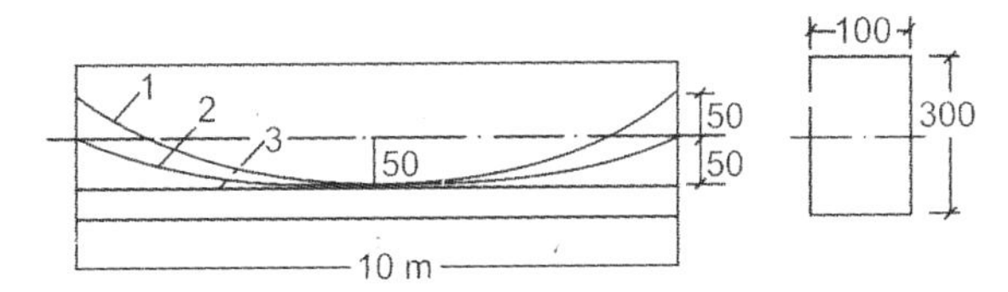

Fig. 3.3

$$P = 240 \text{ kN (3 cables 1, 2 and 3)}$$

$$e_1 = e_2 = 50 \text{ mm}$$

$$I = \left(\frac{100 \times 300^3}{12}\right) = 2.25 \times 10^8 \text{ mm}^4$$

Self weight of beam $= (0.1 \times 0.3 \times 24) = 0.72$ kN/m

$= 0.00072$ kN/mm

Deflection due to cable 1, 2 and 3

$$a_1 = \frac{PL^2}{48\,EI}[-5e_1 + e_2]$$

$$= \frac{240 \times (10 \times 1000)^2}{48 \times 38 \times 2.25 \times 10^8}[-5 \times 50 + 50]$$

$$= -15.8 \text{ mm (upwards)}$$

$$a_2 = -\left(\frac{5PeL^2}{48\,EI}\right) = -\left[\frac{5 \times 240 \times 50 \times (10 \times 1000)^2}{48 \times 38 \times 2.25 \times 10^8}\right]$$

$$= -14.6 \text{ mm}$$

$$a_3 = -\left(\frac{PeL^2}{8\,EI}\right) = -\left[\frac{240 \times 50 \times (10 \times 1000)^2}{8 \times 38 \times 2.25 \times 10^8}\right]$$

$$= -17.5 \text{ mm}$$

Total upward deflection = –[15.8 + 14.6 + 17.5] = –47.9 mm

Deflection due to self weight

$$a_g = \left[\frac{5 \times 0.00072 \times (10 \times 1000)^4}{384 \times 38 \times 2.25 \times 10^8}\right] = 10.96 \text{ mm}$$

Deflection due to live load

$$a_g = \left(\frac{10.96}{0.00072} \times 0.005\right) = 76.1 \text{ mm}$$

a. Prestress + self weight

$$a_R = (-47.9 + 10.96) = -36.94 \text{ (upwards)}$$

b. Prestress + self weight + live load

$$a_R = (-47.9 + 10.96 + 76.1) = 39.16 \text{ (downwards)}$$

Problem 3.3

A prestressed concrete beam spanning over 8 m is of rectangular section, 150 mm wide and 300 mm deep. The beam is prestressed by a parabolic cable having an eccentricity of 75 mm below the centroidal axis at the centre of span and an eccentricity of 25 mm above the centroidal axis at the support sections. The initial force in the cable is 350 kN. The beam supports 3 concentrated loads of 10 kN each at intervals of 2 m.

$$E_c = 38 \text{ kN/mm}^2$$

a. Neglecting losses of prestress, estimate the short term deflection due to (Prestress + self weight) and
b. Allowing for 20 percent loss in prestress, estimate the long term deflection under (prestress + self weight + live load) assuming creep coefficient as 1.80.

Solution

The concrete beam of rectangular section prestressed by a parabolic cable is shown in Fig. 3.4.

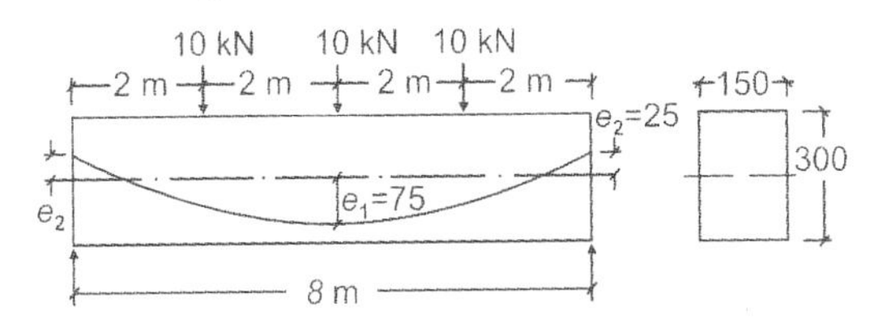

Fig. 3.4

$$P = 350 \text{ kN}$$

$$E_c = 38 \text{ kN/mm}^2$$

$$I = \left(\frac{150 \times 300^3}{12}\right) = 3.375 \times 10^8 \text{ mm}^4$$

Deflection due to prestressing force is

$$a_p = \frac{PL^2}{48\,EI}[-5e_1 + e_2]$$

$$= \frac{350 \times 8000^2}{48 \times 38 \times 3.375 \times 10^8}[-5 \times 75 + 25]$$

$$= -12.7 \text{ mm (upwards)}$$

Self weight of beam

$$= (0.15 \times 0.3 \times 24) = 1.08 \text{ kN/m}$$

$$= 0.00108 \text{ kN/mm}$$

$$a_g = \left(\frac{5 \times 0.00108 \times 8000^4}{384 \times 38 \times 3.375 \times 10^8}\right) = 4.49 \text{ mm}$$

Live load deflections

$$\text{Central load } a_q = \left(\frac{10 \times 8000^3}{48 \times 38 \times 3.375 \times 10^8}\right) = 8.3 \text{ mm}$$

Due to two concentrated loads at

$$a = 2 \text{ m}$$

$$a_q = \frac{Wa}{24\,EI}[3L^2 - 4a^2]$$

$$= \left(\frac{10 \times 2000}{24 \times 38 \times 3.375 \times 10^8}\right)[3 \times 8000^2 - 4(2000)^2]$$

$$= 11.4 \text{ mm}$$

a. Prestress + self weight

$$a_R = (-12.7 + 4.49) = -8.21 \text{ mm (upwards)}$$

b. Prestress + self weight + live load

$$a_R = [-(0.8 \times 12.7) + 4.49 + 8.3 + 11.4]\,(1 + 1.8)$$

$$= 39.28 \text{ mm (downwards)}$$

Problem 3.4

A prestressed beam of rectangular section, 100 mm wide and 200 mm deep, has a straight duct 25 mm by 40 mm with its centre located at 50 mm from the soffit of the beam which is prestressed by 12 wires of 7 mm diameter stressed to 600 N/mm^2. The beam supports an imposed load of 4 kN/m over a span of 6 m. The modulus of elasticity of concrete is 38 kN/mm^2. Estimate the central deflection of the beam under the action of prestress, self weight and live load.

a. Based on net section (beam ungrouted); and
b. Based on transformed section (beam grouted)

Solution

The concrete beam with the cable duct is shown in Fig. 3.5.

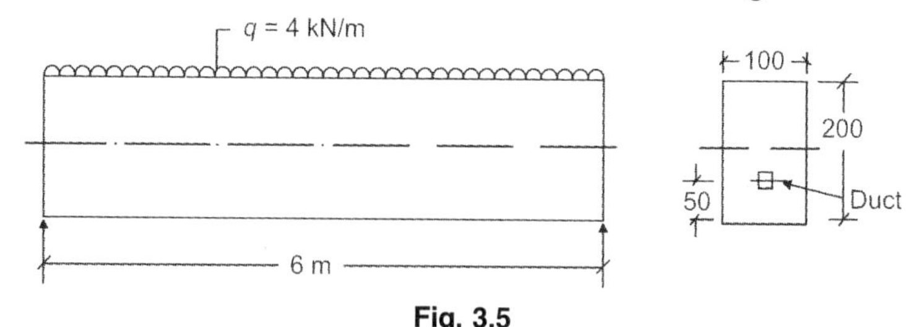

Fig. 3.5

$$P = \left[\frac{12 \times (\pi \times 7^2) \times 600}{4 \times 1000}\right] = 277 \text{ kN}$$

Self weight $= g = (0.1 \times 0.2 \times 24) = 0.48$ kN/m

The cross-section of the beam with the centroidal axis is shown in Fig. 3.6.

$$y = \left[\frac{(1000\times 200\times 100)-(25\times 40\times 150)}{19000}\right]$$

$$= 97 \text{ mm}$$

$$I_{xx} = \left(\frac{100\times 97^3}{3}\right)+\left(\frac{100\times 103^3}{3}\right) - \left[\frac{25\times 40^3}{12}+25\times 40\times 53^2\right]$$

$$= 63.87 \times 10^6 \text{ mm}^4$$

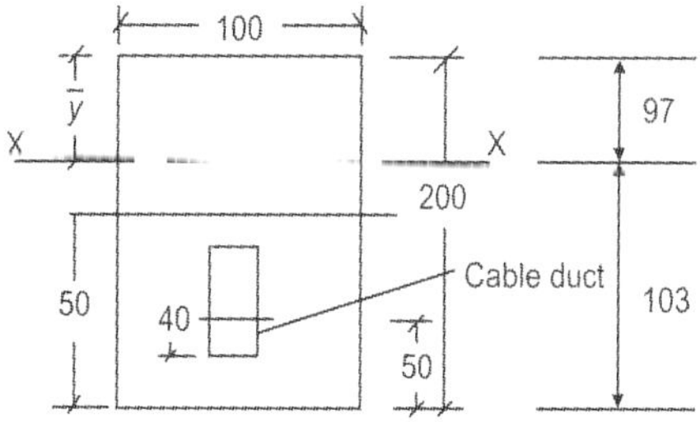

Fig. 3.6

Transformed section

$$= \left[\frac{100\times 200^3}{12}\right] + (6\times 12\times 38.4\times 50^2)$$

$$= 73.5 \times 10^6 \text{ mm}^4$$

$$E_c = 38 \text{ kN/mm}^2$$

Self weight of beam

$$= (0.1 \times 0.2 \times 24) = 0.48 \text{ kN/m}$$

$$= 0.00048 \text{ kN/mm}$$

Live load on beam

$$= 4 \text{ kN/m} = 0.004 \text{ kN/mm}$$

a. *Based on ungrouted section*

$$\text{Deflection due to prestress} = \left(\frac{PeL^2}{8\,EI}\right)$$

$$= \left(\frac{277\times 50\times 6000^2}{8\times 38\times 63.87\times 10^6}\right) = 25.6 \text{ mm (upwards)}$$

Deflection due to $(g + q)$

$$= \left(\frac{5 \times 0.00448 \times 6000^4}{384 \times 38 \times 63.87 \times 10^6}\right) = 31.1 \text{ mm}$$

$\therefore$ Net deflection = (31.1 – 25.6) = 5.5 mm

b. *Based on transformed section*

Deflection due to prestress

$$= \left(\frac{25.6 \times 63.87}{73.5}\right) = 22.2 \text{ mm}$$

Deflection due to dead and live load

$$= \left(\frac{31.1 \times 63.87}{73.5}\right) = 27.0$$

$\therefore$ Net deflection = (27 – 22.2) = 4.8 mm (downwards)

Problem 3.5

A pre-tensioned concrete beam with a cross-section, 120 mm wide and 300 mm deep is used to support a uniformly distributed live load of 3 kN/m over an effective span of 6 m. The beam is prestressed by a straight cable carrying an effective prestressing force of 180 kN at a constant eccentricity of 50 mm. Given E_c = 38 kN/mm^2, the modulus of rupture = 5 N/mm^2, area of the cable = 200 mm^2 and modular ratio = 6, estimate the deflection of the beam at the following stages:

a. Working load
b. Cracking load
c. 1.5 times the cracking load

Solution

The loaded prestressed concrete beam with its rectangular section is shown in Fig. 3.7.

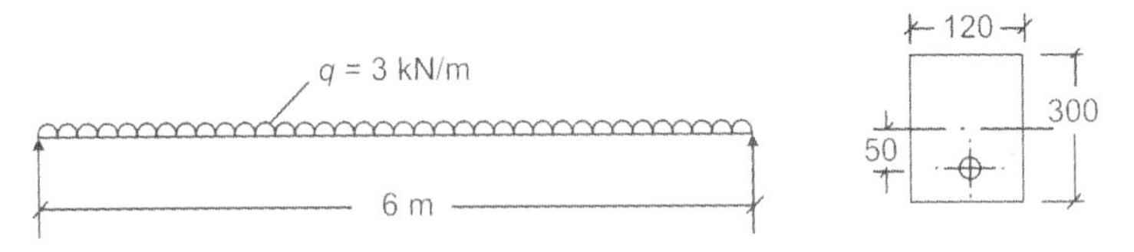

Fig. 3.7

P = 180 kN I = 27 × 10^7 mm^4

e = 50 mm Z = 18 × 10^5 mm^3

E_c = 38 kN/mm^2 Modulus of rupture = 5 N/mm^2

α_e = 6

Self weight of beam g = (0.12 × 0.3 × 24) = 0.864 kN/m

Working load q = 3.000

$(g + q)$ = 3.864

$$a_p = \left(\frac{PeL^2}{8\,EI}\right) = \left(\frac{180 \times 50 \times 6000^2}{8 \times 38 \times 27 \times 10^7}\right) = 3.9 \text{ mm}$$

$$a_{(g+q)} = \left(\frac{5 \times (0.003864)6000^4}{384 \times 38 \times 27 \times 10^7}\right) = 6.3 \text{ mm}$$

a. *At working load*

$$a_R = (6.3 - 3.9) = 2.4 \text{ mm (downwards)}$$

b. *Cracking load*

$$\text{Working moment} = \left(\frac{3.864 \times 6^2}{8}\right) = 17.388 \text{ kN.m}$$

Stress at bottom fibre due to prestress

$$= \left(\frac{180 \times 10^3}{120 \times 300}\right) + \left(\frac{180 \times 10^3 \times 50}{18 \times 10^5}\right) = 10 \text{ N/mm}^2$$

Stress at bottom due to working moment

$$= \left(\frac{17.388 \times 10^6}{18 \times 10^5}\right) = -9.66 \text{ N/mm}^2$$

Resultant stress at bottom fibre = (10 – 9.66) = 0.34 N/mm^2

Extra moment required to cause cracking

$$= \frac{(5 + 0.34)\,18 \times 10^5}{10^6} = 9.6 \text{ kN.m}$$

∴ Cracking moment

$$= (17.388 + 9.6) = 27 \text{ kN.m}$$

$$\therefore \text{ Cracking load} = \left(\frac{8 \times 27}{6^2}\right) = 6 \text{ kN/m}$$

∴ Deflection due to cracking load

$$= \left(\frac{6.3}{3.864} \times 6\right) = 9.78 \text{ mm}$$

∴ Resultant deflection = (9.78 – 3.9) = 5.88 mm

c. *At 1.5 times cracking load*

The cross section of the beam with the eccentric cable and neutral axis is shown in Fig. 3.8. If x = depth of neutral axis

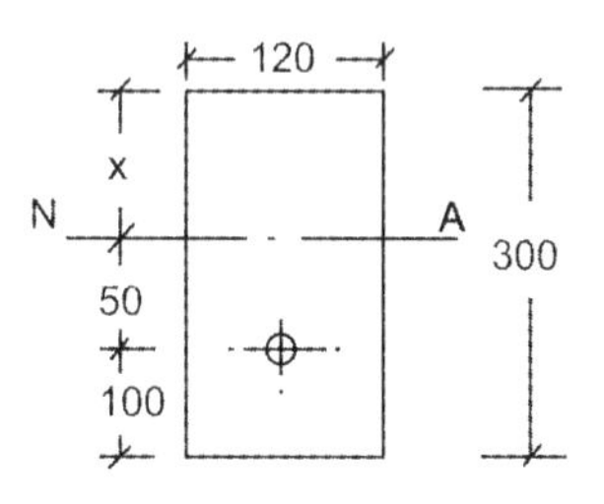

Fig. 3.8

$$\frac{120 \cdot x^2}{2} = 6 \times 200(200 - x)$$

$\therefore$ $x = 54$ mm

M.I. of cracked transformed section

$$I_r = \left[\frac{bx^3}{3}\right] + \alpha_e \,.\, A_s \,.\, r^2$$

$$= \left[\frac{120 \times 54^3}{3} + 6 \times 200 \times 146^2\right]$$

$$= 31.8 \times 10^6 \text{ mm}^4$$

Load at 1.5 times cracking load

$$= (6 + 3) = 9 \text{ kN/m}$$

Cracking moment

$$= \left(\frac{9 \times 6^2}{8}\right) = 40.5 \text{ kN.m}$$

$$\text{Deflection} = \frac{5L^2}{48}\left[\frac{M_{cr}}{E_c I_c} + \frac{(M - M_{cr})}{0.85\, E_c I_r}\right]$$

$$= \frac{5 \times 6000^2}{48}\left[\frac{27 \times 10^3}{38 \times 27 \times 10^7} + \frac{(40.5 - 27)10^3}{0.85 \times 38 \times 31.8 \times 10^6}\right]$$

$$= 59.13 \text{ mm (downwards)}$$

Resultant deflection = (59.13 – 3.9) = 55.2 mm (downwards)

Problem 3.6

A concrete beam with a section, 90 mm wide and 180 mm deep is prestressed by two wires of 7 mm diameter initially stressed to 920 N/mm^2. The wires are located in a parabolic profile with an eccentricity of 36.8 mm at the centre span (3 m) and concentric at the supports. The beam supports two concentrated live loads of 7 kN each spaced 1 m apart. The modulus of elasticity of concrete is 30.9 kN/mm^2. Compute the initial deflection of the beam at the centre of span under (prestress + self weight) and the final deflection, including live loads, assuming 15 percent loss in prestress due to various causes. Compare these deflections with the limits prescribed in the IS: 1343. Assume $\phi = 1.6$

Solution

The concrete beam prestressed by a parabolic cable and supporting concentrated loads at one third points is shown in Fig. 3.9.

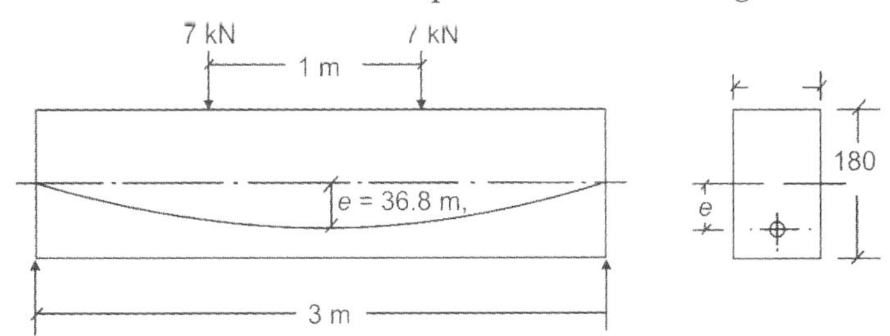

Fig. 3.9

$$E_c = 30.9 \text{ kN/mm}^2$$

$$I = 43.74 \times 10^6 \text{ mm}^4$$

$$P = (2 \times 38.4 \times 920) = 70811 \text{ N} = 70.8 \text{ kN}$$

$$a_p = \left(\frac{5\,PeL^2}{48\,EI}\right) = \left(\frac{5 \times 70.8 \times 36.8 \times 3000^2}{48 \times 30.9 \times 43.74 \times 10^6}\right)$$

$$= 1.8 \text{ mm (upwards)}$$

Self weight of beam

$$g = (0.09 \times 0.18 \times 24) = 0.3888 \text{ kN/m}$$

$$= 0.000388 \text{ kN/mm}$$

$$a_g = \left(\frac{5 \times 0.000388 \times 3000^2}{384 \times 30.9 \times 43.74 \times 10^6}\right)$$

$$= 0.06 \text{ mm (downwards)}$$

∴ Deflection due to prestress + self weight

$$= (1.8 - 0.06) = 1.74 \text{ mm (upwards)}$$

Deflection due to live load

$$a_q = \frac{Wa}{24\,EI}\,[3L^2 - 4a^2]$$

$$= \left(\frac{7 \times 1000}{24 \times 30.9 \times 43.74 \times 10^6}\right)[3 \times 3000^2 - 4 \times 1000^2]$$

$$= 4.96 \text{ mm (downwards)}$$

Final deflection (Prestress + self weight + live load)

$$= [-(0.85 \times 1.8) + 0.06 + 4.96]\,(1 + 1.6)$$

$$= 9.074 \text{ mm (downwards)}$$

According to IS; 1343

Maximum permissible limiting deflection

$$= \left(\frac{\text{Span}}{250}\right) = \left(\frac{3000}{250}\right) = 12 \text{ mm} > 9.074 \text{ mm.}$$

(Hence safe)

Problem 3.7

A concrete beam having a rectangular section, 150 mm wide by 300 mm deep is prestressed by a parabolic cable having an eccentricity of 75 mm at centre of span towards the soffit and an eccentricity of 25 mm towards the top at support sections. The effective force in the cable is 350 kN. The beam supports a concentrated load of 20 kN at the centres of span in addition to the self weight. If the modulus of elasticity of concrete is 38 kN/mm^2 and span is 8 m; calculate,

a. Short term deflection at centre of span under prestress, self weight and live load.
b. Long term deflection assuming the loss ratio as 0.8 and creep coefficient as 1.6.

Solution

Given data:

$P = 35$ kN,	$e_1 = 75$ mm
$e_2 = 25$ mm	$L = 8$ m
$E_c = 38$ kN/mm^2	$\phi = 1.6$
$\eta = 0.8$	$b = 150$ mm
$d = 300$ mm,	$Q = 20$ kN

$$I = \frac{(bd^3)}{12} = \frac{(150 \times 300^3)}{12}$$

$$= 3.375 \times 10^8 \text{ mm}^4$$

Self weight of beam = g = $(0.15 \times 0.3 \times 24) = 1.08$ kN/m

$= 0.00108$ kN/mm

Deflection due to self weight

$$= a_g = \frac{(5gL^4)}{(384\ EI)} = \frac{(5 \times 0.00108 \times 8000^4)}{(384 \times 38 \times 3.375 \times 10^8)}$$

$= 4.49$ mm (downwards)

Deflection due to prestress

$$= a_p = \frac{PL^2(-5e_1 + e_2)}{(48\ EI)}$$

$$= \frac{(350 \times 8000^2)(-5 \times 75 + 25)}{(48 \times 38 \times 3.375 \times 10^8)}$$

$= -12.7$ mm (upwards)

Deflection due to concentrated load

$$= a_q = \frac{(QL^3)}{(48\ EI)} = \frac{(20 \times 8000^3)}{(48 \times 38 \times 3.375 \times 10^8)}$$

$= 16.6$ mm

Short term deflection $= a_{RS} = [-12.7 + 4.49 + 16.6]$

$= 9.39$ mm (downwards)

Long term deflection $= a_{RL} = [0.8(-12.7) + 4.49 + 16.6](1 + 1.6)$

$= 28.41$ mm (downwards)

Problem 3.8

A concrete beam having a rectangular section, 250 mm wide by 500 mm deep is prestressed by a parabolic cable carrying an effective prestressing force of 250 kN. The cable has an eccentricity of 75 mm at centre of span and is concentric at ends. The span of the beam is 9.5 m and it is subjected to a live load of 2.5 kN/m. Estimate the short term and long term deflections of the beam at the centre of span assuming that the dead load are applied simultaneously after the release of prestress. Take $E_c = 40$ kN/mm^2, creer coefficient = 2.0 and loss of prestress = 18%.

Solution

Data: $P = 250$ kN, $e = 75$ mm (parabolic cable)

$q = 2.5$ kN/m, $E_c = 40$ kN/mm^2

$\phi = 2.0$

Loss ratio $= \eta = 0.82$ (18 per cent loss)

$$I = \frac{(bd^3)}{12} = \frac{(250 \times 500^3)}{12} = 2604 \times 10^6 \text{ mm}^4$$

$$g = (0.25 \times 0.5 \times 24) = 3 \text{ kN/m} = 0.003 \text{ kN/mm}$$

$$q = 2.5 \text{ kN/m} = 0.0025 \text{ kN/mm}$$

$$a_p = \left[\frac{5\,PeL^2}{48\,EI}\right] = -\left[\frac{5 \times 250 \times 75 \times (9.5 \times 10^3)^2}{48 \times 40 \times 2604 \times 10^6}\right]$$

$$= -1.692 \text{ mm (upwards)}$$

$$a_{(g+q)} = \left[\frac{5\,(g+q)L^4}{384\,EI}\right] = \left[\frac{5 \times 0.0055 \times (9.5 \times 10^3)^4}{384 \times 40 \times 2604 \times 10^6}\right]$$

$$= 5.6 \text{ mm (downwards)}$$

Short term deflection assuming that dead and live loads are applied simultaneously is given by

$$a_{RS} = \left[a_p + a_{(g+q)}\right] = [-1.692 + 5.6]$$

$$= 3.908 \text{ mm (downwards)}$$

Long term deflection is computed as

$$a_{RL} = [1+\phi]\left[\eta a_p + a_{(g+q)}\right]$$

$$= [1 + 2][0.82 \times (-1.692) + 5.6]$$

$$= 12.639 \text{ mm (downwards)}$$

Problem 3.9

A concrete beam rectangular in section, 150 mm wide by 300 mm deep is prestressed by a cable which is parabolic in shape with an eccentricity of 75 mm towards the soffit at centre of span and 25 mm towards the top at supports with an effective force of 350 kN. Assuming the modulus of elasticity of concrete as 38 kN/mm^2 and creep coefficient as 1.6 and loss ratio is 0.8, compute, (a) short term deflection at centre of span under (prestress + self weight + live load of 20 kN at centre of span of 8 m) (b) long term deflection.

Solution

Data:

$$P = 350 \text{ kN}, \quad e_1 = 75 \text{ mm},$$
$$e_2 = 25 \text{ mm}, \quad \phi = 1.6$$
$$E = 38 \text{ kN/mm}^2, \quad L = 8 \text{ m}$$
$$Q = 20 \text{ kN} \quad \eta = 0.8$$

$$I = \frac{(bd^3)}{12} = \frac{(150 \times 300^3)}{12} = 3375 \times 10^5 \text{ mm}^4$$

$$a_p = \frac{P.L^2}{48\, EI}[-5e_1 + e_2]$$
$$= \left[\frac{350 \times 8000^2}{48 \times 38 \times 3375 \times 10^5}\right][(-5 \times 75) + 25]$$
$$= 12.7 \text{ mm (upwards)}$$

$$a_g = \left(\frac{5gL^4}{384\, EI}\right) = \left(\frac{5 \times 0.00108 \times 8000^4}{384 \times 38 \times 3375 \times 10^5}\right)$$
$$= 4.49 \text{ mm (downwards)}$$

$$a_Q = \left(\frac{qL^3}{48\, EI}\right) = \left(\frac{20 \times 8000^3}{48 \times 38 \times 3375 \times 10^5}\right)$$
$$= 16.6 \text{ mm (downwards)}$$

a. *Short term deflection is computed as*

$$a_{RS} = \left[a_p + a_g + a_Q\right] = [-12.7 + 4.49 + 16.6]$$
$$= 9.39 \text{ mm (downwards)}$$

b. *Long term deflection is computed as*

$$a_{RL} = [1 + \phi]\left[\eta a_p + a_g + a_Q\right]$$
$$= [1 + 1.6]\,[(0.8)(-12.7) + 4.49 + 16.6]$$
$$= 28.41 \text{ mm (downwards)}$$

Problem 3.10

A concrete girder of unsymmetrical I-section used for a bridge spans over 30 m and its self weight is 10.8 kN/m. The girder is prestressed by a parabolic cable having an eccentricity of 580 mm at centre of span and 170 mm at supports towards the soffit of the girder. The initial force in the cable is 3200 kN. If loss ratio is 0.85 and the creep coefficient is 1.6, modulus of elasticity of concrete is 34 kN/mm^2, estimate the long term deflection of the bridge girder and compare it with the permissible deflection as per IS: 1343 code specifications.

Assume second moment of area as 72490×10^6 mm^4 and live load is 9 kN/m.

Solution

Data: $P = 3200$ kN, $e_1 = 580$ mm,
$e_2 = 170$ mm $E = 34$ kN/mm^2
$I = 72490 \times 10^6$ mm^4 $\eta = 0.85$,
$\phi = 1.6$ $L = 30$ m

Maximum deflection at centre of span due to prestressing force is

$$a_p = -\left[\frac{PL^2}{48\,EI}(5e_1 + e_2)\right]$$

$$= -\left[\frac{3200\times10^3\times30^2\times10^6}{48\times34\times10^3\times72490\times10^6}\right](5 \times 580 + 170)$$

$$= -74.7 \text{ mm (upwards)}$$

Maximum deflection due to dead load of girder is

$$a_g = \left(\frac{5gL^4}{384\,EI}\right) = \left(\frac{5\times10.8\times(30\times1000)^4}{384\times34\times10^3\times72490\times10^6}\right)$$

$$= 46 \text{ mm (downwards)}$$

Maximum deflection due to live load is

$$a_q = \left(\frac{5qL^4}{384\,EI}\right) = \left(\frac{5\times9\times(30\times1000)^4}{384\times34\times10^3\times72490\times10^6}\right)$$

$$= 38.5 \text{ mm (downwards)}$$

Long term deflection including effects of creep and loss of prestress is

$$a_{RL} = [1+\phi]\left[-(\eta\times a_p)+a_g+a_q\right]$$

$$= [1 + 1.6][-(0.85 \times 74.7) + 38.5]$$

$$= 54.61 \text{ mm (downwards)}$$

Maximum permissible long term deflection as per IS: 1343 is

$$= \left(\frac{\text{Span}}{250}\right) = \left(\frac{30\times10^3}{250}\right) = 120 \text{ mm}$$

Problem 3.11

A simply supported concrete beam of span 8 m and rectangular cross-section 125 mm wide and 250 mm deep, is prestressed by a cable in which the total tensile force is 220 kN. The centre line of the cable is parallel to the axis of the beam and 75 mm above the soffit over the middle third of the span and is curved upward in a parabola over the outer thirds of the span to a distance of 175 mm above the soffit at the supports. If the modulus of elasticity of concrete is 35 kN/mm^2 and the density of concrete is 24 kN/m^3, calculate the deflection when the beam is supporting its own weight.

Solution

The beam with the cable profile is shown in Fig. 3.10

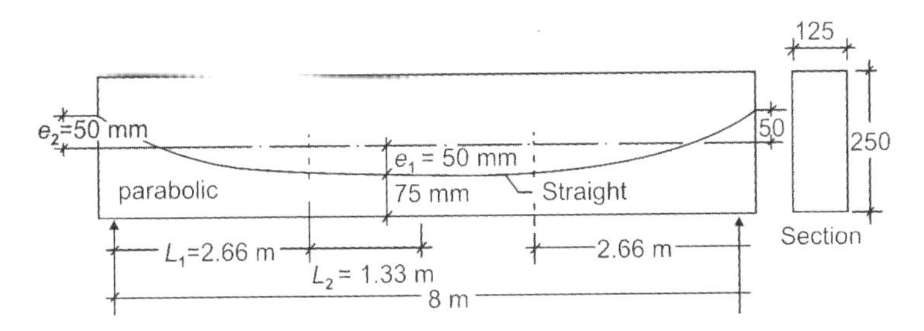

Fig. 3.10

Given data: $P = 220$ kN, $L_1 = 2.66$ m, $I = \left(\dfrac{bD^3}{12}\right)$

$L = 8$ m $L_2 = 1.33$ m,

$$I = \frac{(125 \times 250^3)}{12} = 1.63 \times 10^8 \text{ mm}^4$$

$e_1 = 50$ mm $\quad E = 35$ kN/mm^2

$e_2 = 50$ mm

$g = 0.75$ kN/m $= 0.00075$ kN/mm

The cable is parabolic towards the supports and straight in the mid third span. Deflection due to prestressing force is given by the relation

$$a_p = \frac{P(e_1 + e_2)}{12\,EI}[L_1^2 + 12L_1L_2 + 6L_2^2] + \left[\frac{Pe_2L^2}{8\,EI}\right]$$

$$= \left(\frac{220(50+50)}{12 \times 35 \times 1.63 \times 10^8} \right)$$

$$[(2660)^2 + (12 \times 2660 \times 1330) + (6 \times 1330^2)]$$

$$+ \left[\frac{220 \times 50 \times 8000^4}{8 \times 35 \times 1.63 \times 10^8} \right]$$

$= 12.59$ mm (downwards)

Deflection due to self weight of the beam is given by

$$a_g = \left(\frac{5gL^4}{384\,EI} \right) = \left(\frac{5 \times 0.00075 \times 8000^4}{384 \times 35 \times 1.63 \times 10^8} \right)$$

$= 7.01$ mm (downwards)

Final deflection $= (a_p + a_g) = (12.59 + 7.01)$

$= 19.60$ mm (downwards)

Problem 3.12

A concrete beam is prestressed by a linearly varying cable having eccentricities as shown in the Fig. 3.11. The force in the cable is 350 kN. The beam supports a concentrated load of 20 kN at centre of span. If $E = 38$ kN/mm^2, loss ratio = 0.8 and creep coefficient is 1.6, compute (a) short term deflection under prestress + self weight and (b) long term deflection under prestress + self weight + live load. The cross section of beam is 150 mm wide by 300 mm deep.

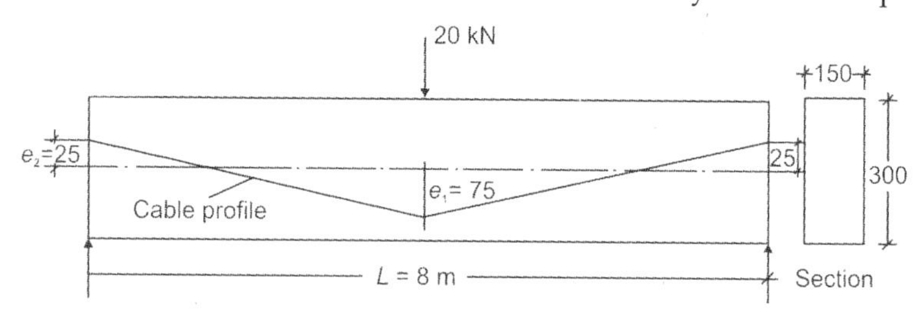

Fig. 3.11

Solution

The beam with the cable profile is shown in Fig. 3.11.

Given data: $P = 350$ kN, $e_1 = 75$ mm,

$e_2 = 25$ mm $L = 8$ m

$$E = 38 \text{ kN/mm}^2 \qquad \phi = 1.6,$$
$$\eta = 0.8 \qquad b = 150 \text{ mm}$$
$$d = 300 \text{ mm}, \qquad Q = 20 \text{ kN}$$

$$I = \frac{(bd^3)}{12} = \frac{(150 \times 300^3)}{12} = 3.375 \times 10^8 \text{ mm}^4$$

Self weight of beam = g = $(0.15 \times 0.3 \times 24) = 1.08$ kN/m
$= 0.00108$ kN/mm

Deflection due to self weight

$$a_g = \frac{(5gL^4)}{(384\ EI)} = \frac{(5 \times 0.00108 \times 8000^4)}{(384 \times 38 \times 3.375 \times 10^8)}$$
$$= 4.49 \text{ mm (downwards)}$$

Deflection due to prestress

$$a_p = P.L^2 \frac{(-2e_1 + e_2)}{24\ EI}$$

$$= \frac{(350 \times 8000^2)(-2 \times 75 + 25)}{(384 \times 38 \times 3.375 \times 10^8)}$$
$$= -9.09 \text{ mm (upwards)}$$

Deflection due to live load

$$a_Q = \frac{(QL^3)}{(48\ EI)} = \frac{(20 \times 8000^2)}{(48 \times 38 \times 3.375 \times 10^8)}$$
$$= 16.6 \text{ mm (downwards)}$$

Short term deflection $= a_{RS} = (a_g + a_P)$
$= (4.49 - 9.09) = -4.60$ mm (upwards)

Long term deflection $= a_{RL} = (1 + \phi)(\eta a_p + a_g + a_Q)$
$= [1 + 1.6][0.8(-9.09) + 4.49 + 16.6]$
$= 35.92$ mm (downwards)

Problem 3.13

A prestressed concrete beam of rectangular section, 120 mm wide by 300 mm deep is prestressed by a linearly varying cable as shown in Fig. 3.12. The beam supports a uniformly distributed live load of 4 kN/m. The effective force in the cable is 200 kN. Estimate the short term deflection of the beam under (prestress + self weight + live load). Assume $E = 38$ kN/mm^2.

Solution

The beam with the linearly varying cable is shown in Fig. 3.12

Given data: $P = 200$ kN, $b = 120$ mm,

$L = 6$ mm $e = 50$ mm,

$d = 300$ mm, $E = 38$ kN/mm^2

$$I = \frac{(bd^3)}{12} = 27 \times 10^7 \text{ mm}^4$$

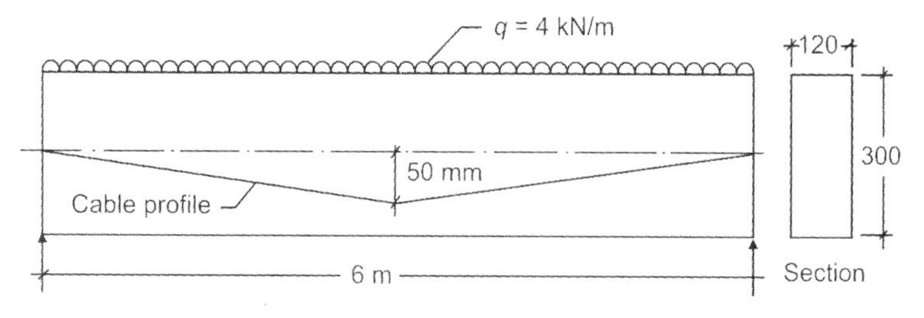

Fig. 3.12

Self weight of beam = g = (0.12 × 0.3 × 24) = 0.864 kN/m

= 0.000864 kN/mm

Deflection due to self weight

$$a_g = \frac{(5gL^4)}{(384\,EI)} = \frac{(5 \times 0.000864 \times 6000^4)}{(384 \times 38 \times 27 \times 10^7)}$$

= 1.421 mm (downwards)

Deflection due to prestress

$$a_p = -\frac{(PeL^2)}{12\,EI} = -\frac{(200 \times 50 \times 6000^2)}{(12 \times 38 \times 27 \times 10^7)}$$

= –2.92 mm (upwards)

Deflection due to live load

$$a_q = \left[\frac{(1.421 \times 4)}{0.864}\right] = 6.57 \text{ mm (downwards)}$$

Short term deflection $= a_{RS} = [a_p + a_g + a_q]$

= [–2.92 + 1.421 + 6.57]

= 5.071 mm (downwards)

Problem 3.14

A concrete beam with a cross-sectional area of 32×10^3 mm^2 and radius of gyration of 72 mm is prestressed by a parabolic cable carrying an effective stress of 1 kN/mm^2.

The span of the beam is 8 m. The cable contains 6 high tensile wires of 7 mm diameter and it has an eccentricity of 50 mm towards the soffit at the centre of span and concentric at the supports.

Neglecting all losses, estimate the deflection at the centre of span for the following conditions:

a. Self weight + prestress
b. Self weight + prestress + live load of 2 kN/m over the beam.

Adopt $E_c = 38$ kN/mm^2 and density of concrete as 24 kN/m^3.

Solution

$$A = (32 \times 10^3)\ \text{mm}^2$$

Radius of gyration $= k = 72$ mm

$$L = 8000\ \text{mm},$$

Eccentricity $= e = 50$ mm

$$I = A k^2 = (32 \times 10^3)\ 72^2 = (166 \times 10^6)\ \text{mm}^4$$

$$P = (6 \times 38.5 \times 1000) = 231000\ \text{N} = 231\ \text{kN}$$

$$g = \left[\frac{32 \times 10^3 \times 24}{10}\right] = 0.77\ \text{kN/m} = 0.00077\ \text{kN/mm}$$

Downwards deflection due to self weight

$$= \left[\frac{5\ gL^4}{384EI}\right] = \left[\frac{5 \times 0.00077 \times 8000^4}{384 \times 38 \times 166 \times 10^6}\right] = 6.5\ \text{mm}$$

Upward deflection due to prestressing force

$$= \left[\frac{5\ PeL^4}{48EI}\right] = \left[\frac{5 \times 231 \times 50 \times 800^2}{48 \times 38 \times 166 \times 10^6}\right]$$

$$= 12.2\ \text{mm (negative)}$$

Downward deflection due to live load

$$= \left[\frac{6.5}{0.77}\right] \times 24 = 16.9\ \text{mm}$$

Resultant deflections:

Case a: Deflection due to self weight + pressure

$$= (6.5 - 12.2) = -5.7\ \text{mm (upwards)}$$

Case b: Deflection due to self weight + pressure + live load

$$= (6.5 - 12.2 + 16.9) = 11.2\ \text{mm (downwards)}$$

Problem 3.15

A rectangular concrete beam of cross-section 150 mm wide by 300 mm deep is simply supported over a span of 8 m. The beam is prestressed by a parabolic cable having an eccentricity of 25 mm towards the top at supports and 75 mm towards the soffit at mid span. If the force in the cable is 350 kN and the modulus of elasticity of concrete is 38000 N/mm^2 estimate:

a. The deflection at mid span when the beam is supporting its own self weight

b. The concentrated load which must be applied at mid span to restore it to the level of supports.

Solution

$P = 350$ kN $\qquad I = (3375 \times 10^5)$ mm^4

$E_c = 38$ kN/mm^2 $\qquad e_1 = 75$ mm

$e_2 = 25$ mm

Net deflection due to the prestressing force is computed as

$$a_p = \left[\frac{PL^2}{48EI}\right](-5e_1 + e_2)$$

$$= \left[\frac{350 \times 8000^2}{48 \times 38 \times 3375 \times 10^5}\right](-5 \times 75 + 25)$$

$= -12.7$ mm (upwards)

Self weight of the beam

$= g = (0.15 \times 0.30 \times 24) = 1.08$ kN/m

$= 0.00108$ kN/mm

Downward deflection due to self weight

$$= a_g = \left[\frac{5gL^4}{384EI}\right] = \left[\frac{5 \times 0.00108 \times 8000^4}{384 \times 38 \times 3375 \times 10^5}\right] = 4.5 \text{ mm}$$

Case a: Deflection due to prestress + self weight

$= [-12.7 + 4.5] = -8.2$ mm (upwards)

Case b: If Q = Concentrated load required at the centre of span, then we have the relation,

$$\left[\frac{QL^3}{48EI}\right] = 8.2$$

$$\text{Hence } Q = \left[\frac{8.2 \times 48 \times EI}{L^3}\right] = \left[\frac{8.2 \times 48 \times 38 \times 3375 \times 10^5}{8000^3}\right] = 9.9 \text{ kN.}$$

Problem 3.16

A simply supported concrete beam spanning over 6 m is post-tensioned by two cables, both of which have an eccentricity of 100 mm towards the soffit at mid span. The first cable is parabolic and is anchored at an eccentricity of 100 mm towards the top of beam at supports. The second cable is straight and is parallel to the longitudinal axis of the beam with constant eccentricity. Each of these cables carry a force of 120 kN. The cross-sectional area of the beam is (2×10^4) mm^2 and the radius of gyration is 120 mm.

The beam supports two concentrated loads of 20 kN each at the third point of the span. Modulus of elasticity = 38 kN/mm^2. Calculate using Lin's simplified method,

a. The instantaneous deflection at mid span
b. The deflection at mid span after 2 years, assuming 20 percent loss in prestress and the effective modulus of elasticity to be one-third of the short term modulus of elasticity.

Solution

$$P = 120 \text{ kN} \qquad e_1 = e_2 = 100 \text{ mm}$$
$$L = 6 \text{ m} = 6000 \text{ mm} \qquad A = (2 \times 10^4) \text{ mm}^4$$
$$k = 120 \text{ mm}$$
$$I = A\,k^2 = (2 \times 10^4) \times 120^2 = (288 \times 10^6) \text{ mm}^4$$
$$\text{Self weight} = g = [2 \times 10^4 \times 24 \times 10^{-9}] = 0.00048 \text{ kN/mm}$$

Concentrated load at third span point = Q = 20 kN

Downward deflection due to self weight of the beam is computed as,

$$a_g = \left[\frac{5gL^4}{384EI}\right] = \left[\frac{5 \times 0.00048 \times 6000^4}{384 \times 38 \times 288 \times 10^6}\right] = 0.74 \text{ mm}$$

Downward deflection due to concentrated loads is evaluated as,

$$\left[\frac{\text{span}}{300}\right] = \left[\frac{8000}{300}\right] = 26.6 \text{ mm}$$

Deflection due to prestressing force (parabolic cable) is calculated as,

$$a_p = \left[\frac{PL^2}{48EI}\right](-5e_1 + e_2)$$
$$= \left[\frac{120 \times 6000^2}{48 \times 38 \times 288 \times 10^6}\right](-5 \times 100 + 100)$$
$$= -3.27 \text{ mm (upwards)}$$

Deflection due to prestressing force (straight cable) is calculated as,

$$a_p = \left[\frac{-PeL^2}{8EI}\right] = -\left[\frac{120 \times 100 \times 6000^2}{8 \times 38 \times 288 \times 10^6}\right]$$

$$= -4.92 \text{ mm (upwards)}$$

a. Instantaneous deflection due to prestress + self weight + live loads is

$$a_i = [-3.27 - 4.92 + 0.74 + 14.10]$$
$$= 6.65 \text{ mm (downwards)}$$

b. Long term deflection at the end of two years is computed by using

$$E_{ce} = \left[\frac{E}{3}\right] \text{ and loss of prestress} = 20\%$$

Net upward deflection

$$= 3[0.8\ (3.27 + 4.92) = 19.65 \text{ mm}$$

Net downward deflection

$$= 3[0.74 + 14.10] = 44.52 \text{ mm}$$

Long term final downward deflection

$$= [44.52 - 19.65] = 24.87 \text{ mm.}$$

Problem 3.17

A concrete beam having a symmetrical I-section has flange width and depth of 200 mm and 60 mm respectively. The thickness of the web is 80 mm and the overall depth is 400 mm. The beam is prestressed by a cable carrying a force of 1000 kN. The span of the beam is 8 m. The centre line of the cable is 150 mm from the soffit of the beam at the centre of span, linearly varying to 250 mm at the supports. Compute the initial deflection at mid span due to prestress and the self weight of the beam, assuming the modulus of elasticity of concrete as 38 kN/mm^2. Compare the deflection with the limiting deflection permitted in IS: 1343 code.

Solution

$P = 1000$ kN $I = (847 \times 10^6)$ mm^4

$e_1 = 50$ mm $e_2 = 50$ mm

$L = 8000$ mm

Self weight of the beam $= g = 1.12$ kN/m $= 0.00112$ kN/mm

Deflection due to self weight

$$= \left[\frac{5gL^4}{384EI}\right] = \left[\frac{5 \times 0.00112 \times 8000^4}{384 \times 38 \times 847 \times 10^6}\right]$$

$$= 1.86 \text{ mm (downwards)}$$

Deflection due to the prestressing force

$$= \left[\frac{Pe_2 L^2}{8EI} - \frac{P(e_1 + e_2)L^2}{12EI}\right] = \left[\frac{PL^2}{EI}\right]\left[\frac{e^2}{8} - \frac{(e_1 + e_2)}{12}\right]$$

$$= \left[\frac{1000 \times 8000^2}{38 \times 847 \times 10^6}\right]\left[\frac{50}{8} - \frac{(50 + 50)}{12}\right]$$

$$= -4.1 \text{ mm (upwards)}$$

Deflection due to prestress + self weight

$$= -4.1 + 1.86 = -2.24 \text{ mm (upwards)}$$

Maximum permissible upward deflection according to IS: 1343 code is limited to

$$a_{\max} = \left[\frac{\text{span}}{300}\right] = \left[\frac{8000}{300}\right] = 26.6 \text{ mm}$$

Hence, the actual deflection is well within the permissible limits.

Problem 3.18

A post-tensioned roof girder spanning over 30 m has an unsymmetrical I-section with a second moment of area of section of (72490×10^6) mm^4 and an overall depth of 1300 mm. The effective eccentricity of the group of cables at the centre of span and support sections is 580 and 170 mm, respectively towards the soffit of the beam. The cables carry an initial prestressing force of 3200 kN.

The self weight of the girder is 10.8 kN/m and the live load on the girder is 9 kN/m. The modulus of elasticity of concrete is 34 kN/mm^2. If the creep coefficient is 1.6, and the total loss of prestress is 15 percent, estimate the deflections at the following stages and compare them with the permissible values according to the Indian Standard Code (IS: 1343) limits:

a. Instantaneous deflection due to prestress + self weight
b. Resultant maximum long term deflection allowing for loss of prestress and creep of concrete.

Solution

$P = 3200$ kN $I = (72490 \times 10^6)$ mm^4

$e_1 = 580$ mm $e_2 = 170$ mm

$E = 34$ kN/mm^2

Creep coefficient = $\phi = 1.6$

$g = 10.8$ kN/m = 0.0108 kN/mm

$q = 9$ kN/m = 0.009 kN/mm

Loss ratio = $\eta = 0.85$

Deflection due to initial prestress = $a_p = \left[\dfrac{PL^2}{48EI}\right](-5e_1 + e_2)$

$$a_p = \left[\frac{3200\times10^3\times30000^2}{48\times38\times72490\times10^6}\right](-5 \times 580 + 170)$$

= –74.7 mm (upwards)

Deflection due to self weight

$$= \left[\frac{5gL^4}{384EI}\right] = \left[\frac{5\times0.0108\times30000^4}{384\times34\times72490\times10^6}\right]$$

= 46.2 mm (downwards)

a. Instantaneous deflection due to prestress + self weight

= (–74.7 + 46) = –28.7 mm (upwards)

Permissible upward deflection according to IS: 1343

$$= \left[\frac{\text{span}}{300}\right] = \left[\frac{30000}{300}\right] = 100 \text{ mm}$$

Deflection due to live load

$$= \left[\frac{5qL^4}{384EI}\right] = \left[\frac{5\times0.009\times30000^4}{384\times38\times72490\times10^6}\right]$$

= 38.5 mm (downwards)

b. If creep coefficient = $\phi = 1.6$.

Long term modulus of elasticity of concrete is given by

$$E = \left[\frac{E}{1+\phi}\right] = \left[\frac{E}{1+1.6}\right] = \left[\frac{E}{2.6}\right]$$

∴ Resultant maximum long term deflection

= [(2.6 × 46) + 38.5 – (0.85 × 74.7)] = 95 mm

which is less than the IS: 1343 code limit of

$$\left[\frac{\text{span}}{250}\right] = \left[\frac{30000}{250}\right] = 120 \text{ mm.}$$

Problem 3.19

A prestressed concrete beam of rectangular section, 120 mm wide by 300 mm deep spans over 6 m. The beam is prestressed by a straight cable carrying an effective force of 200 kN at an eccentricity of 50 mm. The modulus of elasticity of concrete is 38 kN/mm^2.

Compute the deflection at centre of span for the following cases:

a. Deflection under prestress + self weight
b. Find the magnitude of the uniformly distributed live load which will nullify the deflection due to prestress and self weight.

Solution

$$P = 200 \text{ kN} \qquad I = (27 \times 10^7) \text{ mm}^4$$
$$g = 0.86 \text{ N/mm} \qquad e = 50 \text{ mm}$$
$$L = 6000 \text{ mm} \qquad E = 38 \text{ kN/mm}^2$$

Deflection due to the prestressing force is computed as,

$$a_p = \left[\frac{-PeL^2}{8EI}\right] = -\left[\frac{200 \times 50 \times 6000^2}{8 \times 38 \times 2 \times 10^7}\right]$$
$$= -4.38 \text{ mm (upwards)}$$

Deflection due to self weight of the beam is

$$a_g = \left[\frac{5gL^4}{384EI}\right] = \left[\frac{5 \times 0.86 \times 6000^4}{384 \times 38 \times 10^3 \times 27 \times 10^7}\right]$$
$$= 1.40 \text{ mm (downwards)}$$

a. Deflection due to prestress + self weight

$$= [1.40 - 4.38] = -2.98 \text{ mm (upwards)}$$

b. If q = uniformly distributed live load on the beam which neutralizes the deflection due to prestress and self weight, its magnitude is calculated as,

$$q = \left[\frac{a \times 384EI}{5 \times L^4}\right] = \left[\frac{2.98 \times 384 \times 38 \times 10^3 \times 27 \times 10^7}{5 \times 6000^4}\right]$$
$$= 1.81 \text{ N/mm} = 1.81 \text{ kN/m.}$$

Problem 3.20

The deck of a prestressed concrete culvert is made up of a slab 500 mm thick. The slab is spanning over 10.4 m and supports a total uniformly distributed load comprising the dead and live loads of 23.5 kN/m^2. The modulus of elasticity of concrete is 38 kN/mm^2. The concrete slab is prestressed by straight cables each containing 12 high tensile wires of 7 mm diameter stressed to 1200 N/mm^2 at a constant eccentricity of 195 mm. The cables are spaced at 328 mm intervals in the transverse direction. Estimate the instantaneous deflection of the slab at centre of span under prestress and the imposed loads.

Solution

Considering 1 m width of cable, the properties of the cross-section are computed.

Thickness of the slab = d = 500 mm

Width of the slab = b = 1000 mm

Span of the slab = L = 10.4 m

Second moment of area

$$= I = \left[\frac{bd^3}{12}\right] = \left[\frac{1000 \times 500^3}{12}\right] = (1041 \times 10^7)\ \text{mm}^4$$

$$\text{Force in each cable} = \left[\frac{12 \times 38.5 \times 1200}{1000}\right] = 554\ \text{kN}$$

Spacing of cables in the transverse direction = 328 mm

Hence, the prestressing force per metre width of the slab is computed as,

$$P = \left[\frac{1000 \times 554}{328}\right] = 1689\ \text{kN}$$

Eccentricity = e = 195 mm

Total uniformly distributed load on the beam

$$= w = 33.5\ \text{kN/m} = 0.0335\ \text{kN/mm}$$

Deflection due to prestressing force

$$= a_p = \left[\frac{-PeL^2}{8EI}\right] = -\left[\frac{1689 \times 195 \times (10.4 \times 1000)^2}{8 \times 38 \times 1041 \times 10^7}\right]$$

$$= -11.25\ \text{mm (upwards)}$$

Deflection due to loads

$$= a_w = \left[\frac{5wL^4}{384EI}\right] = \left[\frac{5 \times 0.0335 \times (10.4 \times 1000)^4}{384 \times 38 \times 1041 \times 10^7}\right]$$

$$= 12.90\ \text{mm (downwards)}$$

Resultant deflection = (12.90 – 11.25) = 1.65 mm (downwards)

4

Flexural Strength of Prestressed Concrete Sections

Problem 4.1

A pretensioned beam of rectangular section 400 mm wide by 1000 mm overall depth is prestressed by 800 mm^2 of high tensile steel wires at an eccentricity of 300 mm. If f_{ck} = 40 N/mm^2, f_p = 1600 N/mm^2, estimate the ultimate flexural strength of the section as per IS: 1343 code provisions.

Solution

Given: b = 400 mm, f_{ck} = 40 N/mm^2

d = 800 mm, f_p = 1600 N/mm^2

A_p = 800 mm^2

The effective reinforcement ratio is

$$\left(\frac{A_p \cdot f_p}{f_{ck} \cdot b \cdot d}\right) = \left(\frac{800 \times 1600}{40 \times 400 \times 800}\right) = 0.1$$

Referring to (Table 11 of IS: 1343–2012)

$$\left(\frac{f_{pu}}{0.87\, f_p}\right) = 1.0$$

$\therefore$ $f_{pu} = 0.87, \quad f_p = (0.87 \times 1600) = 1392 \text{ N/mm}^2$

and $\left(\dfrac{x_u}{d}\right) = 0.217$

$\therefore$ $x_u = (0.217 \times 800) = 173.6$ mm

$$\therefore \quad M_u = f_{ru} \cdot A_p(d - 0.42\, x_u)$$
$$= 1392 \times 800(800 - 0.42 \times 173.6)$$
$$= 809.6 \times 10^6 \text{ N.mm} = 809.6 \text{ kN.m}$$

Table 4.1: Conditions at the ultimate limit state for rectangular beams with pre-tensioned tendons or with post-tensioned tendons having effective bond (Table 11 of IS: 1343–2012)

$\left(\frac{A_p f_p}{b.d.f_{ck}}\right)$	Stress in tendons as a proportion of the design strength $(f_{pu}/0.87\,f_p)$		Ratio of the depth of neutral axis to that of the centroid of the tendon in the tension zone (x_u/d)	
	Pre-tensioning	Post-tensioning with effective bond	Pre-tensioning	Post-tensioning with effective bond
0.025	1.0	1.0	0.054	0.054
0.05	1.0	1.0	0.109	0.109
0.10	1.0	1.0	0.217	0.217
0.15	1.0	1.0	0.326	0.316
0.20	1.0	0.95	0.435	0.414
0.25	1.0	0.90	0.542	0.488
0.30	1.0	0.85	0.655	0.558
0.40	0.9	0.75	0.783	0.653

Problem 4.2

A pre-tensioned concrete girder of box section 1 m by 1 m overall dimensions has a uniform wall thickness of 200 mm. The girder is post-tensioned by high tensile wires of area 2250 mm^2 located at an effective depth of 900 mm. If f_{ck} = 40 N/mm^2 and f_p = 1600 N/mm^2. Calculate the ultimate flexural strength of the box girder section.

Solution

The box girder section is shown in Fig. 4.1

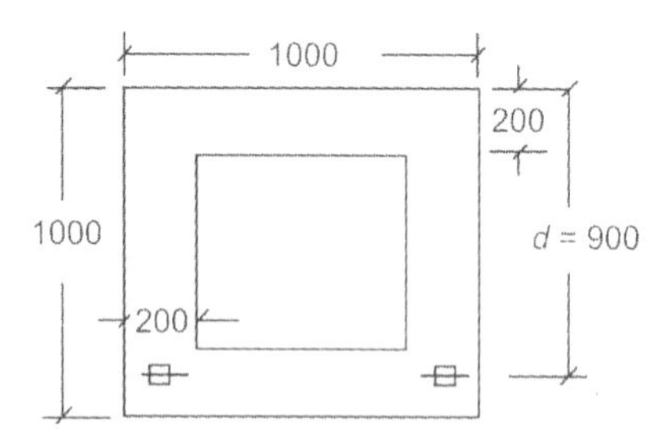

Fig. 4.1

Given: $b = 1000$ mm $\quad f_{ck} = 40\ \text{N/mm}^2$

$d = 900$ mm $\quad f_p = 1600\ \text{N/mm}^2$

$A_p = 2250\ \text{mm}^2$

The ratio $\left(\dfrac{A_p.f_p}{f_{ck}.b.d}\right) = \left(\dfrac{2250 \times 1600}{40 \times 1000 \times 900}\right) = 0.10$

Referring to Table 4.1,

$$\left(\frac{f_{pu}}{0.87\, f_p}\right) = 0.1$$

$$\therefore \quad f_{pu} = (0.87 \times 1600) = 1392\ \text{N/mm}^2$$

$$\left(\frac{x_u}{d}\right) = 0.217$$

$$\therefore \quad x_u = (0.217 \times 900) = 195.3\ \text{mm} < 200\ \text{mm}$$

$$\therefore \quad M_u = f_{pu} \cdot A_p(d - 0.42\, x_u)$$

$$= 1392 \times 2250(900 - 0.42 \times 195.3)$$

$$= 2561 \times 10^6\ \text{N.mm} = 2561\ \text{kN.m}$$

Problem 4.3

A post-tensioned bonded prestressed concrete beam of unsymmetrical tee-section has a flange width of 1500 mm and thickness of flange is 200 mm, and thickness of rib is 300 mm. The area of high tensile steel is 5000 mm^2 located at an effective depth of 1800 mm. If the characteristic strength of concrete and steel are 40 and 1600 N/mm^2, respectively, calculate the flexural strength of the tee-section.

Solution

The cross-section of the post-tensioned prestressed concrete beam is shown in Fig. 4.2.

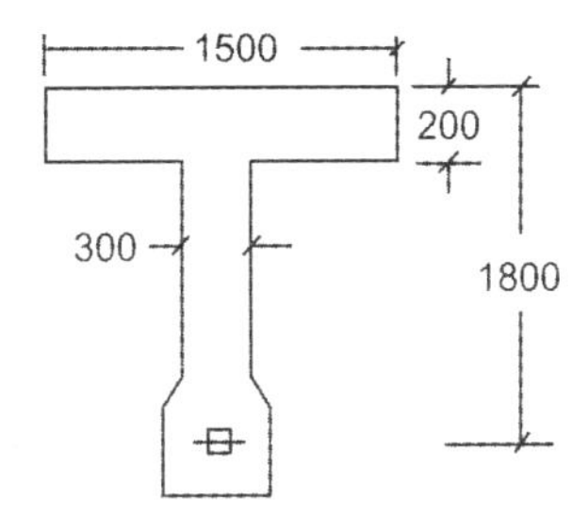

Fig. 4.2

Given:
$$A_p = 5000 \text{ mm}^2 \qquad f_{ck} = 40 \text{ N/mm}^2$$
$$b = 1500 \text{ mm} \qquad f_p = 1600 \text{ N/mm}^2$$
$$b_w = 300 \text{ mm} \qquad t = 200 \text{ mm}$$
$$A_p = (A_{pf} + A_{pw})$$
$$A_{pf} = 0.44 f_{ck}(b - b_w)(t/f_p)$$
$$= 0.44 \times 40(1500 - 300)(200/1600)$$
$$= 2640 \text{ mm}^2$$
$$\therefore \quad A_{pw} = (A_p - A_{pf}) = (5000 - 2640) = 2360 \text{ mm}^2$$

$$\text{Ratio} \left(\frac{A_{pw} \cdot f_p}{b_w \cdot d \cdot f_{ck}}\right) = \left(\frac{2360 \times 1600}{300 \times 1800 \times 40}\right) = 0.1748$$

From Table 4.1, interpolating we have
$$(f_{pu}/0.87 f_p) = 0.975$$
$$\therefore \quad f_{pu} = (0.975 \times 0.87 \times 1600) = 1357.2 \text{ N/mm}^2$$
$$(x_u/d) = 0.36$$
$$\therefore \quad x_u = (0.36 \times 1800) = 648 \text{ mm} > 200 \text{ mm}$$

The ultimate moment capacity of the tee-section is expressed as
$$M_u = f_{pu} \cdot A_{pw}(d - 0.42\, x_u) + 0.44 f_{ck}(b - b_w)\, t\,(d - 0.5t)$$
$$= [1357.2 \times 2360\,(1800 - 0.42 \times 648) + 0.44 \times 40 \times 1200 \times 200\,(1800 - 100)]$$
$$= [(4890 \times 10^6) + (7180 \times 10^6)]$$
$$= (12070 \times 10^6) \text{ N.mm} = 12070 \text{ kN.m}$$

Problem 4.4

A double tee-section having a flange 1200 mm wide and 150 mm thick is prestressed by 4700 mm^2 of high tensile steel located at an effective depth of 1600 mm. The ribs have a thickness of 150 mm each. If the cube strength of concrete is 40 N/mm^2 and tensile strength of steel is 1600 N/mm^2, determine the flexural strength of the double tee-girder using IS: 1343 code provisions.

Solution

The cross-section of the double tee-girder is shown in Fig. 4.3.

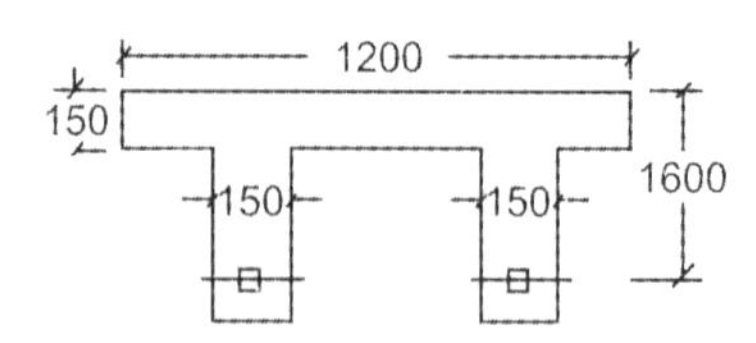

Fig. 4.3

Given: $b = 1200$ mm $\quad f_{ck} = 40\ \text{N/mm}^2$

$$b_w = (150 + 150) = 300 \text{ mm}$$
$$f_p = 1600\ \text{N/mm}^2 \qquad t = 150 \text{ mm}$$
$$A_p = 4700\ \text{mm}^2$$
$$A_p = (A_{pw} + A_{pf})$$
$$A_{pf} = 0.44 f_{ck}(b - b_w)(t/f_p)$$
$$= 0.44 \times 40(1200 - 300)(150/1600)$$
$$= 1485\ \text{mm}^2$$

$$\therefore \quad A_{pw} = (A_p - A_{pf}) = (4700 - 1485) = 3215\ \text{mm}^2$$

$$\text{Ratio} \left(\frac{A_{pw} \cdot f_p}{b_w \cdot d\, f_{ck}}\right) = \left(\frac{3215 \times 1600}{300 \times 1600 \times 40}\right) = 0.268$$

From Table 4.1, interpolating we have

$$\left(\frac{f_{pu}}{0.87 f_p}\right) = 0.93$$

$$\therefore \quad f_{pu} = (0.87 \times 1600 \times 0.93) = 1225\ \text{N/mm}^2$$

and
$$\left(\frac{x_u}{d}\right) = 0.51$$

$$\therefore \quad x_u = (0.51 \times 1600) = 816 \text{ mm} > 150 \text{ mm}$$

The ultimate flexural strength of the section is expressed by the relation:

$$M_u = f_{pu} \cdot A_{pw}(d - 0.42\, x_u) + 0.44 f_{ck}(b - b_w)\, t(d - 0.5t)$$
$$= [1225 \times 3215(1600 - 0.42 \times 816)$$
$$+ 0.44 \times 40(1200 - 300)\, 150\, (1600 - 0.5 \times 150)]$$
$$= [(4951 \times 10^6) + (3623 \times 10^6)]$$
$$= (8574 \times 10^6) \text{ N.mm} = 8574 \text{ kN.m}$$

Problem 4.5

A post-tensioned prestressed concrete tee-beam with unbonded tendons is made up of a flange 300 mm wide by 150 mm thick and the thickness of rib is 150 mm. The beam is prestressed by 24 H.T. wires of 5 mm diameter with an effective stress of 65 per cent of the ultimate tensile strength of wires. The wires are located at an effective depth of 320 mm. Loss ratio is 0.8. If $f_{ck} = 56\ \text{N/mm}^2$ and $f_p = 1650\ \text{N/mm}^2$, estimate the flexural strength of the section assuming the span/depth ratio of the beam as 20.

Solution

Given:

$b = 300$ mm $\qquad f_{ck} = 56$ N/mm^2

$d = 320$ mm $\qquad f_p = 1650$ N/mm^2

$t = 150$ mm $\qquad f_{pe} = (0.65 \times 1650 \times 0.8)$

$\qquad = 858$ N/mm^2

$(L/d) = 20 \qquad A_p = (24 \times 20)$

$\qquad = 480$ mm^2

The effective reinforcement ratio is

$$\left(\frac{A_p \cdot f_{pe}}{f_{ck} \cdot b \cdot d}\right) = \left(\frac{480 \times 858}{56 \times 300 \times 320}\right) = 0.076$$

Referring to Table 4.2,

Table 4.2: Conditions at the ultimate limit state for post-tensioned rectangular beams having unbonded tendons (IS: 1343)

$\left(\frac{A_p f_{pe}}{bdf_{ck}}\right)$	Stress in tendons as a proportion of the effective prestress (f_{pu}/f_{pc}) for values of (L/D) $\left(\frac{\text{Effective span}}{\text{Effective depth}}\right)$			Ratio of depth of neutral axis to that of the tendons in the tension zone (x_u/d) for values of (L/d) $\left(\frac{\text{Effective span}}{\text{Effective depth}}\right)$		
	30	20	10	30	20	10
0.025	1.23	1.34	1.45	0.10	0.10	0.10
0.05	1.21	1.32	1.45	0.16	0.16	0.18
0.10	1.18	1.26	1.45	0.30	0.32	0.36
0.15	1.14	1.20	1.36	0.44	0.46	0.52
0.20	1.11	1.16	1.27	0.56	0.58	0.64

$(f_{pu}/f_{pe}) = 1.29$

$\therefore \quad f_{pu} = (1.29 \times 858) = 1106$ N/mm^2

$(x_u/d) = 0.24$

$\therefore \quad x_u = (0.24 \times 320) = 76.8$ mm < 150 mm

$\therefore \quad M_u = f_{pu} A_p (d - 0.42\, x_u)$

$= 1106 \times 480\,(320 - 0.42 \times 76.8)$

$= 152 \times 10^6$ N.mm

$= 152$ kN.m

Problem 4.6

A pre-tensioned tee-section has a flange width of 300 mm and thickness of flange is 200 mm. The rib is 150 mm wide by 350 mm deep. The effective depth at which high tensile steel of area 200 mm^2 is provided is 500 mm. Given $f_{ck} = 50$ N/mm^2 and $f_p = 1600$ N/mm^2. Estimate the flexural strength of the tee-section.

Solution

Given:

$$b = 300 \text{ mm} \qquad f_{ck} = 50 \text{ N/mm}^2$$
$$d = 500 \text{ mm} \qquad f_p = 1600 \text{ N/mm}^2$$
$$A_p = 200 \text{ mm}^2$$

Assuming that the neutral axis falls within the flange, the value of $b = 300$ mm for computations of the effective reinforcement ratio.

$$\left(\frac{f_p \cdot A_p}{f_{ck} \cdot b \cdot d}\right) = \left(\frac{1600 \times 200}{50 \times 300 \times 500}\right) = 0.04$$

From Table 4.1, the corresponding values of the ratios are

$$(f_{pu}/0.87\, f_p) = 1.0 \quad \text{and} \quad (x_u/d) = 0.09$$

$$\therefore \quad f_{pu} = (0.87 \times 1600) = 1392 \text{ N/mm}^2$$
$$x_u = (0.09 \times 500) = 45 \text{ mm} < 200 \text{ mm}$$
$$\therefore \quad M_u = f_{pu}\, A_p(d - 0.42\, x_u)$$
$$= [1392 \times 200\,(500 - 0.42 \times 45)]$$
$$= [134 \times 10^6] \text{ N.mm} = 134 \text{ kN.m}$$

Problem 4.7

A pre-tensioned prestressed concrete beam having rectangular section 150 mm wide by 350 mm deep has an effective cover of 50 mm. If $A_p = 461$ mm^2, $f_{ck} = 40$ N/mm^2, and $f_p = 1600$ N/mm^2. Calculate the ultimate flexural strength of the section using IS: 1343 code specifications.

Solution

Data:

$$f_{ck} = 40 \text{ N/mm}^2, \qquad b = 150 \text{ mm}$$
$$A_p = 461 \text{ mm}^2 \qquad f_p = 1600 \text{ N/mm}^2$$
$$d = 300 \text{ mm}$$

Effective reinforcement ratio is given by

$$\left(\frac{A_p f_p}{f_{ck} b d}\right) = \left(\frac{461 \times 1600}{40 \times 150 \times 300}\right) = 0.40$$

From Table 11 of IS: 1343 code, read out

$$\left[\frac{f_{pu}}{0.87 f_p}\right] = 0.9 \quad \therefore \quad f_{pu} = (0.9 \times 0.87 \times 1600)$$

$$= 1253 \text{ N/mm}^2$$

$$\left[\frac{x_u}{d}\right] = 0.783 \qquad x_u = (0.783 \times 300) = 234.9 \text{ mm}$$

The ultimate flexural strength is given by the relation

$$M_u = A_p f_{pu} [d - 0.42\, x_u]$$
$$= (461 \times 1253) [300 - 0.42 \times 234.9]$$
$$= 116 \times 10^6 \text{ N.mm} = 116 \text{ kN.m}$$

Problem 4.8

A post-tensioned beam having a rectangular section, 150 mm wide by 350 mm deep has an effective cover of 50 mm. If $A_p = 461$ mm^2, $f_p = 40$ N/mm^2 and $f_p = 1600$ N/mm^2, estimate the ultimate moment capacity of the section assuming the ratio of effective depth to span as 20 and the effective stress in tendons after all losses as 800 N/mm^2.

Solution

Data:

$$b = 150 \text{ mm} \qquad \left(\frac{L}{d}\right) = 20$$
$$f_{ck} = 40 \text{ N/mm}^2, \qquad d = 300 \text{ mm}$$
$$f_p = 1600 \text{ N/mm}^2 \qquad f_{pe} = 800 \text{ N/mm}^2$$
$$A_p = 461 \text{ mm}^2$$

Effective reinforcement ratio is computed as

$$\left[\frac{A_p f_{pe}}{bd\, f_{ck}}\right] = \left[\frac{461 \times 800}{150 \times 300 \times 40}\right] = 0.204$$

From Table 4.2, read out for unbonded beams

$$\left[\frac{f_{pu}}{f_{pe}}\right] = 1.16 \quad \therefore \quad f_{pu} = (1.16 \times 800) = 928 \text{ N/mm}^2$$

$$\left[\frac{x_u}{d}\right] = 0.58 \quad \therefore \quad x_u = (0.58 \times 300) = 174 \text{ mm}$$

The ultimate moment of resistance is computed as

$$M_u = A_p f_p \, [d - 0.42\, x_u]$$
$$= (461 \times 928)\,[300 - 0.42 \times 174]$$
$$= 97.708 \times 10^6 \text{ N.mm} = 97.708 \text{ kN.m}$$

Problem 4.9

A pre-tensioned concrete tee-section having a flange width of 1200 mm and flange thickness of 200 mm, thickness of web being 300 mm is prestressed by 2000 mm^2 of high tensile steel located at an effective depth of 1600 mm. If $f_{ck} = 40$ N/mm^2, $f_p = 1600$ N/mm^2, estimate the ultimate flexural strength of the flanged section.

Solution

Data: $b = 1200$ mm, $f_{ck} = 40$ N/mm^2

$b_w = 300$ mm, $f_p = 1600$ N/mm^2

$d = 1600$ mm, $t = 200$ mm

Assuming the neutral axis to fall within the flange

$$\text{Ratio}\left(\frac{A_p f_{pe}}{bd\, f_{ck}}\right) = \left(\frac{2000 \times 1600}{1200 \times 1600 \times 40}\right) = 0.04$$

From Table 4.1, read out the ratio of

$$\left(\frac{f_{pu}}{0.87\, f_p}\right) = 1.0 \quad \therefore \quad f_{pu} = (0.87 \times 1600) = 1392 \text{ N/mm}^2$$

$$\left(\frac{x_u}{d}\right) = 0.087 \quad \therefore \quad x_u = (0.087 \times 1600)$$
$$= 139.2 \text{ mm} < t = 150 \text{ mm}$$

The ultimate moment of resistance is computed as

$$M_u = A_p f_{pu} \, [d - 0.42\, x_u]$$
$$= (2000 \times 1392)\,[1600 - 0.42 \times 139.2]$$
$$= 4291.6 \times 10^6 \text{ N.mm} = 4291.6 \text{ kN.m}$$

Problem 4.10

A post-tensioned prestressed concrete tee-section having a flange width of 1200 mm and flange thickness of 200 mm, thickness of web being 300 mm is prestressed by 2000 mm^2 of high tensile steel located at an effective depth of 1600 mm. If $f_{ck} = 40$ N/mm^2 and $f_p = 1600$ N/mm^2, estimate the ultimate moment capacity of the unbonded tee-section, assuming (L/d) ratio as 20 and f_{pe} = 1000 N/mm^2.

Solution

Data:

$b = 1200$ mm $f_{ck} = 40\ \text{N/mm}^2$

$b_w = 300$ mm $f_p = 1600\ \text{N/mm}^2$

$d = 1600$ mm $\left(\frac{L}{d}\right) = 20$

$t = 200$ mm $f_{pe} = 1000\ \text{N/mm}^2$

Assuming the neutral axis to fall within the flange

$$\text{Ratio}\left(\frac{A_p f_{pe}}{bd\, f_{ck}}\right) = \left(\frac{2000 \times 1000}{1200 \times 1600 \times 40}\right) = 0.026$$

From Table 4.2, corresponding to $\left(\frac{L}{d}\right) = 20$

$$\left(\frac{f_{pu}}{f_{pe}}\right) = 1.34 \quad \therefore \quad f_{pu} = (1.34 \times 1000) = 1340\ \text{N/mm}^2$$

and
$$\left(\frac{x_u}{d}\right) = 0.10 \quad \therefore \quad x_u = (0.1 \times 1600)$$
$$= 160\ \text{mm} < t = 200\ \text{mm}$$

Hence, the neutral axis falls within the flange.

The ultimate flexural strength is computed as

$$\begin{aligned} M_u &= A_p f_{pu}\,[d - 0.42\,x_u] \\ &= [(2000 \times 1340)]\,[1600 - 0.42 \times 160] \\ &= (4107 \times 10^6)\ \text{N.mm} = 4107\ \text{kN.m} \end{aligned}$$

Problem 4.11

A pre-tensioned prestressed concrete tee-section having a flange width of 1200 mm and thickness of flange 150 mm, thickness of web being 300 mm is prestressed by 4700 mm^2 of high tensile steel located at an effective depth of 1600 mm. If $f_{ck} = 40\ \text{N/mm}^2$, $f_p = 1600\ \text{N/mm}^2$, estimate the ultimate moment capacity of the pre-tensioned tee-section.

Solution

Given data:

$b = 1200$ mm $f_{ck} = 40\ \text{N/mm}^2$

$b_w = 300$ mm $f_p = 1600\ \text{N/mm}^2$

$d = 1600$ mm $t = 150$ mm

$A_p = 4700\ \text{mm}^2$

$$A_p = A_{pw} + A_{pf}$$

But $$A_{pf} = \left[\frac{0.45 f_{ck}(b - b_w)t}{f_p}\right]$$

$$= \left[\frac{0.45 \times 40(1200 - 300)150}{1600}\right]$$

$$= 1518.75 \text{ N/mm}^2$$

$\therefore$ $$A_{pw} = (A_p - A_{pf})$$

$$= (4700 - 1518.75) = 3181.25 \text{ N/mm}^2$$

$$\text{Ratio}\left(\frac{A_{pw} f_p}{b_w d\, f_{ck}}\right) = \left(\frac{3181.25 \times 1600}{300 \times 1600 \times 40}\right) = 0.265$$

From Table 4.1, read out the ratio

$$\left(\frac{f_{pu}}{0.87\, f_p}\right) = 1.0$$

$\therefore$ $$f_{pu} = (0.87 \times 1600) = 1392 \text{ N/mm}^2$$

$$\left(\frac{x_u}{d}\right) = 0.575$$

$\therefore$ $$x_u = (0.575 \times 1600) = 921.4 \text{ mm} > t = 150 \text{ mm}$$

$$M_u = f_{pw} A_{pu}(d - 0.42\, x_u) + 0.45 f_{ck}\,(b - b_w)\, t(d - 0.5t)$$

$$= 1392 \times 3181.25(1600 - 0.42 \times 921.4)$$

$$+ 0.45 \times 40(1200 - 300)\, 150(1600 - 0.5 \times 150)$$

$$= 9007 \times 10^6 \text{ N.mm} = 9007 \text{ kN.m}$$

Problem 4.12

A post-tensioned unbonded prestressed concrete beam of tee-section having a flange width of 1200 mm and thickness of flange 150 mm, thickness of web being 300 mm is prestressed by 4700 mm^2 of high tensile steel located at an effective depth of 1600 mm. If $f_{ck} = 40$ N/mm^2 and $f_p = 1600$ N/mm^2, span to effective depth ratio is 20 and the effective prestress after all losses is 1000 N/mm^2 estimate the ultimate flexural strength of the unbonded section.

Solution

Given data:

$b = 1200$ mm	$f_{ck} = 40$ N/mm^2
$b_w = 300$ mm	$f_p = 1600$ N/mm^2
$d = 1600$ mm	$t = 200$ mm

$$A_p = 4700 \text{ mm}^2 \qquad \left(\frac{L}{d}\right) = 20$$

$$f_{pe} = 1000 \text{ N/mm}^2$$

Referring to problem 4.11, the data being the same, therefore, the values of A_{pw} and A_{pf} will be the same.

Hence, we have, $A_{pw} = 3181.25 \text{ mm}^2$ and $A_{pf} = 1518.75 \text{ N/mm}^2$

$$\text{Ratio of } \left(\frac{A_{pw} f_{pe}}{b_w d\, f_{ck}}\right) = \left(\frac{3181.25 \times 1000}{300 \times 1600 \times 40}\right) = 0.165$$

From Table 4.2 for (L/d) ratio of 20, read out the ratio

$$\left(\frac{f_{pu}}{f_{pe}}\right) = 1.185 \quad \therefore \quad f_{pu} = (1.185 \times 1000) = 1185 \text{ N/mm}^2$$

$$\left(\frac{x_u}{d}\right) = 0.49 \quad \therefore \quad x_u = (0.49 \times 1600) = 784 \text{ mm} > t$$

Hence, the ultimate moment of resistance of the unbonded tee-section is computed as,

$$\begin{aligned} M_u &= f_{pu} A_{pw} [d - 0.42\, x_u] + A_{pf} f_p (d - 0.5t) \\ &= (1185 \times 3181.25)(1600 - 0.42 \times 784) \\ &\qquad + (1518.75 \times 1600)(1600 - 75) \\ &= 8495 \times 10^6 \text{ N.mm} = 8495 \text{ kN.m} \end{aligned}$$

Problem 4.13

A pre-tensioned prestressed concrete girder of box section 1 m by 1 m overall dimensions has a uniform wall thickness of 200 mm. The girder is prestressed by high tensile wires of area 2250 mm^2, located at an effective depth of 900 mm. If $f_{ck} = 40 \text{ N/mm}^2$, $f_p = 1600 \text{ N/mm}^2$, calculate the ultimate flexural strength of the box girder.

Solution

Given data:

$$b = 1000 \text{ mm} \qquad f_{ck} = 40 \text{ N/mm}^2$$

$$b_w = 400 \text{ mm} \qquad f_p = 1600 \text{ N/mm}^2$$

$$d = 900 \text{ mm} \qquad A_p = 2250 \text{ N/mm}^2$$

$$t = 200 \text{ mm}$$

Assuming the neutral axis to fall within the flange, the ratio

$$\left(\frac{A_p f_p}{f_{ck} b d}\right) = \left(\frac{2250 \times 1000}{40 \times 1000 \times 900}\right) = 0.10$$

Refer Table 4.1 and read out the values of

$$\left(\frac{f_{pu}}{0.87 f_p}\right) = 0.1 \quad \therefore \quad f_{pu} = (0.87 \times 1600) = 1392 \text{ N/mm}^2$$

$$\left(\frac{x_u}{d}\right) = 0.217 \quad \therefore \quad x_u (0.217 \times 900) = 195.3 \text{ mm} < t$$

Hence the assumption that neutral axis falls within the flange is correct. Using the relation of M_u, we have

$$\begin{aligned} M_u &= A_p f_{pw} (d - 0.42\, x_u) \\ &= 2250 \times 1392\, (900 - 0.42 \times 195.3) \\ &= 2561 \times 10^6 \text{ N.mm} = 2561 \text{ kN.m} \end{aligned}$$

Problem 4.14

A pre-tensioned beam of rectangular section 400 mm wide by 1000 mm, overall depth is prestressed by 800 mm^2 of high tensile steel at an eccentricity of 300 mm. If $f_{ck} = 40$ N/mm^2, $f_p = 400$ N/mm^2. Estimate the ultimate flexural strength of the section.

Solution

Given data: $A_p = 800 \text{ mm}^2$ $\quad f_{ck} = 40 \text{ N/mm}^2$

$b = 400$ mm $\quad f_p = 1600 \text{ N/mm}^2$

$d = 800$ mm

$$\text{Ratio } \left(\frac{A_p f_p}{f_{ck} bd}\right) = \left(\frac{800 \times 1600}{40 \times 400 \times 800}\right) = 0.1$$

Refer Table 4.1 and read out the values of

$$\left(\frac{f_{pu}}{0.87 f_p}\right) = 1.0 \quad \therefore \quad f_{pu} = (0.87 \times 1600) = 1392 \text{ N/mm}^2$$

and

$$\left(\frac{x_u}{d}\right) = 0.217 \quad \therefore \quad x_u = (0.217 \times 800) = 173.6 \text{ mm}$$

Hence, the ultimate moment of resistance of rectangular section is given by the relation,

$$\begin{aligned} M_u &= A_p f_p\, [d - 0.42\, x_u] \\ &= (800 \times 1600)\, [800 - 0.42 \times 173.6] \\ &= 809 \times 10^6 \text{ N.mm} = 809 \text{ kN.m} \end{aligned}$$

Problem 4.15

A pre-tensioned prestressed concrete beam of symmetrical I-section has flanges 160 mm wide by 70 mm thick. The thickness of the web is 50 mm. The overall depth of the I-section is 320 mm. The beam is prestressed by four high tensile wires of 7 mm diameter at an effective depth of 265 mm. If f_{ck} = 50 N/mm^2, and f_p = 1600 N/mm^2, compute the ultimate flexural strength of the I-section.

Solution

Given data:

$$A_p = (4 \times 38.5) = 154 \text{ mm}^2$$
$$b = 160 \text{ mm}$$
$$f_{ck} = 50 \text{ N/mm}^2$$
$$d = 265 \text{ mm}$$
$$f_p = 1600 \text{ N/mm}^2$$

The ratio $\left(\dfrac{A_p f_p}{f_{ck} b d}\right) = \left(\dfrac{154 \times 1600}{50 \times 160 \times 265}\right) = 0.116$

Referring to Table 4.1, read out

$$\left(\frac{f_{pu}}{0.87 f_p}\right) = 1.00 \quad \therefore \quad f_{pu} = (0.87 \times 1600) = 1392 \text{ N/mm}^2$$

$$\left(\frac{x_u}{d}\right) = 0.24 \quad \therefore \quad x_u = (0.24 \times 265) = 64 \text{ mm}$$

Hence, the ultimate moment of resistance is computed as

$$M_u = A_p f_{pu} [d - 0.42\, x_u]$$
$$= (154 \times 1392)\,[265 - 0.42 \times 64)$$
$$= 51.04 \times 10^6 \text{ N.mm} = 51.04 \text{ kN.m}$$

Problem 4.16

A simply supported post-tensioned prestressed concrete deck slab of a road bridge is 500 mm thick spanning over 10 m. The slab is prestressed by Freyssinet cables each containing 12 high tensile wires of 8 mm diameter. The cable are spaced at 500 mm centres at an effective depth of 450 mm. If f_{ck} = 40 N/mm^2, f_p = 1600 N/mm^2, estimate the

a. Ultimate flexural strength of the slab for 1 m width
b. Maximum permissible uniformly distributed ultimate live load on slab assuming a load factor of 1.5 for dead load.

Solution

Given data: $b = 1000$ mm

$A_p = (12 \times 50) = 600$ mm^2/cable

$d = 450$ mm

$L = 10$ m

$f_{ck} = 40$ N/mm^2

For 1 m width of slab

$A_p = (2 \times 600) = 1200$ mm^2/cable

$f_p = 1600$ N/mm^2

L.F. = 1.5 for dead load

$$\text{Ratio} \left(\frac{A_p f_p}{f_{ck} b d}\right) = \left(\frac{1200 \times 1600}{40 \times 1000 \times 450}\right) = 0.106$$

From Table 4.1, read out the ratio

$$\left(\frac{f_{pu}}{0.87 f_p}\right) = 1.0 \quad \therefore \quad f_{pu} = (0.87 \times 1600) = 1302 \text{ N/mm}^2$$

and $$\left(\frac{x_u}{d}\right) = 0.218 \quad \therefore \quad x_u = (0.218 \times 450) = 98.1 \text{ mm}$$

Hence, the ultimate flexural strength is computed as

$$M_u = A_p f_p [d - 0.42 x_u]$$
$$= 1200 \times 1600 [450 - 0.42 \times 98.1]$$
$$= 785 \times 10^6 \text{ N.mm} = 785 \text{ kN.m}$$

If w_u = Ultimate load,

$$w_u = (g_u + q_u) = \left(\frac{8M_u}{L^2}\right) = \left(\frac{8 \times 785}{10^2}\right) = 62.8 \text{ kN/m}$$

$$g_u = [1.5 \times 0.5 \times 1 \times 24] = 18 \text{ kN/m}$$

$$\therefore \quad q_u = (w_u - g_u) = (62.8 - 18) = 44.8 \text{ kN/m}$$

Problem 4.17

A tensioned prestressed concrete beam which is bonded is of rectangular section 300 mm wide by 650 mm overall depth. It is prestressed by 800 mm^2 area of high tensile steel at an effective depth of 600 mm. The section is also reinforced with two HYSD bars of 25 mm diameter on the tension side at an effective cover of 50 mm. If $f_{ck} = 40$ N/mm^2, $f_p = 1600$ N/mm^2, $f_y = 415$ N/mm^2, estimate the ultimate flexural strength of the section.

Solution

Given data: $b = 300$ mm $\quad f_p = 1600$ N/mm^2

$d = 600$ mm $\quad A_{p1} = 800$ mm^2

$f_{ck} = 40\ \text{N/mm}^2 \qquad A_s = \dfrac{(2 \times 3.14 \times 25^2)}{4}$

$= 156.25\ \text{N/mm}^2$

$f_y = 415\ \text{N/mm}^2$

The supplementary reinforcement of area A_s is converted into an equivalent area of high tensile steel given by the relation

$$A_{pe} = \frac{(A_s \cdot f_y)}{f_p} = \frac{(156.25 \times 415)}{1600} = 40.52\ \text{mm}^2$$

$\therefore$ Total area of high tensile steel $= (A_{p1} + A_{pe})$

$= (800 + 40.52) = 840.52\ \text{mm}^2$

$$\text{Ratio} \left(\frac{A_p f_p}{f_{ck} bd}\right) = \left(\frac{840.52 \times 1600}{40 \times 300 \times 600}\right) = 0.186$$

From Table 4.1, read out the ratio

$$\left(\frac{f_{pu}}{0.87 f_p}\right) = 1.0 \quad \therefore \quad f_{pu} = (1.0 \times 0.87 \times 1600)$$

$$= 1392\ \text{N/mm}^2$$

and $$\left(\frac{x_u}{d}\right) = 0.39 \quad \therefore \quad x_u = (0.39 \times 600) = 234\ \text{mm}$$

The ultimate flexural strength of the section is computed as

$$M_u = A_p f_{pu} [d - 0.42\, x_u]$$
$$= 840.52 \times 1392 [600 - 0.42 \times 234]$$
$$= (587 \times 10^6)\ \text{N.mm} = 587\ \text{kN.m}$$

Problem 4.18

A composite tee-section is made up of a precast pre-tensioned rib 100 mm wide by 350 mm deep and a cast *in situ* slab 400 mm wide by 100 mm thick. The precast rib is prestressed by 8 high tensile wires of 5 mm diameter with an effective cover of 50 mm. The characteristic compressive strength of concrete in rib and cast *in situ* slab are 40 and 20 mm^2, respectively. If the tensile strength of tendons is 1600 N/mm^2, estimate the ultimate flexural strength of the composite section.

Solution

Given data: $b = 400$ mm $\qquad f_{ck}\,(\text{rib}) = 40\ \text{N/mm}^2$

$d = 400$ mm $\qquad f_{ck}\,(\text{slab}) = 20\ \text{N/mm}^2$

$$f_p = 1600\ \text{N/mm}^2$$
$$A_p = (8 \times 20) = 160\ \text{N/mm}^2$$

Assuming the neutral axis to fall within the depth of cast *in situ* slab, the effective reinforcement ratio is given by

$$\left(\frac{A_p f_p}{f_{ck} bd}\right) = \left(\frac{160 \times 1600}{20 \times 400 \times 400}\right) = 0.08$$

From Table 4.1, read out the ratio

$$\left(\frac{f_{pu}}{0.87 f_p}\right) = 1.0 \quad \therefore \quad f_{pu} = (0.87 \times 1600) = 1392\ \text{N/mm}^2$$

and $$\left(\frac{x_u}{d}\right) = 0.16 \quad \therefore \quad x_u = (0.16 \times 400)$$
$$= 64\ \text{mm} < t = 100\ \text{mm}$$

Hence, the assumption of neutral axis falling within the slab is correct.

The ultimate moment of resistance is computed as

$$M_u = A_p f_{pu} [d - 0.42\, x_u]$$
$$= 160 \times 1392\, (400 - 0.42 \times 64)$$
$$= 83.1 \times 10^6\ \text{N.mm} = 83.1\ \text{kN.m}$$

Problem 4.19

A pre-tensioned prestressed concrete beam having a rectangular section 150 mm wide and 350 mm deep has an effective cover of 50 mm. If $f_{ck} = 40\ \text{N/mm}^2$, $f_p = 1600\ \text{N/mm}^2$ and the area of high tensile wires is 461 mm^2, estimate the flexural strength of the section using the (IS: 1343-2000) Indian standard code provisions.

Solution

$$f_{ck} = 40\ \text{N/mm}^2 \qquad b = 150\ \text{mm}$$
$$f_p = 1600\ \text{N/mm} \qquad d = 300\ \text{mm}$$
$$A_p = 461\ \text{mm}^2$$

The effective reinforcement ratio is calculated as,

$$\left[\frac{f_p A_p}{f_{ck} b d}\right] = \left[\frac{1600 \times 461}{40 \times 150 \times 300}\right] = 0.40$$

From Table 4.1, the corresponding value of

$$\left[\frac{f_{pu}}{0.87 f_p}\right] = 0.9 \text{ and } \left(\frac{x_u}{d}\right) = 0.783$$

Hence, we have

$$f_{pu} = (0.87 \times 0.9 \times 1600) = 1253 \text{ N/mm}^2$$
$$x_u = (0.783 \times 300) = 234.9 \text{ mm}$$
$$M_u = f_{pu} A_p (d - 0.42 x_u)$$
$$= (1253 \times 461)(300 - 0.42 \times 234.9)$$
$$= (116 \times 10^6) \text{ N.mm} = 116 \text{ kN.m}$$

Problem 4.20

A pre-tensioned beam of rectangular section 300 mm wide by 700 mm deep is stressed by 800 mm^2 of high tensile steel located at effective depth of 600 mm. The beam is also reinforced with supplementary reinforcements consisting of 4 bars of 25 mm diameter of Fe-415 grade HYSD steel, located 100 mm from the soffit. Estimate the flexural strength of the section. Assume the ultimate tensile strength of tendons as 1600 N/mm^2 and the characteristic cube strength of concrete as 40 N/mm^2.

Solution

$b = 300$ mm $\qquad d = 600$ mm

$f_{ck} = 40$ N/mm^2 $\qquad f_p = 1600$ N/mm^2

$f_y = 415$ N/mm^2 $\qquad A_p = 800$ mm^2

$A_s = 1964$ mm^2

The untensioned supplementary reinforcement is replaced by an equivalent area of prestressing steel given by the relation

$$\left[\frac{A_s f_y}{f_p}\right] = \left[\frac{1964 \times 415}{1600}\right] = 509 \text{ mm}^2$$

Total area of prestressing steel = A_p = (800 + 509) = 1309 mm^2

$$\text{Ratio} \left[\frac{f_p A_p}{f_{ck}\, b d}\right] = \left[\frac{1600 \times 1309}{40 \times 300 \times 600}\right] = 0.29$$

From Table 4.1, the corresponding value of

$$\left[\frac{f_{pu}}{0.87 f_p}\right] = 0.86 \text{ and } \left(\frac{x_u}{d}\right) = 0.63$$

Hence, we have

$$f_{pu} = (0.86 \times 0.87 \times 1600) = 1197 \text{ N/mm}^2$$
$$x_u = (0.63 \times 600) = 378 \text{ mm}$$
$$M_u = f_{pu} A_p (d - 0.42 x_u)$$
$$= (1197 \times 1309)(600 - 0.42 \times 378)$$
$$= (691.36 \times 10^6) \text{ N.mm} = 691.36 \text{ kN.m}$$

5

Shear and Torsional Resistance of PSC Sections

Problem 5.1

The horizontal prestress at the centroid of a concrete beam of rectangular cross-section, 120 mm by 250 mm, is 7 N/mm^2 and the maximum shearing force on the beam is 70 kN. Calculate the maximum principal tensile stress. What is the minimum vertical prestress required to eliminate this principal tensile stress?

Solution

The cross-section of the beam is shown in Fig. 5.1.

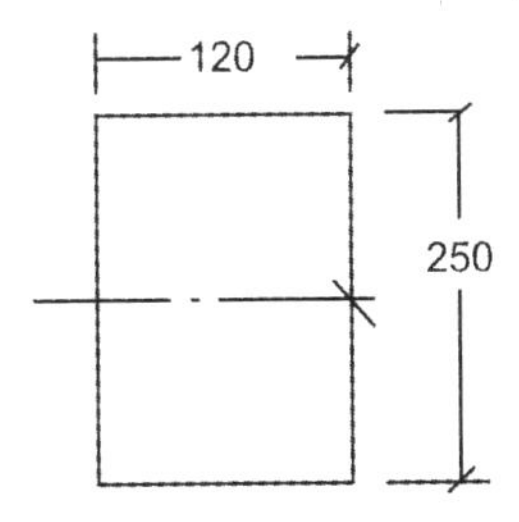

Fig. 5.1

$$V = 70 \text{ kN} \qquad b = 120 \text{ mm}$$

$$D = 250 \text{ mm} \qquad \sigma_x = 7 \text{ N/mm}^2$$

$$\tau = \frac{3}{2}\left(\frac{V}{bD}\right) = \frac{3}{2}\left(\frac{70 \times 10^3}{120 \times 250}\right) = 3.5 \text{ N/mm}^2$$

$$\sigma_{\substack{max \\ min}} = \frac{\sigma_x}{2} \pm \frac{1}{2}\sqrt{\sigma_x^2 + 4\tau^2}$$

$$= \frac{7}{2} \pm \frac{1}{2}\sqrt{7^2 + 4 \times 3.5^2}$$

$$= 3.5 \pm 4.9$$

Maximum principal tensile stress = 3.5 – 4.9 = –1.4 N/mm²

If σ_y = Vertical prestress

$$0 = \frac{7+\sigma_y}{2} - \frac{1}{2}\sqrt{(7-\sigma_y)^2 + 4\times 3.5^2}$$

Solving $\sigma_y = 1.75\ \text{N/mm}^2$

Problem 4.2

A pre-tensioned concrete beam, having an unsymmetrical I-section, has a fibre stress distribution of 13 N/mm² (compression) at the top edge linearly reducing to zero at the bottom. The top flange width and thickness are 2400 and 400 mm respectively, the bottom flange width and thickness are 1200 and 900 mm respectively, and the depth and thickness of web are 1000 and 600 mm respectively. The total vertical service load shear in the concrete at the section is 2350 kN, compute and compare the principal tensile stress at the centroidal axis and the junction of the web with the lower flange.

Solution

The unsymmetrical cross-section of the beam with the horizontal prestress distribution is shown in Fig. 5.2.

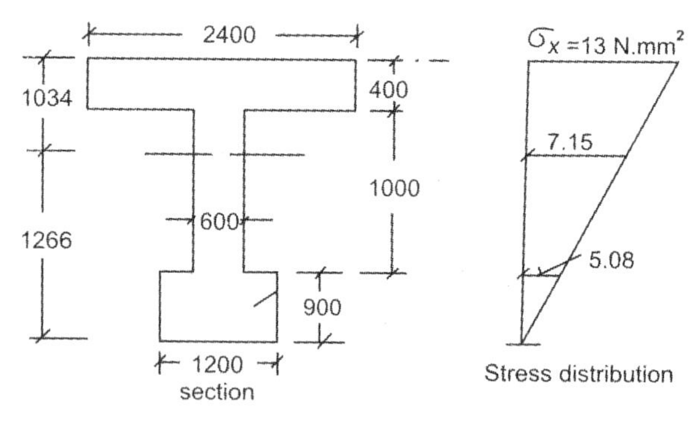

Fig. 5.2

$$V = 2350\ \text{kN} \qquad I_{xx} = 1.54\times 10^{12}\ \text{mm}^4$$

$$\tau_{xx} = \frac{V}{Ib}[A\bar{y}]$$

$$= \frac{2350\times 10^3}{1.54\times 10^{12}\ 600}[(2400\times 400\times 834) + (634\times 600\times 317)]$$

$$= 2.33\ \text{N/mm}^2$$

τ (Junction of web and bottom flange)

$$= \left[\frac{2350 \times 10^3}{1.54 \times 10^{12} \times 600}\right](1200 \times 900)816$$

$$= 2.23 \text{ N/mm}^2$$

Principal tensile stress

$$\text{At the centroid} = \left(\frac{7.15}{2}\right) - \frac{1}{2}\sqrt{7.15^2 + 4 \times 2.33^2}$$

$$= 0.7 \text{ N/mm}^2$$

At junction of web and lower flange

$$= \left(\frac{5.08}{2}\right) - \frac{1}{2}\sqrt{5.08^2 + 4 \times 2.23^2}$$

$$= 0.83 \text{ N/mm}^2$$

Problem 5.3

A concrete beam of rectangular section, 200 mm wide and 600 mm deep, is prestressed by a parabolic cable located at an eccentricity of 100 mm at mid-span and zero at the supports. If the beam has a span of 10 m and carries a uniformly distributed live load of 4 kN/m, find the effective force necessary in the cable for zero shear stress at the support section. For this condition calculate the principal stresses. The density of concrete is 24 kN/m³.

Solution

The longitudinal elevation and cross-section of the beam is shown in Fig. 5.3.

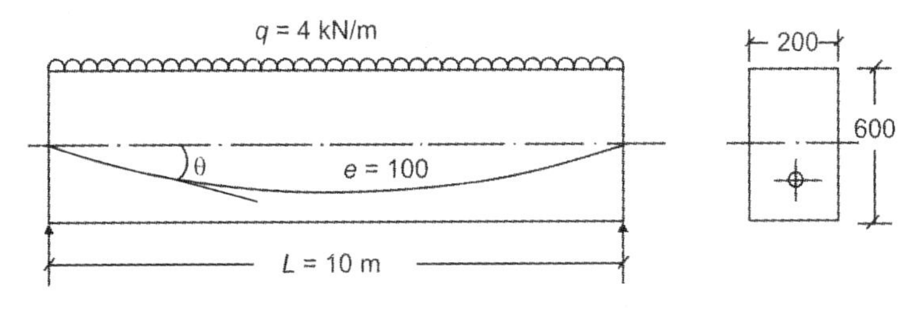

Fig. 5.3

Self weight	= (0.2 × 0.6 × 24)	= 2.88 kN/m
Live load		= 4.00
Total load		= 6.88 kN/m

$$V = \left(\frac{wL}{2}\right) = \left(\frac{6.88 \times 10}{2}\right) = 34.40 \text{ kN}$$

If $\quad P$ = Prestressing force

Slope of cable = $\sin\theta$

$$= \theta = \left(\frac{4e}{L}\right) = \left(\frac{4 \times 0.1}{10}\right)$$

$$= 0.04$$

$$P \sin\theta = 34.40$$

$$\therefore \quad P = 860 \text{ kN}$$

$$\text{Direct stress} = \left(\frac{P}{A}\right) = \left(\frac{860 \times 10^3}{200 \times 600}\right)$$

$$= 7.2 \text{ N/mm}^2$$

Problem 5.4

The shear stress due to imposed load at the centre of the web in an I-section is 3 N/mm^2 and the horizontal prestress at this point is 8.4 N/mm^2. The details of the cross-section are:

Width of top and bottom flange = 250 mm, average thickness of top and bottom flange = 120 and 80 mm, respectively. Overall depth = 750 mm, and thickness of web = 80 mm. Find the increase in the principal tensile stress, if due to eccentricity of the load, a torque of 5 kN.m is applied on the section.

Solution

The symmetrical I-section of the concrete beam is shown in Fig. 5.4.

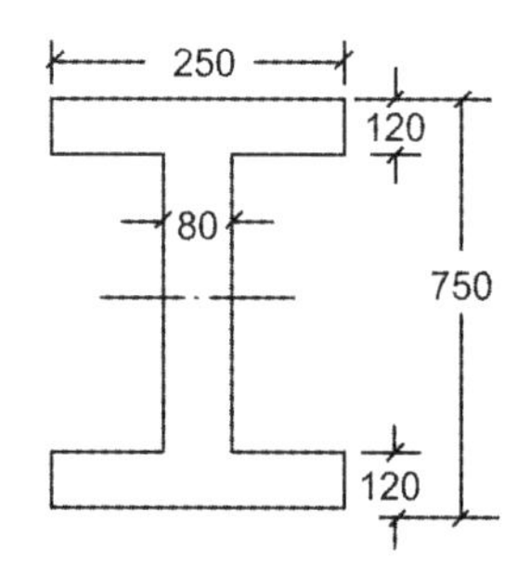

Fig. 5.4

Torque $\quad T$ = 5 kN.m (web)

τ = 3 N/mm^2 at the centre of web

Shear stress at centre of web due to torque is given by

$$\tau_T = \left(\frac{3T\,t_i}{b_i\,t_i^3}\right) = \left(\frac{3\times5\times10^6\times80}{510\times80^3}\right)$$

$$= 4.28\ \text{N/mm}^2$$

Total shear stress at centre of web

$$= (3 + 4.28) = 7.28\ \text{N/mm}^2$$

Horizontal prestress (σ_x) = 8.4 N/mm^2

Principal tensile stress without torque

$$= \left(\frac{8.4}{2}\right) - \frac{1}{2}\sqrt{8.4^2 + 4\times3^2} = -0.9\ \text{N/mm}^2$$

Principal tensile stress with torque

$$= \left(\frac{8.4}{2}\right) - \frac{1}{2}\sqrt{8.4^2 + 4\times7.28^2} = -1.35\ \text{N/mm}^2$$

Increase in tensile stress = (1.35 – 0.9) = 0.45 N/mm^2

Problem 5.5

A concrete beam of rectangular section, 250 mm wide and 650 mm overall depth, is subjected to a torque of 20 kN.m and a uniform prestressing force of 150 kN. Calculate the maximum principal tensile stress. Assuming 15 percent loss of prestress, calculate the prestressing force necessary to limit the principal tensile stress to a value of 0.4 N/mm^2.

Solution

The rectangular cross-section of the beam is shown in Fig. 5.5.

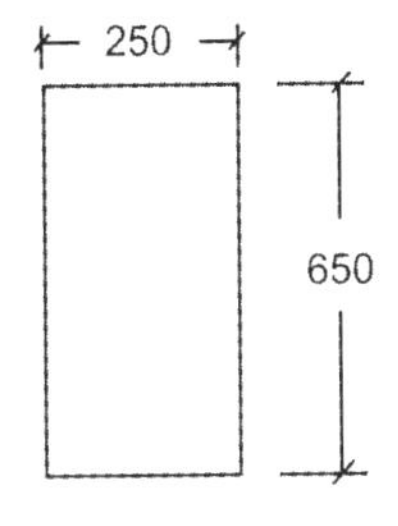

Fig. 5.5

T = 20 kN.m $\quad$ P = 150 kN

b = 250 mm $\quad$ h = 650 mm

$$\tau_T = \frac{T\left(3+2\frac{b}{h}\right)}{b^2 h} = \frac{20\times10^6\left[3+2\left(\frac{250}{650}\right)\right]}{250^2\times650}$$

$$= 1.8 \text{ N/mm}^2$$

$$\sigma_x = \left(\frac{P}{A}\right) = \left(\frac{150\times10^3}{250\times650}\right) = 0.92 \text{ N/mm}^2$$

$$\sigma_{min} = \left(\frac{0.92}{2}\right) - \frac{1}{2}\sqrt{0.92^2 + 4\times1.8^2}$$

$$= -1.40 \text{ N/mm}^2$$

If P = Prestressing force required to limit tensile stress to a value of 0.4 N/mm²

$$\sigma_x = \left(\frac{0.85P\times10^3}{250\times650}\right) - 0.4$$

$$= \left(\frac{1.85P\times10^3}{250\times650\times2}\right) - \frac{1}{2}\sqrt{\left(\frac{0.85P\times10^3}{250\times650\times2}\right)^2 + 4\times1.8^2}$$

Solving $P = 1570$ kN

Problem 5.6

A concrete beam of rectangular section, 300 mm wide and 800 mm deep is subjected to a twisting moment of 30 kN.m and a prestressing force of 150 kN acting at an eccentricity of 220 mm. Calculate the maximum principal tensile stress. If the beam is subjected to a bending moment of 100 kN.m in addition to the twisting moment, calculate the maximum principal tensile stress.

Solution

The rectangular cross-section of the beam is shown in Fig. 5.6.

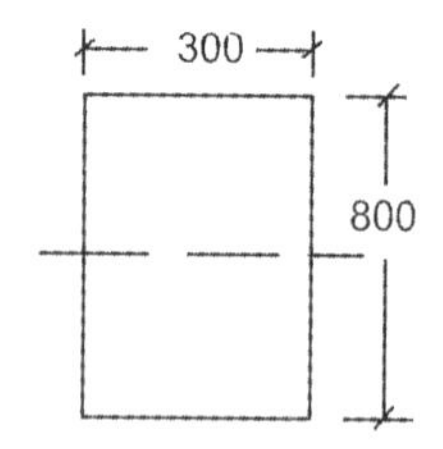

Fig. 5.6

$$T = 30 \text{ kN.m} \qquad P = 150 \text{ kN}$$

$$e = 220 \text{ mm}$$

$$Z = \left(\frac{300 \times 800^2}{6}\right) = 32 \times 10^6 \text{ mm}^2$$

Shear stress at soffit of the beam

$$\tau_T = \frac{30 \times 10^6 \left[3 + 2\left(\frac{800}{300}\right)\right]}{800^2 \times 300} = 1.3 \text{ N/mm}^2$$

Compressive or tensile stress due to P

$$= \left[\frac{150 \times 10^3}{300 \times 800} \pm \frac{150 \times 10^3 \times 220}{32 \times 10^6}\right]$$

$$= -0.405 \text{ N/mm}^2 \ (1.655 \text{ N/mm}^2)$$

$$\therefore \qquad \sigma_{min} = -\left(\frac{0.405}{2}\right) - \frac{1}{2}\sqrt{(0.405)^2 + 4 \times 1.3^2}$$

$$= -1.52 \text{ N/mm}^2$$

Tensile stress at bottom due to B.M.

$$= \left(\frac{100 \times 10^6}{32 \times 10^6}\right) = -3.125 \text{ N/mm}^2$$

Prestress + tensile stress = (1.655 – 3.125) = –1.47 N/mm^2

$$\therefore \qquad \sigma_{min} = -\left(\frac{1.47}{2}\right) - \frac{1}{2}\sqrt{1.47^2 + 4 \times 1.3^2}$$

$$= -2.22 \text{ N/mm}^2$$

Problem 5.7

A concrete box section girder has an overall depth and width of 800 and 600 mm, respectively. The concrete walls are 100 mm thick on both the horizontal and vertical parts of the box. Determine the maximum permissible torque, if the section is uniformly prestressed by a force of 200 kN. Assume the maximum permissible diagonal tensile stress as 0.7 N/mm^2.

Solution

The cross-section of the concrete box girder is shown in Fig. 5.7.

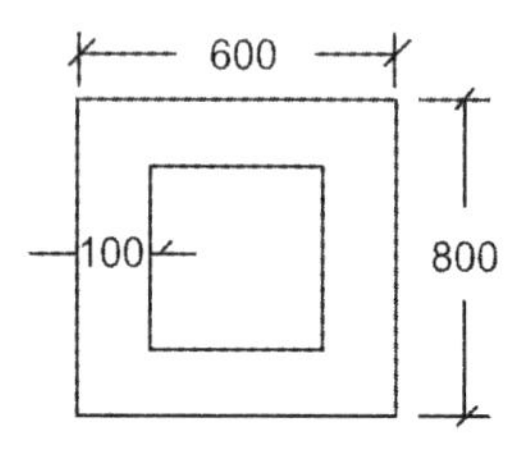

Fig. 5.7

$$P = 200 \text{ kN}$$

$$\tau_t = \left(\frac{T}{2At_i}\right)$$

where, $A = [(600 \times 800) - (400 \times 600)] = 24 \times 10^4 \text{ mm}^2$

$$\text{Direct stress} = \left(\frac{200 \times 10^3}{2400 \times 100}\right) = 0.83 \text{ N/mm}^2$$

Maximum permissible tensile stress = 0.7 N/mm²

$$\tau_T = \left[\frac{T}{2b_w(h - h_f)(b - b_w)}\right]$$

$$= \left[\frac{T}{2 \times 100(800 - 100)(600 - 100)}\right]$$

$$= \left(\frac{T}{7 \times 10^7}\right) \qquad \therefore \quad T = (\tau_T \times 7 \times 10^7)$$

$$\sigma_{min} = -7 = \left(\frac{0.83}{2}\right) - \frac{1}{2}\sqrt{0.83^2 + 4\tau^2}$$

$$\therefore \quad (-7 - 0.42)^2 = \frac{1}{4}(0.83^2 + 4\,\tau^2)$$

$$\therefore \quad \tau_T = 7.41 \text{ N/mm}^2$$

$$\therefore \quad T = \tau_T.7 \times 10^7$$

$$= 7.41 \times 7 \times 10^7 \text{ N.mm}$$

$$= 518 \times 10^6 \text{ N.mm} = 518 \text{ kN.m}$$

Problem 5.8

An unsymmetrical I-section bridge girder has the following sectional properties:

Area of cross-section = 777×10^3 mm^2, second moment of area = 22×10^{10} mm^4, width and thickness of top flange = 1200 and 360 mm, respectively, and thickness of web = 240 mm. The centroid of the section is located at 580 mm from the top. The girder is used over a span of 40 m. The tendons with a cross-section of 700 mm^2 are parabolic with an eccentricity of 1220 mm at the centre of span and zero at the supports. The effective prestress in the wires is 800 N/mm^2. If the tensile strength of concrete is 4.5 N/mm^2, estimate the ultimate shear resistance of the section, assuming failure to take place when the principal tensile stress reaches a value equal to the tensile strength of concrete. Overall depth is 2000 mm.

Solution

The unsymmetrical I-section of the prestressed concrete beam is shown in Fig. 5.8.

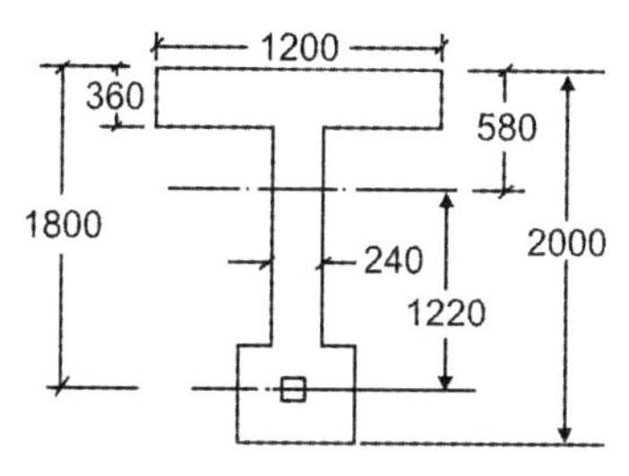

Fig. 5.8

$$I = 22 \times 10^{10} \text{ mm}^4 \qquad A = 777 \times 10^3 \text{ mm}^2$$

$$f_t = 4.5 \text{ N/mm}^2$$

$$S = A\bar{y} = [(1200 \times 360 \times 400) + (240 \times 220 \times 110)]$$

$$= 17.78 \times 10^7 \text{ mm}^3$$

$$\therefore \quad e = (1220 - 580) = 640 \text{ mm}$$

$$V_{cw} = b\left(\frac{I}{S}\right)\sqrt{f_t^2 + f_{cp} \cdot f_t} + P \cdot \sin\theta$$

$$\therefore \quad P = \frac{(700 \times 800)}{100} = 560 \text{ kN}$$

$$f_{cp} = \left(\frac{560 \times 10^3}{777 \times 10^3}\right) = 0.72 \text{ N/mm}^2$$

$$\text{Slope of cable at support} = \left(\frac{4e}{L}\right) = \frac{(4\times1.22)}{40} = 0.122$$

$$V_{cw} = 240\left(\frac{22\times10^{10}}{17.78\times10^{7}}\right)\sqrt{4.5^2+(0.72\times4.5)} + \left(560\times10^3\times0.122\right)$$

$$= 1507396 \text{ N} = 1507 \text{ kN}$$

According to IS: 1343 code, shear strength is computed as

$$V_{cw} = 0.67b_w D\sqrt{f_t^2+0.8f_{cp}f_t} + P\cdot\sin\theta$$

$$= (0.67\times240\times2000)\sqrt{4.5^2+(0.8\times0.72\times4.5} + \left(560\times10^3\times0.122\right)$$

$$= 1605600 \text{ N} = 1605 \text{ kN}$$

Problem 5.9

A prestressed concrete beam of rectangular section, 300 mm wide by 600 mm deep is prestressed by two post-tensioned cables of area 600 mm^2 each initially stressed to 1600 N/mm^2. The cables are located at a constant eccentricity of 100 mm. The span of the beam is 10 m. If f_{ck} = 40 N/mm^2, estimate the ultimate shear resistance of support section uncracked in flexure.

Solution

Given data:

b = 300 mm A_s = (2 × 600) = 1200 mm^2

D = 600 mm A = (300 × 600) = 18 × 10^4 mm^2

e = 100 mm P = (1200 × 1600) = 192 × 10^4 N

Compressive prestress at centroid is given by the relation

$$f_{cp} = \left(\frac{P}{A}\right) = \frac{\left(1992\times10^4\right)}{\left(18\times10^4\right)} = 10.66 \text{ N/mm}^2$$

Tensile strength of concrete

$$= f_t = 0.24\sqrt{f_{ck}} = 0.24\sqrt{40} = 1.517 \text{ N/mm}^2$$

According to IS: 1343 code, the ultimate shear strength of support section uncracked in flexure is given by the relation

$$V_{co} = 0.67\,bD\sqrt{f_t^2 + 0.8 f_{cp} f_t}$$

$$= (0.67 \times 300 \times 600)\sqrt{1.517^2 + 0.8 \times 10.66 \times 1.517}$$

$$= 475164 \text{ N} = 475.164 \text{ kN}$$

Problem 5.10

The support section of a prestressed concrete beam, 120 mm wide and 250 mm deep, is required to support an ultimate shear force of 60 kN. The compressive prestress at the centroidal axis is 5 N/mm^2. The characteristic cube strength of concrete is 40 N/mm^2. The cover to the tension reinforcement is 50 mm. If the characteristic tensile strength of steel in stirrups is 250 N/mm^2, design suitable reinforcements at the section using IS: 1343 code specifications.

Solution

Given data: $V_u = 60$ kN $\quad b = 120$ mm

$f_y = 250$ N/mm^2 $\quad f_{ck} = 40$ N/mm^2

$D = 250$ mm $\quad f_t = 0.24\sqrt{40} = 1.517$ N/mm^2

$f_{cp} = 5$ N/mm^2 $\quad d = 200$ mm

For the section uncracked in flexure, the shear strength is

$$V_{co} = 0.67\,bD\sqrt{f_t^2 + 0.8 f_{cp} f_t}$$

$$= (0.67 \times 120 \times 250)\sqrt{1.517^2 + (0.8 \times 5 \times 1.517)}$$

$$= 58088 \text{ N} = 58 \text{ kN}$$

Balance shear force = $(V_u - V_{co}) = (60 - 58) = 2$ kN

Using 6 mm diameter two legged stirrups,

$$\text{Spacing of stirrups is } = s_v = \left[\frac{A_{sv} \cdot 0.87 f_y d}{(V_u - V_{co})}\right]$$

$$= \left[\frac{(2 \times 28.2 \times 0.87 \times 250 \times 250)}{2 \times 10^3}\right]$$

$$= 1500 \text{ mm}$$

Maximum spacing of stirrups not to exceed 0.75 d

$\therefore$ Spacing $s_v \not> 0.75\,d \not> (0.75 \times 250) = 187$ mm

Hence, adopt 6 mm diameter two legged stirrups at 180 mm centres.

Problem 5.11

A double tee-section used as bridge girder is made up of a flange 1200 mm wide by 150 mm thick. The ribs are 150 mm wide by 1500 mm deep. The section is prestressed by high tensile wires of area 4700 mm^2 located at 850 mm from the soffit of the girder. The initial stress in the wires is 1200 N/mm^2. If f_{ck} = 40 N/mm^2, estimate the ultimate shear strength of the support section of the girder.

Solution

Given data: $b = 1200$ mm

Overall depth $= D = 1650$ mm

$b_w = (2 \times 150) = 300$ mm

$A = (1200 \times 150) + (1500 \times 300) = 63 \times 10^4$ mm^2

$f_x = 1200$ N/mm^2

$f_{ck} = 40$ N/mm^2

$f_t = 0.24\sqrt{f_{ck}} = 0.24\sqrt{40} = 1.517$ N/mm^2

$P = (4700 \times 1200) = 564 \times 10^4$ N

$$f_{cp} = \left(\frac{P}{A}\right) = \frac{(564 \times 10^4)}{(63 \times 10^4)} = 8.95 \text{ N/mm}^2$$

The ultimate shear strength of double tee-section is

$$V_{co} = 0.67\, b_w D\sqrt{f_t^2 + 0.8 f_{cp} f_t}$$

$$= (0.67 \times 300 \times 1650)\sqrt{1.517^2 + (0.8 \times 8.95 \times 1.517)}$$

$$= 1203117 \text{ N} = 1203.117 \text{ kN}$$

Problem 5.12

A prestressed concrete bridge girder of unsymmetrical I-section has an overall depth of 1300 mm with a cross-sectional area of 328500 mm^2. Thickness of web is 150 mm, span of the girder is 25 m. The girder is prestressed by a parabolic cable having an eccentricity of 650 mm at centroid and 285 mm at support sections. The effective prestressing force in the cable is 2070 kN. If f_t = 1.6 N/mm^2, estimate the ultimate shear resistance of the support section.

Solution

Given data: b_w = 150 mm e_1 = 650 mm

A = 328500 mm^2 e_2 = 285 mm

P = 2070 kN L = 25 m

f_t = 1.6 N/mm^2

$$f_{cp} = \left(\frac{P}{A}\right) = \frac{(2070 \times 10^3)}{(328500)} = 6.3 \text{ N/mm}^2$$

$$\text{Slope of cable at support} = \Theta = \left(\frac{4e}{L}\right) = 4\frac{(e_1 - e_2)}{L}$$

$$= \frac{4(650 - 285)}{(25 \times 10^3)} = 0.0585$$

The ultimate shear strength of the support section is given by

$$V_{co} = 0.67\, b_w D \sqrt{f_t^2 + 0.8 f_{cp} f_t} + P \sin\Theta$$

$$= (0.67 \times 150 \times 1300)\sqrt{1.6^2 + (0.8 \times 6.3 \times 1.6)} + (2070 \times 10^3 \times 0.0585)$$

$$= 546 \times 10^3 \text{ N} = 546 \text{ kN}$$

Problem 5.13

The cross section of a prestressed concrete beam is an unsymmetrical I-section with an overall depth of 1300 mm. Thickness of web is 150 mm. Distance of top and bottom fibres from the centroid are 545 and 755 m, respectively. At a particular section, the beam is subjected to an ultimate moment M = 2130 kN.m and an ultimate shear force of 237 kN. Effective depth d = 1100 mm, f_{ck} = 45 N/mm^2. Effective prestress at tensile face f_{pt} = 19.3 N/mm^2, I = 665 × 10^8 mm^4, A_p = 2310 mm^2 and f_p = 1500 N/mm^2. Effective prestress in tendons after lossess f_{pe} = 890 N/mm^2. Estimate the ultimate shear resistance of the section cracked in flexure using IS: 1343 code specifications.

Solution

$$M_o = \left[\frac{0.8 f_{pt} I}{y_b}\right] = \left[\frac{0.8 \times 19.3 \times 665 \times 10^8}{755}\right]$$

$$= 138 \times 10^7 \text{ N.mm}$$

$$\left(\frac{100A_p}{b_w d}\right) = \left(\frac{100 \times 2310}{150 \times 1100}\right) = 1.40$$

From Table 8 of IS: 1343-2012 code for $f_{ck} = 45$ N/mm², read out the value of $\xi_c = 0.77$

The ultimate shear resistance of section cracked in flexure is

$$V_{cr} = \left[1 - 0.55\left(\frac{f_{pe}}{f_p}\right)\right]\xi_c\, b d + M_o\left(\frac{V}{M}\right)$$

$$= \left[1 - 0.55\left(\frac{890}{1500}\right)\right](0.77 + 150 \times 1100)$$

$$+ \left(138 \times 10^7\right)\left[\frac{\left(237 \times 10^3\right)}{\left(2130 \times 10^6\right)}\right]$$

$$= (240 \times 10^3)\text{ N} = 240\text{ kN}$$

But $$V_{cr} \nless 0.1 b \cdot d\sqrt{f_{ck}}$$

$$\nless (0.1 \times 150 \times 1100\sqrt{45} \nless 110685\text{ N}$$

$$\nless 110.685\text{ kN}$$

Problem 5.14

The support section of a prestressed concrete beam 100 mm wide by 250 mm deep is required to support an ultimate shear force of 60 kN. The compressive prestress at centroid is 5 N/mm², $f_{ck} = 40$ N/mm², effective cover to reinforcement = 50 mm. If $f_y = 415$ N/mm², design suitable shear reinforcement in the section using IS: 1343 code recommendations.

Solution

Given data: $b = 100$ mm $\quad f_y = 145$ N/mm²

$D = 200$ mm

$$f_t = 0.24\sqrt{f_{ck}} = 0.24\sqrt{40} = 1.517\text{ N/mm}^2$$

$V = 60$ kN $\quad f_{ck} = 40$ N/mm²

For the support section uncracked in flexure

$$V_{co} = 0.67\, bD\sqrt{f_t^2 + 0.8 f_{cp} f_t}$$

$$= (0.67 \times 100 \times 250)\sqrt{1.517^2 + (0.8 \times 5 \times 1.517)}$$

$$= 48407\text{ N} = 48.407\text{ kN}$$

Balance shear force $V_s = (V_u - V_{co}) = (60 - 48.4) = 11.6$ kN

Using 6 mm diameter two legged stirrups, spacing is given by

$$s_v = \left[\frac{A_{sv} \cdot 0.87 f_y d}{(V_u - V_{co})}\right]$$

$$= \left[\frac{(2 \times 28.2 \times 0.87 \times 415 \times 200)}{11.6 \times 10^3}\right] = 351 \text{ mm}$$

Maximum permissible spacing = $s_v \not> 0.75\ d$

$$\not> (0.75 \times 200) \not> 150 \text{ mm}$$

Adopt 6 mm diameter two legged stirrups at 150 mm centres.

Problem 5.15

A prestressed concrete beam of symmetrical I-section has an overall depth of 2 m and the thickness of the web is 200 mm. The effective span of the beam is 40 m. The beam is prestressed by cables which are concentric at supports and have an eccentricity of 750 mm at centre of span. The force in the cable is 12000 kN at transfer stage. $f_{ck} = 60$ N/mm². Estimate the ultimate shear resistance of the support section. If the ultimate shear force at support due to loads is 2834 kN and the loss ratio is 0.8, design suitable shear reinforcements using Fe-415 HYSD bars. Area of section = 0.88×10^6 mm².

Solution

Given data: $b_w = 200$ mm

$$f_{cp} = \frac{(0.8 \times 12000 \times 10^3)}{(0.88 \times 10^6)} = 10.9 \text{ N/mm}^2$$

$D = 2000$ mm

$f_{ck} = 60$ N/mm²

$f_t = 0.24\sqrt{60} = 1.85$ N/mm²

$A = 0.88 \times 10^6$ mm², $P = 2834$ kN

$$\text{Slope of cable at support} = \theta = \left(\frac{4e}{L}\right) = \frac{(4 \times 750)}{(40 \times 10^3)} = 0.075$$

$$V_{co} = 0.67\, b_w D \sqrt{f_t^2 + 0.8 f_{cp} f_t} + \eta\, P \,.\, \sin\theta$$

$$= (0.67 \times 200 \times 2000)\sqrt{1.85^2 + (0.8 \times 10.9 \times 1.85) + \left(0.8 \times 12000 \times 10^3 \times 0.075\right)}$$

$$= 1900 \times 10^3 \text{ N} = 1900 \text{ kN} < 2834 \text{ kN}$$

Balance shear force = (2834 – 1900) = 934 kN

Using 12 mm diameter two legged stirrups, the spacing is

$$s_v = \left[\frac{0.87 \times 415 \times 2 \times 113 \times 1900)}{934 \times 10^3}\right] = 165 \text{ mm}$$

Hence, adopt 12 mm diameter two legged stirrups at 160 mm centres.

Problem 5.16

A concrete beam of rectangular section has a width of 250 mm and depth of 600 mm. The beam is prestressed by a parabolic cable carrying an effective force of 1000 kN. The cable is concentric at supports and has a maximum eccentricity of 100 mm at the centre of span. The beam spans over 10 m and supports an uniformly distributed live load of 20 kN/m. Assuming the density of concrete as 24 kN/m^3, estimate:

a. The maximum principal stress developed in the section of the beam at a distance of 300 mm from the support.
b. The prestressing force required to nullify the shear force due to dead and live loads at the support section.

Solution

$$A = (250 \times 600) = (15 \times 10^4) \text{ mm}^2$$

$$L = 10 \text{ m}$$

$$e = 100 \text{ mm}$$

$$I = \left[\frac{250 \times 600^3}{12}\right] = (45 \times 10^8) \text{ mm}^4$$

$$P = 1000 \text{ kN}$$

$$Z = \left[\frac{250 \times 600^2}{6}\right] = (15 \times 10^6) \text{ mm}^3$$

$$b = 250 \text{ mm and } h = 600 \text{ mm}$$

Self weight of the beam

$$= g = (0.25 \times 0.6 \times 24) = 3.6 \text{ kN/m}$$

Live load on the beam = q = 20 kN/m

Total load on the beam

$= w = (g + q) = (3.6 + 20) = 23.6$ kN/m

Shear force at support section

$= 0.5\,(23.6 \times 10) = 118$ kN

Shear force at a section 300 mm from the support

$= [118 - (0.3 \times 23.60) = 110.92$ kN

Slope of the cable at support

$$= \left[\frac{4e}{L}\right] = \left[\frac{4 \times 100}{10 \times 1000}\right] = 0.04$$

Slope of the cable at 300 mm from support is computed as

$$= \frac{4e}{L^2}\,[(L - 2x)] = \frac{4 \times 100}{10000^2}\,[(10000 - 2 \times 300)]$$

$= 0.0376$

Vertical component of prestressing force at support

$= (118 \times 0.04) = 4.72$ kN

Vertical component of prestressing force at 300 mm from support is obtained as

$= (118 \times 0.0376) = 4.43$ kN

Net shear force at 300 mm from support

$= V = (110.92 - 4.43) = 106.49$ kN

a. The maximum shear stress at 300 mm from support at neutral axis level is given by

$$\tau_v = \frac{3}{2}\left[\frac{V}{bh}\right] = \frac{3}{2}\,\frac{[106.49 \times 10^3]}{(250 \times 600)} = 1.06\ \text{N/mm}^2$$

Direct stress due to prestressing force

$$= \left[\frac{P}{A}\right] = \left[\frac{1000 \times 10^3}{15 \times 10^4}\right] = 6.67\ \text{N/mm}^2$$

Maximum principal stresses

$$= \left(\frac{6.67}{2}\right) \pm \frac{1}{2}\,[6.67^2 + 4 \times 1.06^2]^{0.5}$$

$= 6.835$ N/mm^2 (compression)

$= -0.165$ N/mm^2 (tension)

b. If P = prestressing force required to nullify the shear force at support due to dead and live loads and $\sin\theta = \theta$ = slope of the cable at support,

we have the relation $P \sin\theta = 118$

$$\therefore \quad P = \left[\frac{118}{0.04}\right] = 2950\ \text{kN}.$$

Problem 5.17

A prestressed concrete box girder of overall dimensions 300 mm by 600 mm has a uniform wall thickness of 50 mm. The section is subjected to an uniform compressive prestress of 12 N/mm^2. The maximum shear stress at the section due to transverse loads is 3 N/mm^2. Evaluate the increase in the principal tensile stress when the box section is subjected to a torsional moment of 3 kN.m.

Solution

$$A = \text{area covered by the centre line of the box girder}$$
$$= b_0 h_0$$

Where b_0 = 250 mm and h_0 = 550 mm

Thickness of box walls = t_i = 50 mm

$$f_x = \text{direct stress} = 12 \text{ N/mm}^2 \text{ and } f_y = 0$$
$$\tau = \text{shear stress} = 3 \text{ N/mm}^2$$
$$T = 3 \text{ kN.m}$$

Maximum principal tensile stress is given by the expression

$$f_{min} = \left[\frac{f_x + f_y}{2}\right] - \frac{1}{2}\sqrt{(f_x - f_y)^2 + 4\tau^2}$$
$$= \left[\frac{12}{2}\right] - \frac{1}{2}\sqrt{12^2 + 4 \times 3^2}$$
$$= -0.708 \text{ N/mm}^2 \text{ (tension)}$$

Shear stress due to torque at the centre of web in box section is computed as

$$\tau_{max} = \left[\frac{T}{2At_i}\right] = \left[\frac{3 \times 10^6}{2 \times 250 \times 550 \times 50}\right] = 0.22 \text{ N/mm}^2$$

Total shear stress inclusive of shear due to transverse loads

$$= [3 + 0.22] = 3.22 \text{ N/mm}^2$$

Principal tensile stress with torque is given by

$$f_{min} = \left[\frac{12}{2}\right] - \frac{1}{2}\sqrt{12^2 + 4 \times 3.22^2}$$
$$= -0.809 \text{ N/mm}^2 \text{ (tension)}$$

Increase in principal tensile stress

$$= [0.809 - 0.708] = 0.101 \text{ N/mm}^2.$$

Problem 5.18

A prestressed girder of rectangular section 150 mm wide by 300 mm deep is to be designed to support an ultimate shear force of 130 kN. The uniform prestress across the section is 5 N/mm^2. Given the characteristic cube strength of concrete as 40 N/mm^2 and Fe-415 HYSD bars of 8 mm diameter, design suitable spacing for the stirrups conforming to the Indian standard code IS: 1343-2012 recommendations. Assume cover to the reinforcement as 50 mm.

Solution

$b_w = 150$ mm $\quad f_{ck} = 40\ N/mm^2$

$D = 300$ mm $\quad f_{cp} = 5\ N/mm^2$

$d = 250$ mm $\quad f_y = 415\ N/mm^2$

$V = 130$ kN $\quad f_t = 0.24\sqrt{f_{ck}} = 0.24\sqrt{40} = 1.518\ N/mm^2$

According to the recommendations of the IS: 1343-2012 code, the ultimate shear strength of the section uncracked in flexure is given by

$$V_{cw} = V_c = 0.67\, b_w\, D\, \sqrt{f_t^2 + 0.8\, f_{cp} f_t}$$

$$= 0.67 \times 150 \times 300\, \sqrt{1.518^2 + (0.8 \times 5 \times 1.518)}$$

$$= 87260 \text{ N} = 87.26 \text{ kN}$$

Hence, balance shear $= [V - V_c] = [130 - 87.26] = 42.74$ kN

Using 8 mm diameter two legged stirrups, the spacing of the stirrups is

$$s_v = \left[\frac{A_{Sv}\, 0.87\, f_y d}{(V - V_c)}\right] = \left[\frac{2 \times 50.26 \times 0.87 \times 415 \times 250}{42.74 \times 10^3}\right]$$

$$= 212.28 \text{ mm}$$

Maximum permissible spacing

$$= (0.75\, d) = (0.75 \times 250) = 187.5 \text{ mm}$$

Adopt 8 mm diameter two legged stirrups at 180 mm centres.

Problem 5.19

A concrete girder of rectangular box section has an overall depth of 1200 mm and width of 900 mm. The concrete walls are 150 mm thick throughout the section.

a. Determine the maximum permissible torque, if the section is concentrically prestressed by a force of 450 kN, and the maximum permissible diagonal tensile stress in concrete is 0.63 N/mm^2.

b. Also determine the number of longitudinal and transverse steel reinforcement of Fe-230 grade required for the box section, if the torsional resistance moment of the section is to be increased to 345 kN.m. Allow 50 mm cover to the reinforcements.

Solution

$$\text{Horizontal prestress} = \left[\frac{450 \times 10^3}{(1200 \times 900) - (900 \times 600)}\right] = 0.83\ \text{N/mm}^2$$

If the permissible diagonal tensile stress = 0.63 N/mm² and τ_t is the corresponding shear stress, then we have the relation

$$f_{\min} = -0.63 = \left[\left\{\frac{0.83}{2}\right\} - 0.5\sqrt{0.83^2 + 4\tau_t^2}\right]$$

Solving $\tau_t = 0.97\ \text{N/mm}^2$

a. Maximum permissible torque $T = (2\, A\, t_i\, \tau_t)$

$= [2 \times 150\,(1200 - 150)\,(900 - 150)\,0.97]/10^6$

$= 230$ kN.m

b. Allowing 50 mm cover, we have

$x_1 = [900 - (2 \times 50)] = 800$ mm

$y_1 = [1200 - (2 \times 50)] = 1100$ mm

Permissible torque on the concrete section = T_{tp} = 230 kN.m

Torque to be resisted by reinforcement (spiral and longitudinal steel) is

$$T_{ts} = [(345 \times 10^6) - (230 \times 10^6)] = (115 \times 10^6)\ \text{N.mm}$$

If s is the spacing of the closed stirrups, using 12 mm diameter stirrups, $A_{sv} = 226\ \text{mm}^2$

$$s = \left[\frac{0.8\, f_s A_{sv} x_1 y_1}{T_{ts}}\right] = \left[\frac{0.8 \times 230 \times 226 \times 800 \times 1100}{115 \times 10^6}\right]$$

$= 318$ mm

If A_{s1} = area of longitudinal steel distributed around the hoops, then we have

$$A_{s1} = \left[\frac{A_{sv}\,(x_1 + y_1)}{s}\right] = \left[\frac{226(800 + 1100)}{318}\right]$$

$= 1350\ \text{mm}^2$

Using 12 mm diameter bars, the number of longitudinal bars required are given as (1350/113) = 12 bars. These bars are distributed at regular spacing around the perimeter of the stirrups.

Problem 5.20

A tee-beam has a flange width and thickness of 500 mm and 200 mm respectively. The web is 200 mm thick and 600 mm deep. The beam spanning over 16 m is prestressed using a cable carrying an effective force of 2000 kN. The cable is parabolic with an eccentricity of 600 mm at the centre of span and 300 mm at the supports. Estimate the ultimate shear resistance of the support section assuming M-40 grade concrete. Also evaluate the maximum service load on the beam assuming a load factor of 2 according to the Indian standard code IS: 1343 code specifications.

Solution

$$A = [(500 \times 200) + (200 \times 600)] = (22 \times 10^4)\ \text{mm}^2$$

$$P = 2000\ \text{kN}$$

$$b_w = 200\ \text{mm}$$

$$D = 800\ \text{mm}$$

$$f_t = 0.24\sqrt{f_{ck}} = 0.24\sqrt{40} = 1.517\ \text{N/mm}^2$$

$$e_1 = 600\ \text{mm and } e_2 = 300\ \text{mm}$$

$$L = 16\ \text{m}$$

Load factor = 2

Slope of the cable at support

$$= \theta = \left[\frac{4(e_1 - e_2)}{L}\right] = \left[\frac{4(600-300)}{16\times 10^3}\right] = 0.075$$

$$f_{cp} = \left(\frac{P}{A}\right) = \left(\frac{2000\times 10^3}{22\times 10^4}\right) = 9.09\ \text{N/mm}^2$$

Ultimate shear resistance of support section is expressed as

$$V_{cw} = V_c = 0.67\, b_w D\sqrt{f_t^2 + 0.8 f_{cp} f_t} + P\sin\theta$$

$$= (0.67 \times 200 \times 800)\sqrt{1.517^2 + 0.8\times 9.09\times 1.517} + (2000 \times 10^3 \times 0.075)$$

$$= (541.27 \times 10^3)\ \text{N} = 541.27\ \text{kN}$$

$$\text{Safe working load} = \left(\frac{\text{Ultimate load}}{\text{Load factor}}\right) = \left(\frac{541.27}{2}\right) = 33.83\ \text{kN/m}$$

6

Anchorage Zone Stresses

Problem 6.1

The end block of a prestressed concrete beam, rectangular in section is 120 mm wide and 300 mm deep. The prestressing force of 250 kN is transmitted to concrete by a distribution plate 120 mm wide and 75 mm deep, concentrically located at the ends. Calculate the position and magnitude of the maximum tensile stress on the horizontal section through the centre of the end block using the method of (a) Magnel, (b) Guyon and (c) Rowe. Design the reinforcement for the end block for the maximum transverse tension. Yield stress in steel = 260 N/mm^2.

Solution

The end block is shown in Fig. 6.1.

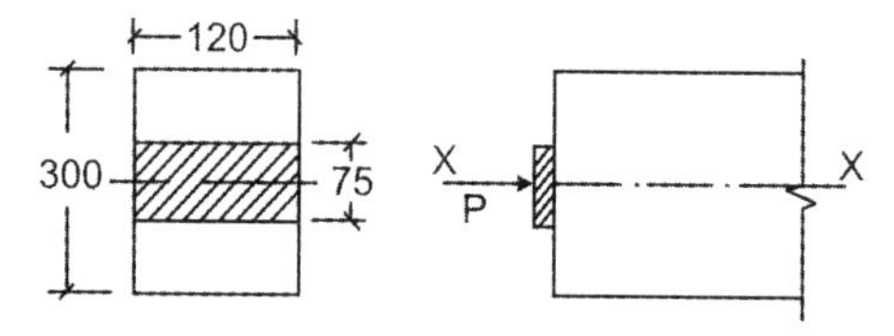

Fig. 6.1

a. *Magnel*

$$P = 250 \text{ kN}$$

$$\text{Direct stress} = f_h = \left(\frac{250\times10^3}{120\times300}\right) = 6.94 \text{ N/mm}^2$$

Vertical stress f_v and principal stress are critical at $x = 0.5\,h$

$$\text{At } \left(\frac{x}{h}\right) = 0.5$$

From Table 6.1 $K_1 = -5$
$K_2 = 2.00$
$K_3 = 1.25$

Table 6.1: Coefficients for stresses in end blocks (Magnel)

Distance from far end, x/h	K_1	K_2	K_3
0	20.00	–2.000	0.000
0.10	9.720	0.000	1.458
0.20	2.560	1.280	2.048
0.30	–1.960	1.960	2.058
0.40	–4.320	2.160	1.728
0.50	–5.000	2.000	1.250
0.60	–4.480	1.600	1.768
0.70	–3.240	1.080	0.378
0.80	–1.760	0.560	0.128
0.90	–0.520	0.160	0.018
1.00	0	0	0

$$M = \left[(6.94\times150\times120)\left(\frac{150}{2}\right)-\left(\frac{250\times10^3}{2}\right)\left(\frac{75}{4}\right)\right]$$

$$= 702650 \text{ N.mm}$$

$$V = 0$$

$$H = 0$$

$$f_v = -5\left(\frac{7026250}{120\times300^2}\right) = -3.25 \text{ N/mm}^2$$

$$f_h = +6.94 \text{ N/mm}^2$$

$$\therefore \quad f_{\min} = \left(\frac{6.94-3.25}{2}\right)-\frac{1}{2}\sqrt{(6.94+3.25)^2+0}$$

$$= -3.25 \text{ N/mm}^2 \text{ (acting at 150 mm) (tension)}$$

b. *Guyon*

$$P = 250 \text{ kN}$$

$$2y_{po} = 75 \text{ mm}$$

$$2y_o = 300 \text{ mm}$$

$\therefore$ Distribution ratio $= \left(\frac{y_{po}}{y_o}\right) = \left(\frac{75}{300}\right) = 0.25$

$\therefore$ Position of maximum stress $= 0.33(2y_o) = 100$ mm (Refer Table 6.2)

Table 6.2: Vertical stresses along axis at ends of prestressed beams (Guyon)

Distribution ratio (y_{po}/y_o)	Position of zero stress ($x/2y_o$)	Position of maximum stress ($x/2y_o$)	Ratio of maximum tensile stress to average stress
0.00	0.00	0.17	0.50
0.10	0.09	0.24	0.43
0.20	0.14	0.30	0.30
0.30	0.16	0.36	0.33
0.40	0.18	0.39	0.27
0.50	0.20	0.43	0.23
0.60	0.22	0.44	0.18
0.70	0.23	0.45	0.13
0.80	0.24	0.46	0.09

$$\text{Maximum tensile stress} = 0.345\left(\frac{P}{A}\right)$$

$$= \left(0.345 \times \frac{250 \times 10^3}{120 \times 300}\right)$$

$$= 2.39 \text{ N/mm}^2$$

c. *Rowe's Method*

$$P_K = 250 \text{ kN}$$

$$f_c = (250 \times 10^3)/(120 \times 120) = 17.36 \text{ N/mm}^2$$

$$f_{y\,(\max)} = f_c\left[0.98 - 0.825\left(\frac{y_{po}}{y_o}\right)\right]$$

$$= 17.36\,[0.98 - 0.825\,(0.25)] = 13.5 \text{ N/mm}^2$$

acting at $(0.5 \times 60) = 30$ mm

Maximum bursting tension by Rowe's method is given by

$$F_{bst} = P_k\left[0.48 - 0.4\left(\frac{y_{po}}{y_o}\right)\right]$$

$$= 250\ [0.48 - 0.4 \times 0.25] = 95 \text{ kN}$$

$$\therefore \quad \text{Area of steel} = A_s = \left(\frac{95 \times 10^3}{260}\right) = 365 \text{ mm}^2$$

Problem 6.2

A prestressing force of 250 kN is transmitted through a distribution plate 120 mm wide and 120 mm deep, the centre of which is located at 100 mm from the bottom of an end block having a section 120 mm wide and 300 mm deep. Evaluate the position and magnitude of the maximum tensile stress on horizontal section passing through the centre of the distribution plate using the methods of (a) Magnel, (b) Guyon and (c) Rowe. Find the area of the steel necessary to resist the largest tensile force resulting from any of these methods. Yield stress in steel = 260 N/mm^2.

Solution

The cross section of the end block with the anchorage plate is shown in Fig. 6.2.

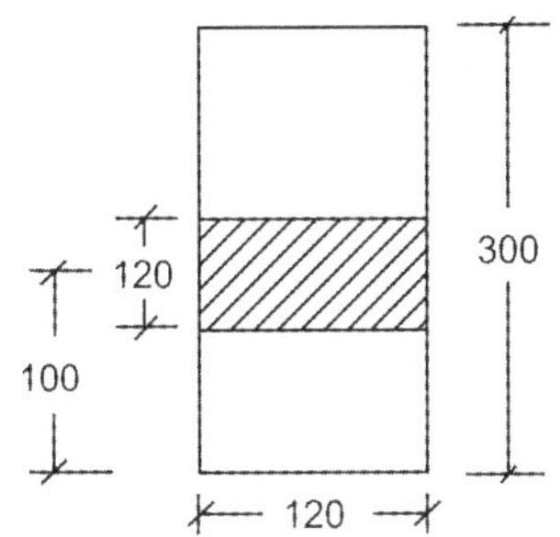

Fig. 6.2

$$P = 250 \text{ kN}$$

$$f_h = (P/A) = (250 \times 10^3)/(120 \times 300)$$

$$= 6.94 \text{ N/ mm}^2$$

a. *Magnel*

Direct stress $f_h = 6.94$ N/mm^2

$$M = \left[\left(6.94 \times 150 \times 120 \times \frac{150}{2}\right) - \left(\frac{250 \times 10^3}{2}\right)\left(\frac{120}{4}\right)\right]$$

$$= 5.61 \times 10^6 \text{ N.mm}$$

$$\therefore \quad f_v = -K_1\left[\frac{M}{bD^2}\right] = -5\left(\frac{5.61\times10^6}{120\times300^2}\right)$$

$$= -2.59\ \text{N/mm}^2$$

$$f_{min} = \left(\frac{6.94-2.59}{2}\right) - \frac{1}{2}\sqrt{(6.94+2.59)^2+0}$$

$$= -2.59\ \text{N/mm}^2,\ \text{acting at } (0.5\times300) = 150\ \text{mm}$$

b. *Guyon*

$$\text{Distribution ratio} = \left(\frac{y_{po}}{y_o}\right) = \left(\frac{120}{300}\right) = 0.4$$

$$\therefore \quad \text{Position of maximum stress} = 0.39(2y_o) = (0.39\times300)$$

$$= 117\ \text{mm}$$

$$\text{Maximum tensile stress} = 0.27\left(\frac{P}{A}\right) = (0.27\times6.94)$$

$$= 1.875\ \text{N/mm}^2$$

c. *Rowe's Method*

$$f_c = \left(\frac{250\times10^3}{120\times300}\right) = 6.94\ \text{N/mm}^2$$

$$f_{v\,(max)} = 6.94\,[0.98 - 0.825\,(0.4)]$$

$$= 4.51\ \text{N/mm}^2,\ \text{acting at } (0.5\times150) = 75\ \text{mm}$$

Bursting tension

$$F_{bst} = P_k\left(0.48 - 0.4\frac{y_{po}}{y_o}\right)$$

$$= 250\,[0.48 - 0.4\times0.4] = 80\ \text{kN}$$

$$\therefore \quad A_{st} = \left(\frac{80\times10^3}{260}\right) = 307\ \text{mm}^2$$

Problem 6.3

The end block of a prestressed concrete beam 200 mm wide and 400 mm deep has two anchor plates, 200 × 50 mm deep at 80 mm from the top and 200 × 80 mm deep located 100 mm from the bottom of the beam, transmitting forces of 250 and 300 kN, respectively.

a. Find the position and magnitude of the maximum tensile stress on a horizontal section passing through the centre of the beam using Guyon's method.
b. Evaluate the maximum tensile stress on section passing through the larger and smaller prestressing forces using Guyon's and Rowe's method.

Solution

The end block with eccentric anchorage plates is shown in Fig. 6.3.

Guyon's method is applied as shown in Example: 10.5 in the text book prestressed concrete Tata Mc Graw Hill Fifth Edition of the same author.

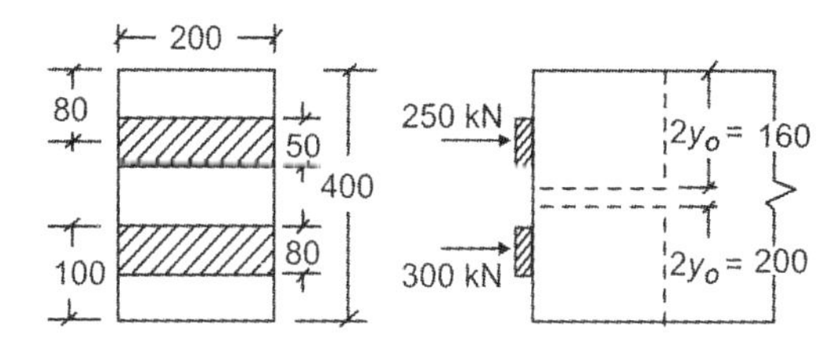

Fig. 6.3

Rowe's method (Larger forces):

$$2y_o = 160 \text{ mm}$$

$$2y_{po} = 80 \text{ mm}$$

$$\left(\frac{y_{po}}{y_o}\right) = 0.4$$

$$f_c = \left(\frac{250 \times 10^3}{200 \times 200}\right) = 6.25 \text{ N/mm}^2$$

$$f_{v\,(\max)} = 6.25\,[0.98 - 0.825 \times 0.4] = 4.06 \text{ N/mm}^2$$

Problem 6.4

The end block of a prestressed beam 250 mm wide and 500 mm deep in section is prestressed by two cables carrying forces of 450 kN each. One of the cables is parabolic, located 125 mm below the centre line at the centre of the span (10 m) and anchored at a point 125 mm above the centre line at the ends. The second cable is straight, located 100 mm from the bottom of the beam. The distribution plates for the cables are 100 mm deep and 250 mm wide. Evaluate the maximum tensile stress on horizontal sections passing through the centre of anchor plates using Rowe's method.

Solution

The end block with the eccentric anchor plates is shown in Fig. 6.4.

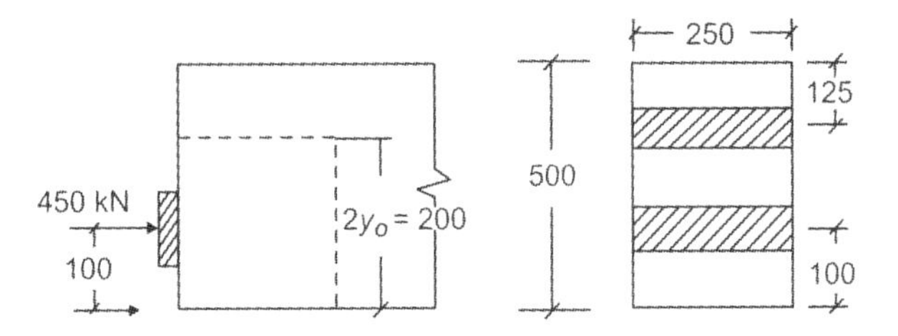

Fig. 6.4

Using Rowe's method, we have the ratio:

$$\left(\frac{2y_{po}}{2y_o}\right) = \left(\frac{100}{200}\right) = 0.5$$

$$f_c = \left(\frac{450 \times 14^3}{250 \times 200}\right) = 9 \text{ N/mm}^2$$

$$f_{v\,(\max)} = 9\,[0.98 - 0.825 \times 0.5)]$$
$$= 5.13 \text{ N/mm}^2$$

Problem 6.5

A Freyssinet anchorage (125 mm diameter) carrying 12 wires of 7 mm diameter, stressed to 950 N/mm^2 is embedded concentrically in the web of an I-section beam at the ends. The thickness of the web is 225 mm. Evaluate the maximum tensile stress and the bursting tensile force in the end block using Rowe's method. Design the reinforcement for the end block.

Solution

The I-section beam with the Freyssinet anchorage is shown in Fig. 6.5.

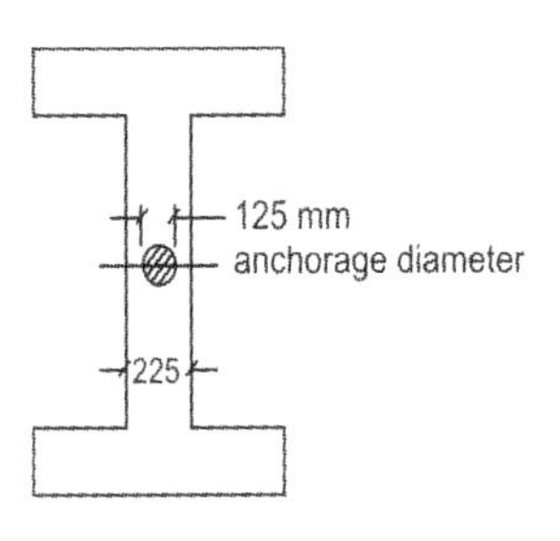

Fig. 6.5

If $\quad 2y_{po}$ = Equivalent side length

$$(2y_{po})^2 = \left(\frac{\pi \times 125^2}{4}\right)$$

$\therefore \quad 2y_{po} = 110$ mm

$$P = \left(\frac{12 \times 38.4 \times 950}{1000}\right) = 438 \text{ kN}$$

$$\left(\frac{2y_{po}}{2y_o}\right) = \left(\frac{110}{225}\right) = 0.48$$

$$f_c = \left(\frac{438 \times 10^3}{225 \times 225}\right) = 8.65 \text{ N/mm}^2$$

$$f_{v\,(\max)} = 8.65\,(0.98 - 0.825 \times 0.48) = 51 \text{ N/mm}^2$$

$$F_{bst} = P_k\left[0.48 - 0.4\frac{y_{po}}{y_o}\right]$$

$$= 438\,[0.48 - 0.4 \times 0.48] = 126 \text{ kN}$$

$$A_{st} = \left(\frac{126 \times 10^3}{0.87 \times 260}\right) = 557 \text{ mm}^2$$

Problem 6.6

The end block of a prestressed beam 500 mm wide and 1050 mm deep contains 6 Freyssinet cables, each carrying a force of 266 kN anchored through 100 mm diameter anchorages, which are spaced 150 mm apart at the end of the beam. Calculate the maximum tensile stress and the bursting tension and design the reinforcement for the end block using Rowe's method. Adopt yield stress in mild steel reinforcement as 260 N/mm^2.

Solution

The end block with the Freyssinet anchorages is shown in Fig. 6.6.

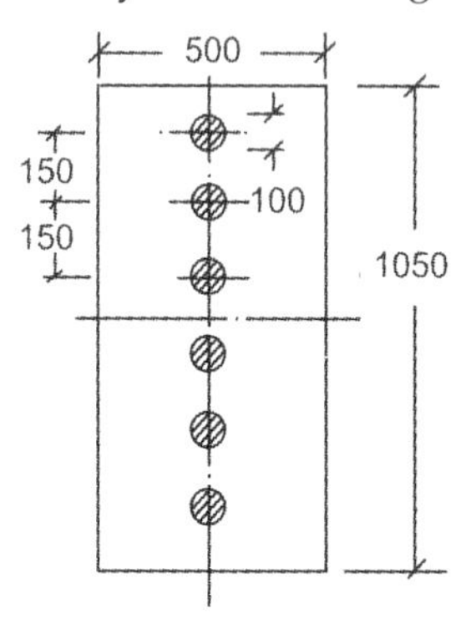

Fig. 6.6

Diameter of anchorage = 100 mm

$$\therefore \quad 2y_{po} = \text{side length of square}$$

$$\left(\frac{\pi\times 100^2}{4}\right) = (2y_{po})$$

$$\therefore \quad 2y_{po} = 89 \text{ mm}$$

$$2y_o = 150 \text{ mm}$$

$$\therefore \quad \text{Ratio} = \left(\frac{2y_{po}}{2y_o}\right) = \left(\frac{89}{150}\right) = 0.59$$

$$f_c = \left(\frac{266\times 10^3}{150\times 150}\right) = 11.8 \text{ N/mm}^2$$

$$f_{v\,(\max)} = 11.8\,[0.98 - 0.825 \times 0.59] = 5.8 \text{ N/mm}^2$$

$$F_{bst} = 266\,[0.48 - 0.4 \times 0.59] = 65 \text{ kN}$$

$$A_{st} = \left(\frac{65\times 10^3}{0.87\times 260}\right) = 287 \text{ mm}^2$$

Problem 6.7

A post-tensioned concrete beam 400 mm wide by 800 mm deep is prestressed by an effective prestressing force of 1100 kN at an eccentricity of 120 mm. The anchor plate is 400 mm wide by 400 mm deep. Calculate the bursting force and design reinforcement to resist this force. Sketch the details of reinforcements.

Solution

Referring to Fig. 6.7.

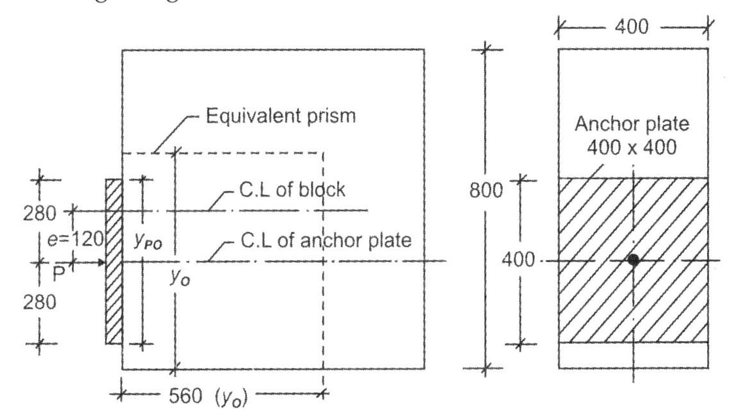

Fig. 6.7

$$y_o = \text{Depth of equivalent prism} = 560 \text{ mm}$$
$$y_{po} = \text{Depth of anchor plate} = 400 \text{ mm}$$

$$\therefore \quad \text{Distribution ratio} = \left(\frac{y_{po}}{y_o}\right) = \left(\frac{400}{560}\right) = 0.714$$

According to IS: 1343-2012 code specifications

$$\text{Bursting tension} = F_{bst} = P_k\left[0.32 - 0.3\left(\frac{y_{po}}{y_o}\right)\right]$$

$$= 1100\,[0.32 - (0.3 \times 0.714] = 121 \text{ kN}$$

$$\text{Anchorage zone reinforcement} = A_{st} = \left(\frac{F_{bst}}{0.87 \times f_y}\right)$$

$$\text{(Using Fe-415 HYSD bars)} = \frac{(121 \times 10^3)}{(0.87 \times 415)} = 335 \text{ mm}^2$$

Using 6 mm diameter bars,

$$\text{Number of bars} = \left(\frac{335}{28}\right) = 12 \text{ numbers in the vertical direction}$$

The details of reinforcements are shown in Fig. 6.8.

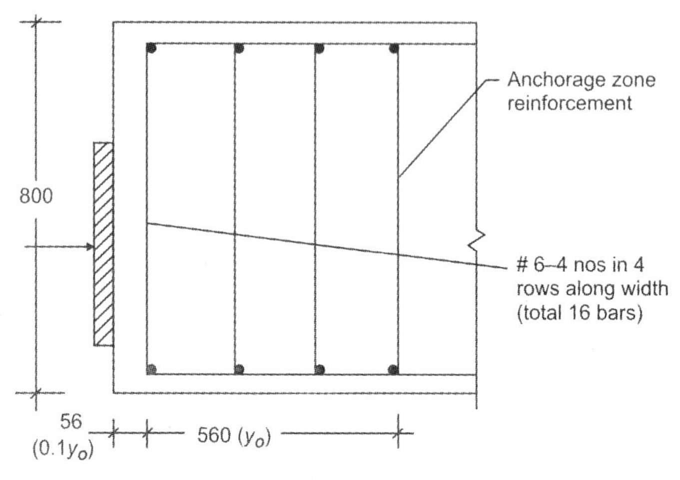

Fig. 6.8

Problem 6.8

The end block of a post-tensioned beam is 300 mm wide by 300 mm deep and is prestressed concentrically by a Freyssinet cylindrical anchorage of 150 mm diameter with a jacking force of 800 kN. Design suitable anchorage zone reinforcement and sketch the details.

Solution

Fig. 6.9 shows the Freyssinet anchorage of diameter 150 mm and the equivalent square for computations.

If y_{po} = side of equivalent square

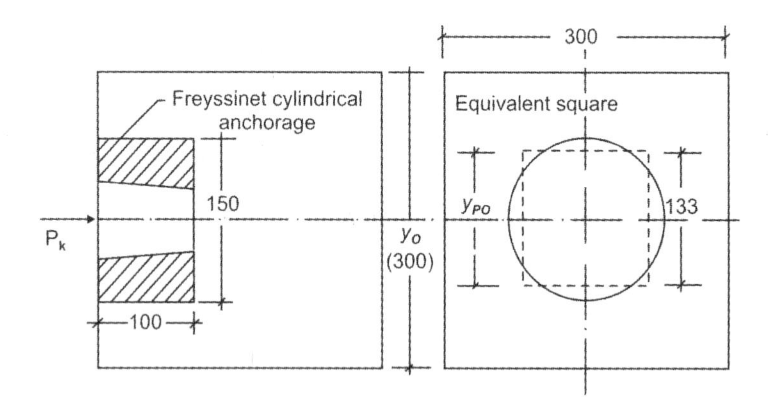

Fig. 6.9

$$y_{po} = \sqrt{\frac{\pi d^2}{4}} = \sqrt{\frac{(\pi \times 150^2)}{4}} = 133 \text{ mm}$$

$$y_o = 300 \text{ mm} \qquad P_k = 800 \text{ kN}$$

$$\text{Bursting tension} = F_{bst} = P_k\left[0.32 - 0.3\left(\frac{y_{po}}{y_o}\right)\right]$$

$$= 800\left[0.32 - 0.3\left(\frac{133}{300}\right)\right] = 152 \text{ kN}$$

$$\text{Area of reinforcement} = A_{st} = \left(\frac{F_{bst}}{0.87 \times f_y}\right)$$

$$= \frac{(152 \times 10^3)}{(0.87 \times 415)} = 421 \text{ mm}^2$$

Using 6 mm diameter bars of Fe-415 grade;

$$\text{number of bars required} = \left(\frac{421}{28}\right) = 15$$

Provide 4 bars in each plane in 4 rows as shown in Fig. 6.10.

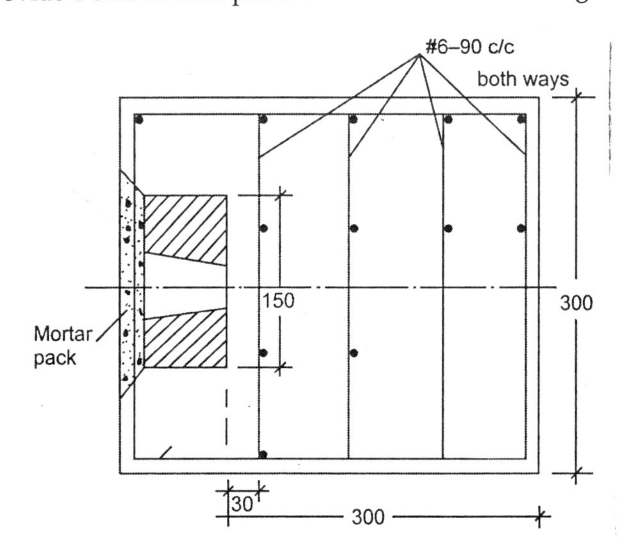

Fig. 6.10

Problem 6.9

The end block of a post-tensioned prestressed concrete bridge girder is of rectangular section 450 mm wide by 1350 mm deep. Freyssinet anchorages of 7K-15 type comprising anchor plates of size 225 mm by 225 mm are used. Three anchorages are provided spaced at 450 mm centres. The jacking force in each anchorage is 1500 kN. Design suitable anchorage zone reinforcement in the end block.

Solution

Referring to Fig. 6.11, the depth of the equivalent prism is obtained as

$$y_o = 450 \text{ mm} \qquad y_{po} = 225 \text{ mm}$$
$$P_k = 1500 \text{ kN}$$

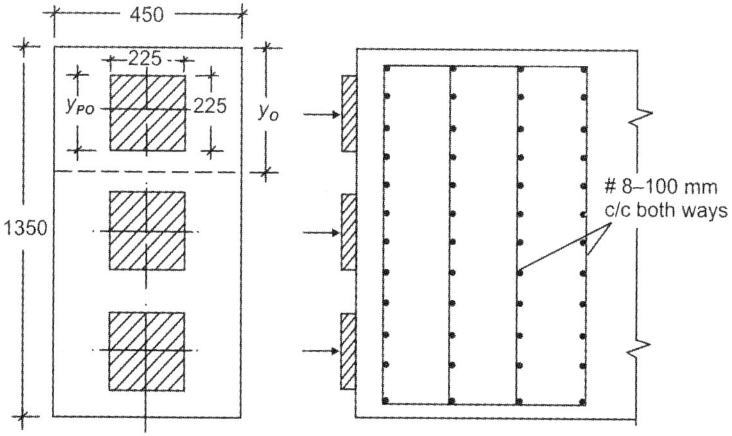

Fig. 6.11

$$\text{Bursting tension } = F_{bst} = P_k\left[0.32 - 0.3\left(\frac{y_{po}}{y_o}\right)\right]$$

$$= 1500\left[0.32 - 0.3\left(\frac{225}{450}\right)\right] = 225 \text{ kN}$$

$$\text{Anchorage zone reinforcement } = A_{st} = \left(\frac{F_{bst}}{0.87 \times f_y}\right)$$

$$= \frac{(255 \times 10^3)}{(0.87 \times 415)} = 706 \text{ mm}^2$$

Using 8 mm diameter Fe-415 grade HYSD bars,

$$\text{Number of bars} = \left(\frac{706}{50}\right) = 14.12$$

Adpot 16 bars in the vertical direction arranged in 4 rows as shown in Fig. 6.11. Similar number of bars are used in the horizontal direction also.

Problem 6.10

A high tensile cable comprising 12 strands of 15 mm diameter (12 K-15 of PSC Freyssinet system) with an effective force of 2500 kN is anchored concentrically in an end block of a post-tensioned beam. The end block is 400 mm wide by 800 mm deep and the anchor plate is 200 mm wide by 260 mm deep. Design suitable anchorage zone reinforcements using Fe-415 HYSD bars using IS: 1343 code specifications.

Solution

Considering the width direction (smaller dimension)

$$y_o = 400 \text{ mm} \qquad y_{po} = 200 \text{ mm}$$

$$\text{Ratio} \quad \left(\frac{y_{po}}{y_o}\right) = \left(\frac{200}{400}\right) = 0.5 \qquad P_k = 2500 \text{ kN}$$

$$\text{Bursting tension} = F_{bst} = P_k\left[0.32 - 0.3\left(\frac{y_{po}}{y_o}\right)\right]$$

$$= 2500\,[0.32 - 0.3 \times 0.5] = 425 \text{ kN}$$

Area of anchorage zone reinforcement is computed as

$$A_{st} = \left(\frac{F_{bst}}{0.87 \times f_y}\right) = \frac{(425 \times 10^3)}{(0.87 \times 415)} = 1177 \text{ mm}^2$$

Using 10 mm diameter Fe-415 HYSD bars;

$$\text{number of bars required} = \left(\frac{1177}{78.5}\right) = 15$$

Adopt 16 number of 10 mm diameter bars in 4 rows with 4 bars in each row both horizontally and vertically.

Problem 6.11

The end block of a post-tensioned beam is 500 mm wide by 1000 mm deep. Two cables each comprising 55 numbers of 7 mm diameter high tensile wires carrying a force of 2800 kN are anchored using the B.B.R.V system anchor plate of side 305 mm. The anchor plate centres are located symmetrically at 250 mm from the top and bottom edges of the beam. Using Fe-415 grade high yield bars design suitable reinforcements in the end block using IS: 1343 code recommendations.

Solution

Data: $P_k = 2800$ kN $\quad f_y = 415$ N/mm^2

$y_o = 500$ mm $\quad y_{po} = 305$ mm

$$\text{Ratio} \left(\frac{y_{po}}{y_o}\right) = \left(\frac{305}{500}\right) = 0.61$$

$$\text{Bursting tension} = F_{bst} = P_k\left[0.32 - 0.3\left(\frac{y_{po}}{y_o}\right)\right]$$

$$= 2800\,[0.32 - 0.3\,(0.61)] = 384 \text{ kN}$$

$$A_{st} = \left(\frac{F_{bst}}{0.87 f_y}\right) = \frac{(384 \times 10^3)}{(0.87 \times 415)} = 1063 \text{ mm}^2$$

Using 10 mm diameter bars;

$$\text{number of bars} = \left(\frac{1063}{78.5}\right) = 14$$

Provide 10 mm diameter bars both ways at 150 mm centres over a length of 500 mm as shown as in Fig. 6.12.

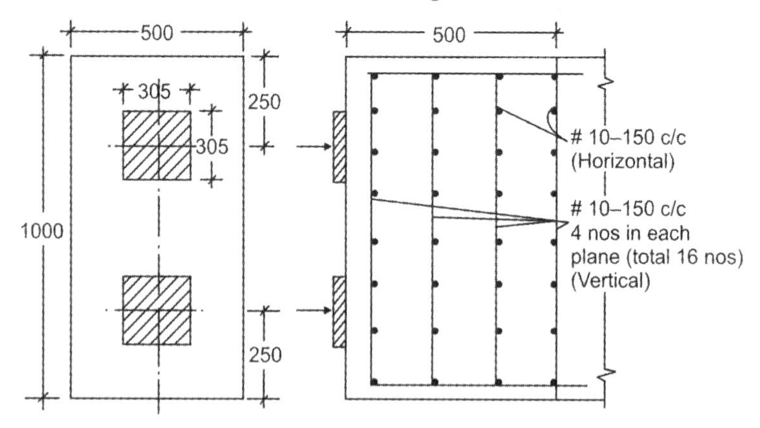

Fig. 6.12

Problem 6.12

The end block of a prestressed concrete girder is 200 mm wide by 300 mm deep. The beam is post-tensioned by two Freyssinet anchorages each of 100 mm diameter with their centres located at 75 mm from the top and bottom of the beam, the force transmitted by each anchorage being 200 kN. Calculate the bursting force according to the Indian standard code IS: 1343 code provisions.

Solution

Anchorge diameter = 100 mm

Equivalent side of the square

$$= 2\,y_{po} = \sqrt{\frac{\pi}{4} \times 100^2} = 89 \text{ mm}$$

Side of the surrounding prism = $2\,y_o$ = 150 mm

$$\text{Distribution ratio} = \left(\frac{y_{po}}{y_o}\right) = \left(\frac{89}{150}\right) = 0.593$$

Bursting tensile force

$$= F_{bst} = P_k\left[0.32 - 0.3\left(\frac{y_{po}}{y_o}\right)\right]$$

$$= 200\,[0.32 - 0.3\,(0.593)] = 28.6 \text{ kN.}$$

Problem 6.13

The end block of a post-tensioned prestressed member is 550 mm wide by 550 mm deep. Four cables, each made up of 7 wires of 12 mm diameter strands, each carrying a force of 1000 kN are anchored 150 by 150 mm plate anchorages, located with their centres at 125 mm from the edges of the end block. The cable duct is of 50 mm diameter. The characteristic compressive strength of concrete is 45 N/mm^2. The cube strength of concrete at transfer is 25 N/mm^2. Permissible bearing stresses behind anchorages should conform to IS: 1343 code specifications. The characteristic tensile strength of steel reinforcement is 260 N/mm^2. Design suitable anchorage reinforcements for the end block and check for bearing stresses.

Solution

P_k = 1000 kN $2\,y_{po}$ = 150 mm

$2\,y_o$ = 250 mm $(y_{po}/y_o) = 0.6$

$f_{ck} = 45 \text{ N/mm}^2$ $\qquad f_{ci} = 25 \text{ N/mm}^2$

$f_y = 260 \text{ N/mm}^2$

Area of the cable duct $= \left[\dfrac{\pi \times 50^2}{4}\right] = 2000 \text{ mm}^2$

Net area of the surrounding prism $= [(250)^2 - 2000] = 60500 \text{ mm}^2$

Average compressive stress $= f_c = \left[\dfrac{1000 \times 10^3}{60500}\right] = 16.5 \text{ N/mm}^2$

According to IS: 1343 code specifications, the bearing stress shall not exceed the value given by the relation $0.48 f_{ci} \sqrt{\dfrac{A_{br}}{A_{pun}}}$ or $0.8 f_{ci}$ whichever is smaller

where $\quad A_{br}$ = bearing area = 60500 mm^2

A_{pun} = punching area = $[150 \times 150] = 22500 \text{ mm}^2$

Ratio $\left[\dfrac{A_{br}}{A_{pun}}\right] = \left[\dfrac{60500}{22500}\right] = 2.70$

Bearing stress is limited to a value $= [0.48 \times 25 \times \sqrt{2.70}\,] = 19.7 \text{ N/mm}^2$ or $(0.8 \times 25) = 20 \text{ N/mm}^2$, whichever is smaller.

Actual bearing stress $= 16.5 \text{ N/mm}^2 < 19.7 \text{ N/mm}^2$ (hence, safe)

Bursting tension in end block

$$= F_{bst} = P_k \left[0.32 - 0.3\left(\frac{y_{po}}{y_o}\right)\right]$$

$$= 1000\,[0.32 - (0.3 \times 0.6)] = 140 \text{ kN}$$

Area of steel reinforcement

$$= A_{st} = \left[\frac{F_{bst}}{0.87\, f_y}\right] = \left[\frac{140 \times 10^3}{0.87 \times 260}\right] = 619 \text{ mm}^2$$

Using 8 mm diameter bars, number of bars required $= \left[\dfrac{619}{50}\right] = 12.38$

Spacing of bars $= \left[\dfrac{275}{13}\right] = 20 \text{ mm}$

Provide a square mesh of 8 mm diameter bars spaced at 20 mm centres both ways in front of the anchorages.

Problem 6.14

The end block of a post-tensioned continuous girder of span 40 m is 800 mm wide by 2000 mm deep. The Freyssinet anchorages used for the strands has plates of size 340 mm by 340 mm. Each of the three strands carry a force of 4000 kN and the three anchorage plates are spaced at intervals 500 mm along the depth of the end block. Design suitable anchorage reinforcement for the end block using Fe-415 HYSD bars according to the IS: 1343 code specifications.

Solution

$$P_k = 4000 \text{ kN} \qquad f_y = 415 \text{ N/mm}^2$$
$$2y_o = 800 \text{ mm} \qquad 2y_{po} = 340 \text{ mm}$$
$$\text{Ratio} \left[\frac{y_{po}}{y_o}\right] = \left[\frac{340}{800}\right] = 0.425$$

According to the Indian standard code (IS : 1343), bursting tension is computed as

$$F_{bst} = P_k \left[0.32 - 0.3\left(\frac{y_{po}}{y_o}\right)\right]$$
$$= 4000\,[0.32 - 0.3 \times 0.425] = 770 \text{ kN}$$

Using Fe-415 HYSD bars, the area of steel required is computed as,

$$A_{st} = \frac{770 \times 10^3}{0.87 \times 415} = 2132 \text{ mm}^2$$

Provide 16 mm diameter bars at 150 mm centres in the horizontal plane distributed in the region from 0.2 y_o to 2 y_o (80 mm to 800 mm). In the vertical plane, the ratio (y_{po}/y_o) being larger, the magnitude of bursting tension is less. However, the same reinforcements are provided in the vertical plane in the form of a mesh to resist bursting tension.

Problem 6.15

The end block of a post-tensioned beam is 500 mm wide by 900 mm deep. The beam is prestressed by a concentric cable with an effective force of 1459 kN. The anchorage plate is 225 mm by 225 mm. Estimate the bursting tension and design suitable end block reinforcements using Indian standard code IS: 1343 specifications. Adopt Fe-415 grade HYSD bars.

Solution

$$P_k = 1459 \text{ kN} \qquad f_y = 415 \text{ N/mm}^2$$
$$2\,y_o = 900 \text{ mm} \qquad 2\,y_{po} = 225 \text{ mm}$$
$$f_y = 415 \text{ N/mm}^2$$

$$\text{Ratio}\left[\frac{y_{po}}{y_o}\right] = \left[\frac{112.5}{450}\right] = 0.25$$

According to the Indian standard code IS: 1343, bursting tension is computed as

$$F_{bst} = P_k\left[0.32 - 0.3\left(\frac{y_{po}}{y_o}\right)\right]$$

$$= 1459\,[0.32 - 0.3 \times 0.25] = 358 \text{ kN}$$

Area of steel required to resist this tension is

$$A_{st} = \left(\frac{358 \times 10^3}{0.87 \times 415}\right) = 991 \text{ mm}^2$$

Provide 10 mm diameter bars at 100 mm centres in the horizontal and vertical planes in the form of a mesh for a distance equal to 900 mm.

7 Design of Prestressed Concrete Sections

Problem 7.1

A prestressed concrete beam of rectangular section, 90 mm wide and 180 mm deep, is to be designed to support two imposed loads of 3.5 kN each located at one-third points over a span of 3 m. If there is to be no tensile stress in the concrete at transfer and service loads, calculate the minimum prestressing force and the corresponding eccentricity.

$$D_c = 24 \text{ kN/m}^3, \quad \text{Loss ratio} = 0.8.$$

Solution

Figure 7.1 shows the loading arrangement and cross-section of the beam.

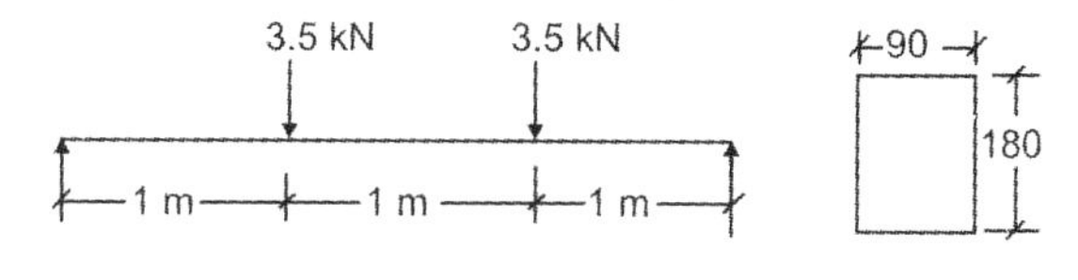

Fig. 7.1

$$Z_b = Z_t = \left(\frac{90 \times 180^2}{6}\right) = 486 \times 10^3 \text{ mm}^3$$

$$A = 90 \times 180 = 16200 \text{ mm}^2$$

$$f_{tt} = f_{tw} = 0 \qquad \text{Loss ratio} = \eta = 0.8$$

$$\text{Self weight} = (0.09 \times 0.18 \times 24) = 0.388 \text{ kN/m}$$

$$M_g = \left(\frac{0.388 \times 3^2}{8}\right) = 0.4365 \text{ kN.m}$$

$$M_q = 3.5 \text{ kN.m}$$

$$\therefore \quad (M_g + M_q) = 3.9365 \text{ kN.m}$$

$$f_t = \left(f_{tt} - \frac{M_g}{Z_t}\right) = \left[0 - \frac{0.4365 \times 10^6}{486 \times 10^3}\right]$$

$$= -0.90 \text{ N/mm}^2$$

$$f_b = \left(\frac{f_{tw}}{\eta} + \frac{M_g + M_q}{\eta Z_b}\right) = \left[0 + \frac{3.9365 \times 10^6}{0.8 \times 486 \times 10^3}\right]$$

$$= 10.12 \text{ N/mm}^2$$

$$\therefore \quad P = \frac{A(f_t Z_t + f_b Z_b)}{Z_t + Z_b}$$

$$= \frac{16200(-0.90 \times 486 \times 10^3 + 10.12 \times 486 \times 10^3)}{2 \times 486 \times 10^3}$$

$$= 74682 \text{ N} = 74.68 \text{ kN}$$

$$e = \frac{Z_t Z_b (f_b - f_t)}{A(f_t Z_t + f_b Z_b)}$$

$$= \frac{(486 \times 10^3)^2 (10.12 + 0.90)}{16200 \times 486 \times 10^3 (-0.90 + 10.12)}$$

$$= 35.6 \text{ mm}$$

Problem 7.2

A prestressed concrete T beam is to be designed to support an imposed load of 4.4 kN/m over an effective span of 5 m. The T beam is made up of a flange 400 mm wide and 40 mm thick. The rib is 100 mm wide and 200 mm deep. The stress in the concrete must not exceed 15 N/mm^2 in compression and zero in tension at any stage. Check for the adequacy of the section provided, and calculate the minimum prestressing force necessary and corresponding eccentricity. Assume 20% loss of prestress.

Solution

The tee-beam section and the live load on beam are shown in Fig. 7.2.

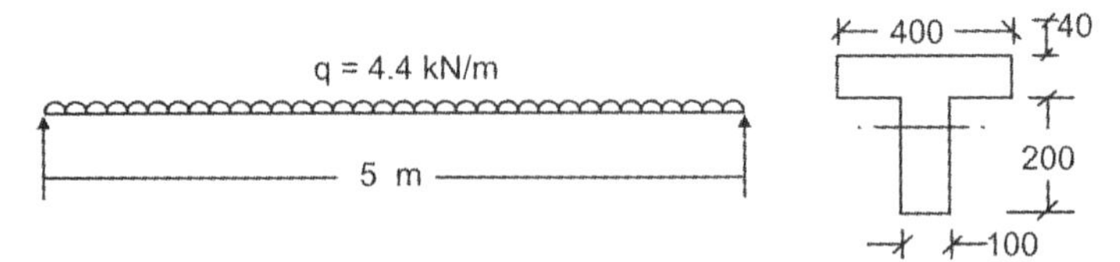

Fig. 7.2

$$I = 19575 \times 10^4 \text{ mm}^4$$
$$A = 36000 \text{ mm}^2$$
$$Z_t = 225 \times 10^4 \text{ mm}^3$$
$$Z_b = 128 \times 10^4 \text{ mm}^3$$
$$M_g = (0.125 \times 0.8 \times 5^2) = 2.7 \text{ N.km}$$
$$M_q = (0.125 \times 4.4 \times 5^2) = 13.75 \text{ N.km}$$

$$\therefore \quad Z_b \geq \frac{M_q + (1-\eta)M_g}{f_{br}} \quad \text{and } f_{br} = (\eta f_{ct} - f_{tw})$$
$$= (0.8 \times 15 - 0) = 12 \text{ N/mm}^2$$
$$\geq \frac{13.75 \times 10^6 + (1-0.8)2.7 \times 10^6}{12}$$
$$\geq 119 \times 10^4 \text{ mm}^2 < 128 \times 10^4 \text{ mm}^3$$

$\therefore$ Section is adequate

$$f_t = \left(f_{tt} - \frac{M_g}{Z_t}\right) = \left(0 - \frac{2.7 \times 10^6}{225 \times 10^4}\right)$$
$$= -1.2 \text{ N/mm}^2$$

$$f_b = \left[\frac{f_{tw}}{\eta} + \frac{M_g + M_q}{\eta Z_b}\right] = 0 + \left[\frac{(2.7 + 13.5)\, 10^6}{0.8 \times 128 \times 10^4}\right]$$
$$= 16 \text{ N/mm}^2$$

$$\therefore \quad P = \left[\frac{36000[(-1.2 \times 225 \times 10^4) + (16 \times 128 \times 10^4)}{(225 + 128)\, 10^4}\right]$$
$$= 1{,}81{,}000 \text{ N} = 181 \text{ kN}$$

$$e = \left[\frac{225 \times 128 \times 10^8\, [16 - (-1.2)]}{36000[(-1.2 \times 225 \times 10^4) + (16 \times 128 \times 10^4)]}\right]$$
$$= 78 \text{ mm}$$

Problem 7.3

A post-tensioned beam of span 15 mm overall depth 900 mm, has a uniform symmetrical cross-section of area 2×10^5 mm^2 and second moment of area being 212×10^8 mm^4 units. The prestress is provided by a cable tensioned to a force of 1450 kN at transfer. If the beam is to support a uniformly distributed live load of 21 kN/m and the minimum load is that due to the self weight of the beam, calculate the vertical limits within which the cable must lie along the beam length. The permissible compressive stresses at transfer and working load are 14 and 16.8 N/mm^2 respectively. The tensile stresses at transfer and working load are zero and 1.75 N/mm^2 respectively. D_c = 24 kN/m^3, loss of prestress = 20%.

Solution

At centre of span

$$e \geq \left[\frac{Z_b \cdot f_{ct}}{p} - \frac{Z_b}{A} + \frac{M_{\min}}{p}\right]$$

$$M_g = M_{\min}$$

$$\text{Self weight} = g = \left(\frac{2\times10^5}{10^6}\times 24\right) = 4.8 \text{ kN/m}$$

$$\therefore \quad M_g = (0.125 \times 4.8 \times 15^2) = 135 \text{ kN.m}$$

Given

$$P = 1450 \times 10^3 \text{ N}$$

$$f_{ct} = 14 \text{ N/mm}^2$$

$$A = 2 \times 10^5 \text{ mm}^2$$

$$I = 212 \times 18^8 \text{ mm}^4$$

$$Z_b = Z_t = 0.47 \times 10^8 \text{ mm}^3$$

At centre of span

$$e \leq \left[\frac{0.47\times10^8\times14}{1450\times10^3} - \frac{0.47\times10^8}{2\times10^5} + \frac{135\times10^6}{1450\times10^3}\right]$$

$$\leq 312 \text{ mm}$$

$$M_q = \left(\frac{21\times15^2}{8}\right) = 590 \text{ kN.m}$$

$$(M_g + M_q) = 725 \text{ kN.m}$$

$$\therefore \quad e \geq \left[\frac{Z_b f_{tw}}{\eta P} - \frac{Z_b}{A} + \frac{(M_g + M_q)}{\eta P}\right]$$

$$\geq \left[\frac{0.47\times10^8(-1.75)}{0.8\times1450\times10^3} - \frac{0.47\times10^8}{2\times10^5} + \frac{725\times10^6}{0.8\times1450\times10^3}\right]$$

$$\geq 249 \text{ mm}$$

At support

$$e \leq \left[\frac{Z_b f_{ct}}{P} - \frac{Z_b}{A}\right] \leq \left[\frac{0.47\times10^8\times14}{1450\times10^3} - \frac{0.47\times10^8}{2\times10^5}\right]$$

$$\leq 219 \text{ mm}$$

$$e \geq \left[\frac{Z_b f_{tw}}{\eta P} - \frac{Z_b}{A}\right]$$

$$\geq \left[\frac{0.47\times10^8\,(-1.75)}{0.8\times1450\times10^3} - \frac{0.47\times10^8}{2\times10^5}\right]$$

$$\geq 376 \text{ mm}$$

Problem 7.4

A prestressed I-section of minimum overall depth 300 mm, is required to have an ultimate flexural strength of 86 kN.m. Find (a) suitable minimum dimensions of the top flange, and (b) the total number of 5 mm wires required in the bottom flange. The cube strength of concrete is 60 N/mm^2 and tensile strength of steel is 1600 N/mm^2.

Solution

The cross-section of the beam is shown in Fig. 7.3.

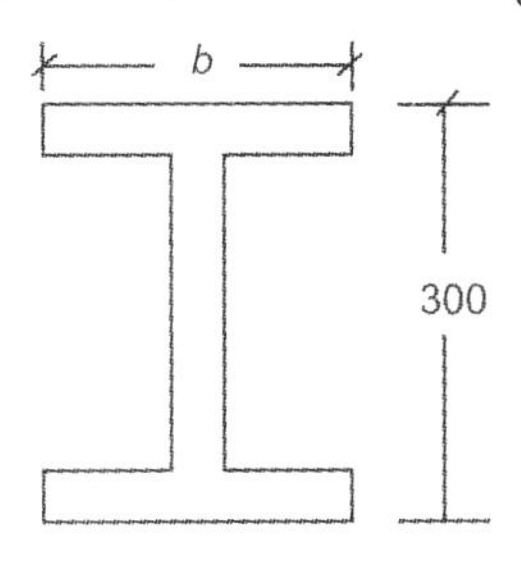

Fig. 7.3

$$M_u = 86 \text{ kN.m}$$
$$f_{cu} = 60 \text{ N/mm}^2$$
$$f_{pu} = 1600 \text{ N/mm}^2$$

Assume effective depth $d = 250$ mm

b = breadth of flange

$$M_u = 0.185\, f_{cu}.b.d^2, \text{ if } b = \frac{d}{2}$$

$$\therefore \quad d^3 = \left(\frac{86 \times 10^6}{0.185 \times 60 \times 0.5}\right)$$

$$d = 250 \text{ mm} \quad \therefore \quad b = 125 \text{ mm}$$

For this case $x = 0.5\ d$ which is the maximum depth of stress block permitted as per British code.

$$A_p = \left[\frac{M_u}{0.87\ f_{pu}(d - 0.5)}\right]$$

$$= \left[\frac{86 \times 10^6}{0.87 \times 1600(250 - 0.5 \times 0.50 \times 250)}\right]$$

$$= 345 \text{ mm}^2$$

Area of 5 mm diameter wire = 19.6 mm^2

$$\therefore \quad \text{No. of wires} = \left(\frac{345}{19.6}\right) = 18.$$

Problem 7.5

Design the prestressing force required for the tie member of reinforced concrete truss. The service load tension in the tie member is 360 kN and the thickness of the member is fixed as 150 mm. The permissible compressive stress in concrete at transfer is 15 N/mm^2 and tension is not permitted under service loads. The loss ratio is 0.8. High tensile wires of 7 mm diameter tensioned to a stress of 1000 N/mm^2 and having an f_{pu} = 1500 N/mm^2 are available for use. The tensile strength of concrete is 2.5 N/mm^2. A load factor of 1.7 at the limit state of collapse and 1.2 against cracking is to be provided in the design.

Solution

Design tensile load $= N_d = 360$ kN

$$f_{ct} = 15 \text{ N/mm}^2 \qquad f_{tw} = 0$$

$$\eta = 0.8$$

Area of concrete section = $(N_d/\eta f_{ct})$

$$= \left(\frac{360 \times 10^3}{0.8 \times 15}\right) = 30{,}000 \text{ mm}^2$$

$\therefore$ Adopt 150 × 200 mm section (Area = 30,000 mm^2)

$$\therefore \quad \text{Compressive stress} = \left(\frac{350 \times 10^3}{0.8 \times 30,000}\right) = 15 \text{ N/mm}^2$$

$$\text{Prestressing force} \qquad P = \left(\frac{15 \times 30,000}{1000}\right) = 450 \text{ kN}$$

$$\text{Number of 7 mm } \phi \text{ H.T. wires} = \left(\frac{450 \times 10^3}{38.4 \times 1000}\right) = 12 \text{ wires}$$

Problem 7.6

Design a simply supported slab for a bridge deck using the following data: Span = 10 m, permissible compressive strength in concrete at transfer = f_{ct} = 16.5 N/mm^2, Type-1 member (No tensile stress at any stage), safe stress in steel = 950 N/mm^2, live load on slab is 10 kN/m^2, loss of stress = 18%. Design the prestressing force and eccentricity.

Solution

Assume thickness of slab = 50 mm per meter of span

Thickness of slab = (50 × 10) = 500 mm

Self weight = g = (0.5 × 24) = 12 kN/m^2

$$M_g = \left(\frac{gL^2}{8}\right) = \frac{(12 \times 10^2)}{8} = 150 \text{ kN.m}$$

$$M_q = \left(\frac{qL^2}{8}\right) = \frac{(10 \times 10^2)}{8} = 125 \text{ kN.m}$$

Section properties

Considering 1 m width of slab, b = 1000 mm

Thickness of slab = h = 500 mm

$$\text{Section modulus} = Z = Z_t = Z_b = \left(\frac{bh^2}{6}\right)$$

$$= \frac{(1000 \times 500^2)}{6} = 41.66 \times 10^6 \text{ mm}^3$$

Cross-sectional area = A = (1000 × 500) = 5 × 10^5 mm^2

$$f_{br} = (\eta f_{ct} - f_{tw}) = (0.82 \times 16.5 - 0) = 13.53 \text{ N/mm}^2$$

Minimum section modulus

$$Z_b \geq \left[\frac{M_q + (1-\eta)M_g}{f_{br}}\right]$$

$$\geq \left[\frac{(125\times10^6) + (1-0.82)150\times10^6}{13.53}\right]$$

$$\geq 11.23 \times 10^6\ \text{mm}^3 \leq 41.66 \times 10^6\ \text{mm}^3$$

... hence adequate

Minimum prestressing force

$$P = \left[\frac{A(f_t Z_t + f_b Z_b)}{Z_t + Z_b}\right]$$

$$f_t = \left[f_{tt} - \frac{M_g}{Z_t}\right] = \left[0 - \frac{150\times10^6}{41.66\times10^6}\right]$$

$$= -3.6\ \text{N/mm}^2$$

$$f_b = \left[\frac{f_{tw}}{\eta} + \frac{(M_g + M_q)}{\eta Z_b}\right]$$

$$= \left[0 - \frac{(125+150)\,10^6}{0.82\times41.66\times10^6}\right] = 8.05\ \text{N/mm}^2$$

$$\therefore\quad P = \left[\frac{5\times10^5\left(-3.6\times41.66\times10^6 + 8.05\times41.66\times10^6\right)}{\left(2\times41.66\times10^6\right)}\right]$$

$$= 1112.5 \times 10^3\ \text{N} = 1112.5\ \text{kN}$$

Corresponding eccentricity is given by the relation

$$e = \left[\frac{Z_t Z_b (f_b - f_t)}{A(f_t Z_t + f_b Z_b)}\right]$$

$$= \frac{\left(41.66\times10^6\right)^2[8.05-(-3.6)]}{\left(5\times10^5\right)\left[-3.6\times41.66\times10^6 = 8.05\times41.66\times10^6\right]}$$

$$= 218.12\ \text{mm}$$

Using Freyssinet cables each containing 12 high tensile wires initially stressed to 950 N/mm^2,

$$\text{Force in each cable} = \left[\frac{(12 \times 50 \times 950)}{1000}\right] = 570 \text{ kN}$$

$$\text{Spacing of cables} = \left[\frac{1000 \times 570}{1112.5}\right] = 512.3 \text{ mm}$$

Adopt cables at 500 mm centres at an eccentricity of 218.12 mm.

Problem 7.7

Design the prestressing force and the eccentricity for a symmetrical I-section beam having flanges of width 260 mm and thickness of flange being 100 mm, thickness of web = 50 mm, overall depth of the I-section = 520 mm, span of the beam = 9 m. The beam supports a uniformly distributed live load of 8 kN/m. Assume compressive strength of concrete at transfer as 15 N/mm^2, loss ratio = 0.8. No tensile stresses are permitted at any stage. If 5 mm diameter high tensile wires initially stressed to 1200 N/mm^2 are used, find the number of wires.

Solution

Section properties

$$A = [(2 \times 260 \times 100) + (320 \times 50)] = 68000 \text{ mm}^2$$

$$I = \left[\frac{(260 \times 520^3)}{12} - \frac{(210 \times 320^3)}{12}\right]$$

$$= 24726 \times 10^5 \text{ mm}^4$$

$$Z = Z_t = Z_b = \left(\frac{I}{Y}\right) = \frac{(24726 \times 10^5)}{260} = 9.51 \times 10^6 \text{ mm}^3$$

$$g = (0.068 \times 24) = 1.632 \text{ kN/m}$$

$$M_g = \left(\frac{gL^2}{8}\right) = \frac{(1.632 \times 9^2)}{8} = 16.52 \text{ kN.m}$$

$$M_q = \left(\frac{qL^2}{8}\right) = \frac{(8 \times 9^2)}{8} = 81 \text{ kN.m}$$

$$f_{ct} = 15 \text{ N/mm}^2, \quad f_{tt} = f_{tw} = 0 \text{ (Type-1 member)}$$

Loss ratio = 0.8

Minimum section modulus is given by

$$Z_b \geq \left[\frac{M_q(1-\eta)M_g}{(\eta f_{ct} - f_{tw})}\right]$$

$$\geq \left[\frac{\left(81\times10^6\right)+(1-0.8)16.52\times10^6}{(0.8\times15-0)}\right]$$

$\geq 7.02 \times 10^6$ mm^3 $\leq 9.51 \times 10^6$ mm^3 (provided)

Hence adequate

Minimum prestressing force

$$P = \left[\frac{A(f_t Z_t + f_b Z_b)}{(Z_t + Z_b)}\right]$$

$$f_t = \left[f_{tt} - \frac{M_g}{Z_t}\right] = \left[0 - \frac{16.52\times10^6}{9.51\times10^6}\right]$$

$= -1.73$ N/mm^2

$$f_b = \left[\frac{f_{tw}}{\eta} + \frac{(M_g + M_q)}{\eta Z_b}\right] = \left[0 + \frac{(16.52+81)\,10^6}{0.8\times9.51\times10^6}\right]$$

$= 12.81$ N/mm^2

$$\therefore \quad P = \left[\frac{68000\left(-1.73\times9.51\times10^6\right)+\left(12.81\times9.51\times10^6\right)}{\left(2\times9.51\times10^6\right)}\right]$$

$= 376000$ N $= 376$ kN

$$\text{Eccentricity} = e = \left[\frac{Z_t Z_b(f_b - f_t)}{A(f_t Z_t + f_b Z_b)}\right]$$

$$= \left[\frac{\left(9.51\times10^6\right)^2[12.81-(-1.73)]}{68000\left[\left(-1.73\times9.51\times10^6\right)+\left(12.81\times9.51\times10^6\right)\right]}\right]$$

$= 184$ mm

Number of 5 mm diameter wires required

$$= \left[\frac{376000}{(20\times1200)}\right] = 16$$

Problem 7.8

A post-tensioned prestressed concrete beam of rectangular section 250 mm wide by 580 mm deep is to be designed to support an imposed load of 12 kN/m uniformly distributed over a span of 12 m. The stress in concrete must not exceed 17 N/mm^2 in compression or 1.4 N/mm^2 in tension at any stage and the loss of prestress is 15%. Determine the minimum prestressing force and the corresponding eccentricity.

Solution

Data:

$$q = 12 \text{ kN/m}$$

$$f_{ct} = 17 \text{ N/mm}^2,$$

$$f_{tw} = -1.4 \text{ N/mm}^2$$

$$A = 250 \times 580 = 145 \times 10^3 \text{ mm}^2$$

$$Z = \frac{250 \times 580^2}{6} = 14 \times 10^6 \text{ mm}^3$$

$$g = (0.25 \times 0.58 \times 24) = 3.48 \text{ kN/m}$$

$$M_g = \frac{(3.48 \times 12^2)}{8} = 62.5 \text{ kN.m}$$

$$M_q = \frac{(12 \times 12^2)}{8} = 216 \text{ kN.m}$$

$$f_t = \left[(-1.4) - \frac{(62.5 \times 10^6)}{(14 \times 10^6)}\right] = -5.9 \text{ N/mm}^2$$

$$f_b = \left[\left(\frac{-14}{0.85}\right) + \frac{(62.5 + 216)\,10^6}{(0.8 \times 14 \times 10^6)}\right] = 22 \text{ N/mm}^2$$

$\therefore$

$$P = \left[\frac{145 \times 10^3 (22 - 5.9) 14 \times 10^6}{28 \times 10^6}\right]$$

$$= 1170000 \text{ N} = 1170 \text{ kN}$$

$$e = \left[\frac{(14 \times 10^6)^2 [22 - (-5.9)]}{145 \times 10^3 (22 - 5.9) 14 \times 10^6}\right] = 167.5 \text{ mm}$$

Hence, prestressing force = 1170 kN

Eccentricity = 167.5 mm.

Problem 7.9

A post-tensioned prestressed concrete beam of rectangular section 400 mm wide by 1000 mm deep spans over 15 m and supports a live load of 25 kN/m. If $f_{ct} = 17$ N/mm² and $f_{tt} = f_{tw} = 0$, $\eta = 0.85$. Design the prestressing force and eccentricity.

Solution

Given data:

$$q = 25 \text{ kN/m}, \quad f_{ct} = 17 \text{ N/mm}^2,$$

$$f_{tt} = 0, \quad \eta = 0.85$$

$$A = (400 \times 1000) = 4 \times 10^5 \text{ mm}^2$$

$$Z = \frac{(400 \times 1000^2)}{6} = 66.66 \times 10^6 \text{ mm}^3$$

$$g = (0.4 \times 24) = 9.6 \text{ kN/m}$$

$$M_g = \frac{(9.6 \times 15^2)}{8} = 270 \text{ kN.m}$$

$$M_q = \frac{(25 \times 15^2)}{8} = 703 \text{ kN.m}$$

$$f_t = \left[0 - \frac{(270 \times 10^6)}{(66.66 \times 10^6)}\right] = -4.05 \text{ N/mm}^2$$

$$f_b = \left[0 + \frac{(703 + 270)\,10^6}{(0.85 \times 66.66 \times 10^6)}\right] = 17.17 \text{ N/mm}^2$$

$$P = \left\{\frac{4 \times 10^3 [-4.05 + 17.17]}{2}\right\}$$

$$= 2624 \times 10^3 \text{ N} = 2624 \text{ kN}$$

$$e = \left[\frac{66.66 \times 10^2 (17.17 + 4.05)}{4 \times 10^3 (-4.05 + 17.17)}\right] = 269.5 \text{ mm}$$

Hence, prestressing force = 2624 kN

Eccentricity = 269.5 mm.

Problem 7.10

A post-tensioned prestressed concrete girder of a bridge spans over 30 mm and is made up of an unsymmetrical I-section with the cross sectional details as shown in Fig. 7.4. The section has to support a dead load bending moment of 4261 kN.m and a live load bending moment of 2070 kN.m. If f_{ct} = 18 N/mm^2 and f_{tw} = 0 and loss ratio is 15%, compute the prestressing force required assuming an effective cover of 200 mm for the cables.

Solution

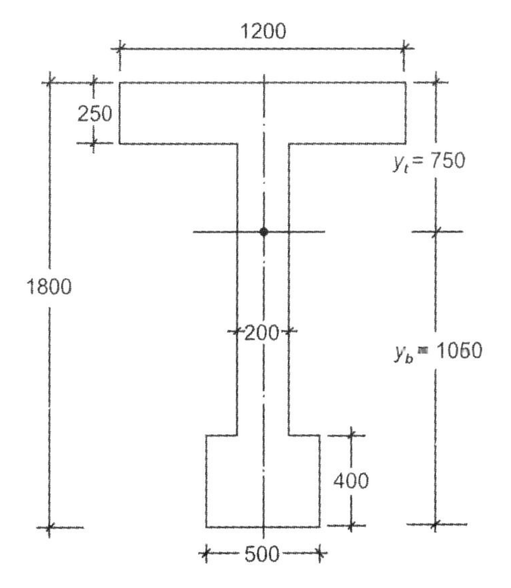

Fig. 7.4

Given data: M_g = 4261 kN.m, M_q = 2074 kN.m,

$A = 73 \times 10^4$ mm^2, f_{ct} = 18 N/mm^2,

f_{tw} = 0, η = 0.85

$I = 2924 \times 10^8$ mm^4, $Z_b = 2.78 \times 10^8$ mm^3

$$Z_b \geq \left[\frac{(2074\times10^6)+(1-0.85)4261\times10^6}{(0.85\times18-0)}\right]$$

$\geq 1.77 \times 10^8 \leq 2.78 \times 10^8$ mm^3; hence adequate.

Prestressing force

Since $M_g > M_q$, the prestressing force is computed by using the modified equation given by

$$P = \left[\frac{(A f_b Z_b)}{Z_b + A.e}\right] \quad \text{and}$$

$$f_b = \left[0 + \left(\frac{(6335 \times 10^6)}{0.85 \times 2.78 \times 10^8}\right)\right] = 26.8 \text{ N/mm}^2$$

In this case $e = (1050 - 200) = 850$ mm

$$P = \left\{\frac{73 \times 10^4 \times 26.80 \times 2.78 \times 10^8}{(2.78 \times 10^8) + (73 \times 10^4 \times 850)}\right\}$$

$$= 6053 \times 10^3 \text{ N} = 6053 \text{ kN}$$

Problem 7.11

Design the thickness and circumferential wire winding for a cylindrical water tank wall subjected to a design tensile force of 600 kN/m. $f_{ct} = 16$ N/mm^2, $f_{tw} = 0$, direct tensile strength of concrete is 3 N/mm^2 and $\eta = 0.85$. High tensile wires of 7 mm diameter ($f_p = 1600$ N/mm^2) initially stressed to 1000 N/mm^2 are available for use. A load factor of 2 against collapse and 1.25 against cracking is required.

Solution

Given data:

$N_d = 600$ kN/m, $f_{ct} = 16$ N/mm^2,

$f_{tw} = 0$, $f_t = 3$ N/mm^2,

$\eta = 0.85$ $f_p = 1600$ N/mm^2

$$A = \left[\frac{N_d}{\eta f_{ct} - f_{tw}}\right] = \left[\frac{600 \times 10^3}{(0.85 \times 16 - 0)}\right]$$

$$= 44118 \text{ mm}^3\text{/m}$$

$$\text{Thickness of wall} = \left(\frac{44118}{1000}\right) = 44.118 \text{ mm}$$

Based on practical considerations adopt a minimum thickness of 100 mm.

$$\text{Prestress required} = \frac{1}{0.85}\left[\frac{600 \times 10^3}{100 \times 10^3} - 0\right] = 7.05 \text{ N/mm}^2$$

$$\text{Prestressing force} = \left[\frac{7.05 \times 100 \times 10^3}{10^3}\right] = 705 \text{ kN}$$

No. of 7 mm diameter H.T. wires $= \left[\dfrac{705 \times 10^3}{38.5 \times 10^3}\right] = 18.3$

Pitch of wires $= \left(\dfrac{1000}{18.3}\right) = 54$ mm

Ultimate tensile force $= \left[\dfrac{(18.3 \times 38.5 \times 0.87 \times 1600)}{1000}\right]$

$= 980$ kN

Required ultimate force $= (2 \times 600) = 1200$ kN

Additional force required $= (1200 - 980) = 220$ kN

Area of HYSD bars $= \left[\dfrac{(220 \times 10^3)}{(0.87 \times 415)}\right] = 610\ \text{mm}^2$

Spacing of 6 mm diameter bars on each face

$= \left[\dfrac{(2 \times 1000 \times 28)}{610}\right] = 90$ mm

Cracking load $= \left[\dfrac{(100 \times 1000)\{0.85 \times 7.05\} + 3}{1000}\right]$

$= 899$ kN

Load factor against cracking $= \left(\dfrac{899}{600}\right) = 1.49$

Problem 7.12

Design a suitable section for the tie member to a truss of support a maximum design tensile force 250 kN, $f_{ct} = 16\ \text{N/mm}^2$. No tensile stresses are permitted. Loss ratio is 0.8. High tensile wires of 7 mm diameter tensioned to 100 N/mm² are available for use, $f_p = 1600\ \text{N/mm}^2$, tensile strength of concrete = 3 N/mm². Load factor against collapse = 1.5 and load factor against cracking is 1.25.

Solution

Given data: $N_d = 250$ kN, $f_t = 3\ \text{N/mm}^2$

$f_{ct} = 16\ \text{N/mm}^2$, $f_{tw} = f_{tt} = 0$

$f_p = 1600\ \text{N/mm}^2$ $\eta = 0.8$

$$\text{Area of concrete section} = \left(\frac{N_d}{\eta f_{ct}}\right) = \frac{(250 \times 10^3)}{(0.8 \times 16)}$$

$$= 19531 \text{ mm}^2$$

Adopt 150 mm by 150 mm section (Area = 22500 mm²)

$$f_c = \frac{(250 \times 10^3)}{(0.8 \times 22500)} = 13.88 \text{ N/mm}^2$$

Prestressing force

$$P = \left[\frac{(13.88 \times 22500)}{1000}\right] = 312 \text{ kN}$$

$$\text{Number of 7 mm diameter H.T. wires} = \left[\frac{(312 \times 10^3)}{(38.5 \times 10^3)}\right] = 8$$

Provide 8 wires in the section;

$$\therefore \text{ ultimate tensile strength of tie} = \left[\frac{(8 \times 38.5 \times 0.87 \times 1600)}{1000}\right]$$

$$= 428 \text{ kN}$$

$$\text{Load factor against collapse} = \left(\frac{428}{250}\right) = 1.712$$

$$\text{Cracking load} = \left[\frac{22500(0.8 \times 13.88 + 3)}{1000}\right] = 317 \text{ kN}$$

$$\text{Load factor against cracking} = \left(\frac{317}{225}\right) = 1.4$$

Problem 7.13

A prestressed girder has to be designed to cover a span of 12 m, and to support an uniformly distributed live load of 15 kN/m. M-45 grade concrete is used for casting the girder. The permissible stress in compression may be assumed as 14 N/mm² and 1.4 N/mm². Assume 15% loss in prestress during service load conditions. The preliminary section proposed for the girder consists of a symmetrical I-section with flanges 300 mm wide and 150 mm thick. The web is 120 mm wide by 450 mm deep.

a. Check the adequacy of the section provided to resist the service loads.

b. Design the minimum prestressing force and the corresponding eccentricity for section.

Solution

$L = 12$ m Loss ratio $= \eta = 0.85$

$f_{ct} = 14$ N/mm^2 $f_{tt} = f_{tw} = -1.4$ N/mm^2

$q = 15$ kN/m $y = y_t = y_b = 375$ mm

Area of section

$$= A = [(2 \times 200 \times 150) + (120 \times 450)] = 144000 \text{ mm}^2$$

Second moment of area

$$= I = \left[\frac{300 \times 750^3}{12} - \frac{180 \times 450^3}{12}\right] = (918 \times 10^7) \text{ mm}^4$$

Second modulus

$$= Z = Z_t = Z_b = \left[\frac{I}{y}\right] = \left[\frac{(918 \times 10^7)}{375}\right]$$

$$= (24.48 \times 10^6) \text{ mm}^3$$

Self weight of girder

$$= g = \left[\frac{144000}{10^6}\right] 24 = 3.456 \text{ kN/m}$$

Dead load moment

$$= M_g = \left[\frac{qL^2}{8}\right] = \left[\frac{3.456 \times 12^2}{8}\right] = 62.208 \text{ kN.m}$$

Live load moment

$$= M_q = \left[\frac{gL^2}{8}\right] = \left[\frac{15 \times 12^2}{8}\right] = 270 \text{ kN.m}$$

$$f_{br} = (\eta f_{ct} - f_{tw}) = [(0.85 \times 14) - (-14)] = 13.3 \text{ N/mm}^2$$

a. Check for adequacy of the section

$$Z_b \geq \left[\frac{M_q + (1 - \eta)\, M_g}{f_{br}}\right]$$

$$= \left[\frac{(270 \times 10^6) + (1 - 0.85)\,(62.208 \times 10^6)}{13.3}\right]$$

$$= (21 \times 10^6) \text{ mm}^3 < Z_b \text{ (provided)}$$

Hence, the section provided is adequate to resist the loads safely.

b. Minimum prestressing force and the corresponding eccentricity:

$$f_t = \left[f_{tt} - \frac{M_g}{Z_t}\right] = \left[-1.4 - \frac{(62.208 \times 10^6)}{(24.48 \times 10^6)}\right]$$

$$= -3.94 \text{ N/mm}^2$$

$$f_b = \left[\frac{f_{tw}}{\eta} + \frac{M_g + M_q}{\eta Z_b}\right] = \left[\frac{-1.4}{0.85} + \frac{(62.208 + 270) \times 10^6}{0.85\,(24.48 \times 10^6)}\right]$$

$$= 14.36 \text{ N/mm}^2$$

Prestressing force is computed using the relation,

$$P = \frac{A(f_t Z_t + f_b Z_b)}{(Z_t + Z_b)} = \frac{144000\,[(-3.94 + 14.36)\,24.48 \times 10^6]}{(2 \times 24.48 \times 10^6)}$$

$$= (747.28 \times 10^6) \text{ N} = 747.28 \text{ kN}$$

Eccentricity is obtained by the relation

$$e = \frac{Z_t Z_b (f_b - f_t)}{A(f_t Z_t + f_b Z_b)}$$

$$= \frac{(24.48^2)\,(10^{12})\,[14.36 - (-3.94)]}{144000\,[(-3.94 + 14.36)\,24.48 \times 10^6]} = 298.5 \text{ mm.}$$

Problem 7.14

Design a suitable section for the tie member of a prestressed concrete truss to carry a design tensile force of 600 kN. Assume the permissible compressive stress in concrete at transfer as 15 N/mm^2 and tension is not allowed under service loads. Loss of prestress is 20%. High tensile wires of 8 mm diameter with an ultimate tensile strength of 1400 N/mm^2 with an initial stress of 800 N/mm^2 are available for use. The direct tensile strength of concrete is 3 N/mm^2. A load factor of 2 against collapse and 1.25 against cracking is to be ensured in the design.

Solution

$$\text{Tensile force} = N_d = 600 \text{ kN}$$

$$f_{ct} = 15 \text{ N/mm}^2$$

$$f_{tw} = 0 \qquad \eta = 0.80$$

Area of concrete section

$$= \left[\frac{N_d}{\eta f_{ct}}\right] = \left[\frac{600 \times 10^3}{0.8 \times 15}\right] = 50000 \text{ mm}^2$$

Adopt a section 200 mm by 250 mm (area = 50000 mm^2)

Compressive prestress

$$= \left[\frac{600 \times 10^3}{0.8 \times 50000}\right] = 15 \text{ N/mm}^2$$

Prestressing force required

$$= P = \left[\frac{15 \times 50000}{1000}\right] = 750 \text{ kN}$$

Number of 8 mm wires

$$= \left[\frac{750 \times 10^3}{50 \times 800}\right]$$

$= 18.75$ (use 20 wires of 8 mm diameter)

Ultimate tensile strength of the tie

$$= \left[\frac{20 \times 50 \times 0.87 \times 1400}{1000}\right] = 1218 \text{ kN}$$

Load factor against collapse

$$= \left[\frac{1218}{600}\right] = 2.03 > 2$$

$$\text{Cracking load} = \left[\frac{50000\,(0.8 \times 15 + 3)}{1000}\right] = 750 \text{ kN}$$

$$\text{Load factor against cracking} = \left(\frac{750}{600}\right) = 1.25.$$

Problem 7.15

The end block of a post-tensioned prestressed concrete beam has a rectangular section 100 mm wide by 200 mm deep. A cable carrying a force of 200 kN is to be anchored against the end block concentrically. If the characteristic cube strength of concrete at transfer is 30 N/mm^2, design the size of a suitable anchorage plate to transmit the force.

Solution

Force in the cable = 200 kN

Assume an anchor plate of size 50 mm by 50 mm

Punching area $= A_{pun} = (50 \times 50) = 2500 \text{ mm}^2$

Bearing area $= A_{br} = (100 \times 100) = 10000 \text{ mm}^2$

$$\text{Actual bearing pressure} = \left[\frac{200 \times 10^3}{10000}\right] = 20\ \text{N/mm}^2$$

According to Indian standard code IS: 1343,
maximum permissible bearing pressure

$$= f_b = 0.48\, f_{ci} \sqrt{\frac{A_{br}}{A_{pun}}}$$

or $0.8\, f_{ci}$ whichever is smaller.

Hence, $$f_b = 0.48 \times 30 \sqrt{\frac{10000}{2500}} \quad \text{or} \quad (0.8 \times 30)$$

$$= 28.8\ \text{N/mm}^2 \ \text{or}\ 24\ \text{N/mm}^2 > 20\ \text{N/mm}^2$$

Hence, the actual bearing pressure behind the anchorages is within the permissible limits.

8

Design of Pretensioned and Post-tensioned Flexural Members

Problem 8.1

Design a pre-tensioned symmetrical I-beam for an effective span of 7 m to support a superimposed load of 6 kN/m. The beam is to be precast in a factory and is to be designed for handling at any point along its length during transport and erection. Load factors against failure by bending and shear:

For dead load = 1.5

For live load = 2.5

Permissible stresses:

At transfer,

Compressive stress = 14 N/mm^2

Tensile stress = 1.4 N/mm^2

At working load,

Compressive stress = 16 N/mm^2

Tensile stress = 1.4 N/mm^2

The specified 28-day cube strength of concrete is 50 N/mm^2. The prestressing force is to be provided by 5 mm high tensile wires having an ultimate tensile strength of 1600 N/mm^2. The loss ratio is 0.8. Design the beam and sketch the cross-section showing the arrangement of wires. Check the safety of the beam for the limit states of cracking, deflection and collapse.

Solution

Given: Effective span = 7 m, q = 6 kN/m

f_{ck} = 50 N/mm^2 f_{tw} = –1.4 N/mm^2

f_{ci} = 35 N/mm^2 f_{pu} = 1600 N/mm^2

η = 0.5 f_{tt} = –1.4 N/mm^2

E_c = 34 kN/mm^2 (assumed)

$f_{cw} = 16 \text{ N/mm}^2 \qquad f_{ct} = 14 \text{ N/mm}^2$

$$\left.\begin{array}{l}\gamma f_1 = 2.5 \\ \gamma f_2 = 1.5\end{array}\right] \text{Load factors}$$

Design calculations

$$\left(\frac{w_{\min}}{w_{ud}}\right) = \left[\frac{K \cdot D_c \cdot g \cdot \beta (L/h) L}{f_{cu} (d/h)^2}\right]$$

$$= \left[\frac{7.5 \times 2400 \times 9.81 \times 0.125 \times 25 \times 7}{50 (0.85)^2 \times 10^6}\right]$$

$$= 0.107$$

But
$$w_{ud} = \left[\frac{\gamma f_1 \cdot q}{1 - y f^2 \left(\dfrac{w_{\min}}{w_{ud}}\right)}\right] = \left[\frac{2.5 \times 6}{1 - 1.5 \times 0.107}\right]$$

$$= 17.87 \text{ kN/m}$$

$\therefore \quad w_{\min} = 0.107 \qquad w_{ud} = (0.107 \times 17.87)$

$= 1.91 \text{ kN/m}$

$M_u = 0.125 \qquad w_{ud} . L^2 = 0.125 \times 17.87 \times 7^2$

$= 109.45 \text{ kN.m}$

$V_u = 0.5 \qquad w_{ud} . L = (0.5 \times 17.87 \times 7)$

$= 62.55 \text{ kN}$

$M_g = 0.125 \times w_{\min} \times L^2 = 0.125 \times 1.91 \times 7^2$

$= 11.69 \text{ kN.m}$

$M_q = 0.125\, q . L^2 = 0.125 \times 6 \times 7^2 = 36.75 \text{ kN.m}$

Cross-sectional dimensions

For symmetrical I-section,

$M_u = 0.10 f_{ck} b d^2 \qquad b = 0.5d$

$$\therefore \quad d = \sqrt[3]{\frac{109.45 \times 10^6}{0.1 \times 50 \times 0.5}} = 360 \text{ mm}$$

Taking $(d/h) = 0.85$

$$h = \left(\frac{360}{0.85}\right) = 425 \text{ mm}$$

Adopt effective depth 'd' = 365 mm

Overall depth 'h' = 430 mm

Width of flange = (0.5 × 430) = 215 mm

Thickness of flange = $0.2d$ = (0.2 × 365) = 73 mm

Since sloping flanges are employed, increase flange thickness by 20%

∴ Average thickness of flange = (73 ± 0.2 × 73) = 90 mm

$$\text{Approximate web thickness} = \left(\frac{0.85\, V_u}{f_t \cdot h}\right)$$

$$= \left(\frac{0.85 \times 62.55 \times 10^3}{1.7 \times 430}\right) = 75 \text{ mm}$$

The assumed section is shown in Fig. 8.1

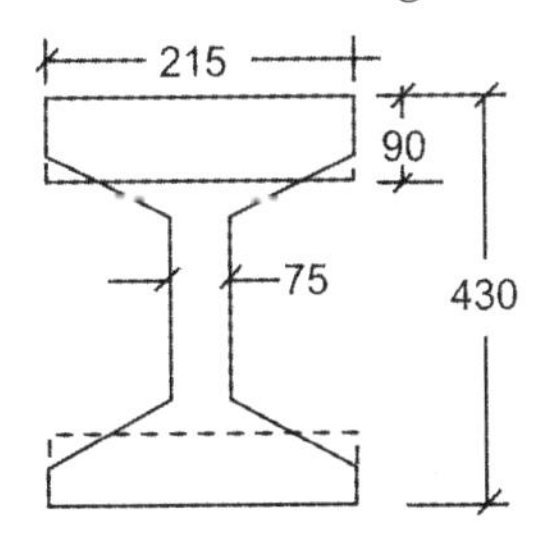

Fig. 8.1

Properties of section

$$A = 57450 \text{ mm}^2$$

$$I = 1.2422 \times 10^9 \text{ mm}^4$$

$$Z_b = Z_t = 5.77 \times 10^6 \text{ mm}^3$$

Self weight = 1.38 kN/m < 1.91 kN/m (assumed)

Minimum section modulus

Range of stress $f_{br} = (\eta f_{ct} - f_{tw})$

$= 0.75 \times 14 - (-1.4) = 12 \text{ N/mm}^2$

$$Z_b \geq \left[\frac{M_g + (1-\eta) M_g}{f_{br}}\right]$$

$$\geq \left[\frac{36.75 \times 10^6 + (1 - 0.75) 11.69 \times 10^6}{12.0}\right]$$

$\geq 2818960 \text{ mm}^3 < 5.77 \times 10^6 \text{ mm}^3$ (provided)

Hence, section is adequate.

Prestressing force and eccentricity

$$P = \frac{A(Z_b f_b + Z_t f_t)}{Z_b + Z_t}$$

$$f_b = \frac{f_{tw}}{\eta} + \frac{(M_q + M_g)}{\eta Z_b}$$

$$= \left(\frac{-1.4}{0.75}\right) + \frac{(11.69 + 36.75)\,10^6}{0.75 \times 5.77 \times 10^6} = 9.31 \text{ N/mm}^2$$

$$f_t = \left(f_{tt} - \frac{M_g}{Z_t}\right), \quad \text{where } f_{tt} = -1.4 \text{ N/mm}^2$$

$$= -1.4 - \left(\frac{11.69 \times 10^6}{5.77 \times 10^6}\right) = -3.42 \text{ N/mm}^2$$

$$\therefore \quad P = \left\{\frac{57450[5.77 \times 10^6 \times 9.31 + 5.77 \times 10^6 (-3.42)]}{2 \times 5.77 \times 10^6}\right\}$$

$$= 169190 \text{ N} = 169.19 \text{ kN}$$

∴ No. of 7 mm ϕ wires stressed to 1200 N/mm²

$$= \left[\frac{169.19 \times 10^3}{(\pi/4)(7^2) \times 1200}\right] = 4$$

$$\text{Eccentricity} = e = \frac{Z_t Z_b (f_b - f_t)}{A(f_t Z_t + f_b Z_b)}$$

$$= \left\{\frac{(5.77 \times 10^6)^2 [9.31 + 3.42]}{57450(-3.42 \times 2 + 9.312)}\right\}$$

$$= 207.4 \text{ mm}$$

Check for ultimate strength (As per revised IS: code) IS: 1343-2012

$$A_p = (38.5 \times 4) = 154 \text{ mm}^2$$

$$f_p = 1600 \text{ N/mm}^2$$

$$b = 215 \text{ mm}$$

$$f_{ck} = 50 \text{ N/mm}^2$$

$$d = (215 + 207.4) = 422.4 \text{ mm}$$

$$\therefore \quad \left(\frac{A_p f_p}{b.d\, f_{ck}}\right) = \left(\frac{154 \times 1600}{215 \times 422.4 \times 50}\right) = 0.054$$

From Table 4.1 or Table 11 of (IS: 1343-2012)

$$\left(\frac{f_{pu}}{0.87 f_p}\right) = 1 \qquad \text{and} \qquad \left(\frac{x_u}{d}\right) = 0.117$$

Hence, $f_{pu} = (0.87 \times 1600) = 1392 \text{ N/mm}^2$

$x_u = (0.117 \times 422.4) = 49.42 \text{ mm}$

$\therefore \quad M_u = f_{pu} \cdot A_p (d - 0.42\, x_u)$

$= [1392 \times 154\, (422.4 - 0.42 \times 49.42)]$

$= (8.6 \times 10^8)$ N.mm > M_u required. Hence safe.

Check for limit state of deflection

$$\text{Permissible deflection} = \left(\frac{\text{span}}{250}\right) = \left(\frac{7000}{250}\right) = 28 \text{ mm}$$

$$\text{Deflection due to prestress} = \left(\frac{P \cdot e \cdot L^2}{8EI}\right)$$

$$a_p = \left(\frac{169.19 \times 10^3 \times 207.4 \times (7000)^2}{8 \times 34 \times 10^3 \times 1.2422 \times 10^9}\right)$$

$= 5.08$ mm (upwards)

Deflection due to live and dead load

$$a_{(g+q)} = \frac{5}{385}\left\{\frac{(g+q)L^4}{EI}\right\}$$

$$= \frac{5}{384}\left[\frac{(1.38 + 6)\,(7000)^4}{34 \times 10^3 \times 1.2422 \times 10^9}\right]$$

$= 5.46$ mm (downwards)

Assuming creep coefficient, $\phi = 1.6$

Long term deflection = $(1 + \phi)$ (short term deflection)

$= (1.6 + 1)\,[5.46 - 5.08]$

$= 0.988 \text{ mm} < 28 \text{ mm}$ $\quad \therefore$ Safe.

The reinforcement details are shown in Fig. 8.2.

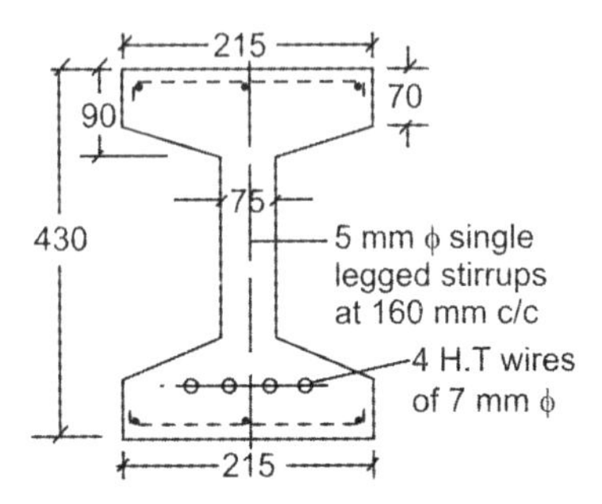

Fig. 8.2

Problem 8.2

A straight, precast, pre-tensioned beam of I-section is to be designed to support a uniformly distributed imposed load of 8 kN/m in addition to the self weight of the member. The effective span of the simply supported beam is to be 9 m. Using concrete of grade M-45 with permissible compressive stress in concrete at transfer and working loads as 15 N/mm^2 and 5 mm diameter high tensile steel wires of U.T.S. = 1600 N/mm^2, and initially stressed to 1200 N/mm^2, design the cross-section of the girder as a class I member without allowing any tension under working loads. Assume the loss of prestress due to elastic deformation, creep, shrinkage and other factors as 20%. Sketch the cross-section of the girder at the centre of span showing the arrangement of wires. Load factors of 1.5 for dead load and 2.5 for live load may be assumed.

Solution

The section is adopted is shown in Fig. 8.3.

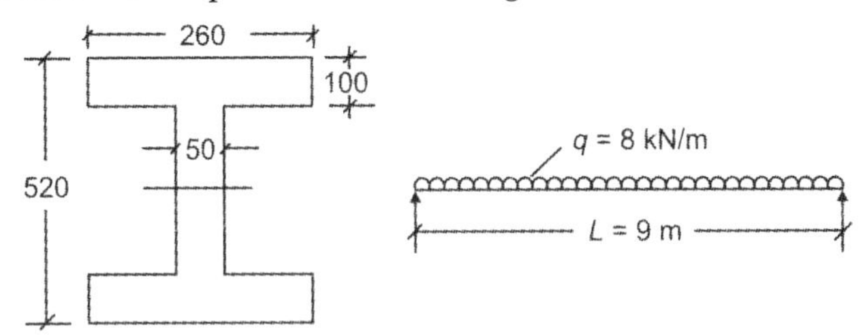

Fig. 8.3

$$f_{ck} = 45 \text{ N/mm}^2$$
$$f_{ct} = f_{cw} = 15 \text{ N/mm}^2$$
$$f_{tw} = 0$$

Initial stress in steel = 1200 N/mm^2

Loss ratio = η = 0.8

Based on preliminary calculations of (w_{min}/w_{ud}) as worked out in example (1), the section adopted is shown in Fig. 8.3.

Properties of section

$$A = 68000 \text{ mm}^2$$

$$Z = 9.51 \times 10^6 \text{ mm}^3$$

Self weight $= 1.632$ kN/m

$$M_g = \left(\frac{1.632 \times 9^2}{8}\right) = 16.52 \text{ kN.m}$$

$$M_q = \left(\frac{8 \times 9^2}{8}\right) = 81 \text{ kN.m}$$

$$f_{br} = (\eta f_{ct} - f_{tw}) = (0.8 \times 15 - 0) = 12 \text{ N/mm}^2$$

$$Z_b \geq \left[\frac{M_q + (1-\eta)M_g}{f_{br}}\right]$$

$$\geq \left[\frac{81 \times 10^6 + (1 - 0.8)\, 16.52 \times 10^6}{12}\right]$$

$$\geq 7.02 \times 10^6 \text{ mm}^3 < 9.51 \times 10^6\, b \text{ mm}^3 \text{ (provided).}$$

Hence, section is adequate.

$$P = \frac{A(f_t Z_t + f_b Z_b)}{Z_t + Z_b}$$

$$f_t = \left(f_{tt} - \frac{M_g}{Z_t}\right) = \left(0 - \frac{16.52 \times 10^6}{9.51 \times 10^6}\right)$$

$$= -1.73 \text{ N/mm}^2$$

$$f_b = \left[\frac{f_{tw}}{\eta} + \frac{M_g + M_q}{\eta Z_b}\right] = \left[0 + \frac{97.52 \times 10^6}{0.8 \times 9.51 \times 10^6}\right]$$

$$= 12.81 \text{ N/mm}^2$$

Prestressing force and eccentricity

$$P = \left\{\frac{68000[-1.73 \times 9.51 \times 10^6 + 1281 \times 9.51 \times 10^6]}{2 \times 9.51 \times 10^6}\right\}$$

$$= 376 \times 10^3 \text{ N} = 376 \text{ kN}$$

Force in each wire of 5 mm ϕ = $\left(\frac{\pi \times 5^2}{4}\right) \times 1200$

$$= 23562 \text{ N} = 23.562 \text{ kN}$$

$$\therefore \text{No. of wires} = \left(\frac{376}{23.562}\right) = 16 \text{ wires}$$

$$e = \frac{Z_t Z_b (f_b - f_t)}{A(f_t Z_t + f_b Z_b)}$$

$$= \left\{\frac{(9.51\times 10^6)^2 (12.81+1.73)}{68000(-1.73\times 9.51\times 10^6 + 12.81\times 9.51\times 10^6}\right\}$$

$$= 184 \text{ mm}$$

Problem 8.3

A post-tensioned, prestressed concrete girder having a span of 40 m between bearings is required for an aircraft hanger. The live load on the girder is 5 kN/m. The specified 28-day cube strength is 50 N/mm^2. The cube strength of concrete at transfer is 30 N/mm^2. Permissible stresses should conform to the provisions of IS: 1343. The prestress is to be provided by seven wire 15 mm strand cables each tensioned to 1200 kN, housed in cable ducts of 64 mm. Ultimate tensile strength of each cable = 1750 kN. Loss ratio = 0.80.

The design has to comply with the various limit states of deflection, cracking and collapse. Design the following particulars:

a. The cross-section of the girder,
b. Prestressing force
c. Eccentricity.

Solution

Design calculations

$$\left(\frac{w_{\min}}{w_{ud}}\right) = \left[\frac{K\cdot D_c \cdot g(L/h)L}{f_{ck}\cdot (d/h)^2}\right]$$

$$= \left[\frac{5\times 2400\times 9.81\times 0.125\times 25\times 40}{50\times 10^6 \times (0.85)^2}\right] = 0.407$$

$$= \frac{\gamma f_1 \cdot q}{1-\gamma f_2\left(\dfrac{w_{\min}}{w_{ud}}\right)}$$

Assuming $\gamma f_1 = 2.4, \quad \gamma f_2 = 1.4$

$$\therefore \quad w_{ud} = \frac{2.4\times 5}{1-1.4\,(0.407)} = 27.9 \text{ kN/m}$$

$\therefore \qquad w_{min} = (0.407 \times 27.9) = 11.355 \text{ kN/m}$

$M_u = 0.125 \times 27.9 \times (40)^2 = 5580 \text{ kN.m}$

$V_u = 0.5 \times 27.9 \times 40 = 558 \text{ kN}$

Cross-sectional dimensions

$$M_u = 0.10 f_{ck} .b.d^2 \qquad \left(\frac{b}{d}\right) = 0.5$$

$$\therefore \quad 5580 \times 10^6 = 0.10 \times 50 \times 0.5\, d \times d^2$$

$$\therefore \quad d = 1300 \text{ mm}$$

Hence $= \left(\dfrac{1300}{0.86}\right) = 1500 \text{ mm}$

$$b = 700 \text{ mm}$$

Thickness of web $= b_w = \left(\dfrac{0.6 V_u}{f_t \cdot h}\right)$

$$= \left(\frac{0.6 \times 558 \times 10^3}{1.7 \times 1500}\right) = 131.3 \text{ mm}$$

Since cables of 64 mm ϕ have to pass through the web, the minimum thickness of web from practical considerations is

$$= [64 + 2\,(64)] = 200 \text{ mm}$$

$$\therefore \quad b_w = 200 \text{ mm}$$

Bottom flange dimensions should be such as to accommodate the cables as well as anchorages at ends of members with suitable minimum cover requirements. So choose bottom flange width and depth as 350 mm and 300 mm respectively.

Dimensions of section selected is shown in Fig. 8.4.

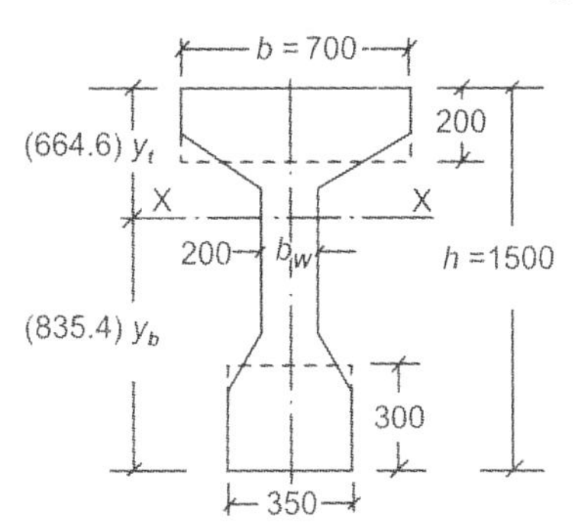

Fig. 8.4

Properties of section

$A = 44.6 \times 10^4$ mm² $\quad \bar{y} = y_t = 664.6$ mm

$y_b = 835.4$ mm $\quad I_{xx} = 1117 \times 10^8$ mm⁴

$Z_t = 1.68 \times 10^8$ mm³ $\quad Z_b = 1.33 \times 10^8$ mm³

Design moments and shear forces

Actual self weight of girder

$$= \left(\frac{44.5 \times 10^4 \times 24}{10^6}\right) = 10.68 \text{ kN/m}$$

$$M_{min} = 0.125 \times 10.68 \times (40)^2 = 2136 \text{ kN/m}$$

Design working load = (10.68 + 5) = 15.68 kN/m

∴ Working moment

$$M_d = (0.125 \times 15.68 \times 40^2) = 3136 \text{ kN.m}$$

Pressing force and eccentricity

$$f_t = \left(f_{tt} - \frac{M_{min}}{Z_t}\right) = \left(0 - \frac{2136 \times 10^6}{1.68 \times 10^8}\right)$$

$$= -12.71 \text{ N/mm}^2$$

$$f_b = \left(\frac{f_{tw}}{\eta} + \frac{M_d}{\eta Z_b}\right) = 0 + \left[\frac{3136 \times 10^6}{0.8 \times 1.33 \times 10^8}\right]$$

$$= 29.47 \text{ N/mm}^2$$

$$\therefore \quad e = \frac{Z_t Z_b (f_b - f_t)}{A(f_t Z_t + f_b Z_b)}$$

$$= \left[\frac{1.68 \times 1.33 \times 10^8 \times 10^8 (29.47 + 12.71)}{44.5 \times 10^4 (-12.71 \times 1.68 + 29.47 \times 1.33) 10^8}\right]$$

$$= 1187 \text{ mm (which is impracticable)}$$

Hence, maximum possible eccentricity

$$= [1500 - 664.6 - 150]$$

$$e = 685 \text{ mm}$$

$$\therefore \quad P = \left(\frac{A f_b Z_b}{Z_b + A_e}\right)$$

$$= \left(\frac{44.5 \times 10^4 \times 29.47 \times 1.33 \times 10^8}{1.33 \times 10^8 + 44.5 \times 10^4 \times 685}\right)$$

$$= 3984 \times 10^3 \text{ N} = 3984 \text{ kN}$$

Force in each cable = 1200 kN

$\therefore$ No. of cables = $\left(\frac{3984}{1200}\right)$ = 4 cables.

Problem 8.4

A post-tensioned prestressed concrete beam for the roof of an industrial structure has a simply supported span of 25 m. The beam has to support a dead load of 2 kN/m together with an imposed load of 15 kN/m in addition to the self weight. The grade of concrete specified is M-40. The compressive strength of concrete at transfer is 35 N/mm^2. The loss ratio is 0.80. The 64 mm cables containing 7–15 mm strands with an ultimate load capacity of 1750 kN are available. Using IS: 1343 provisions, design the cross-section of the girder to comply with various limit states. Sketch the details of cables in the cross-section and the profile of cables along the depth and length of the beam.

Solution

The loaded beam and cross-section assumed is shown in Fig. 8.5 (a) and (b).

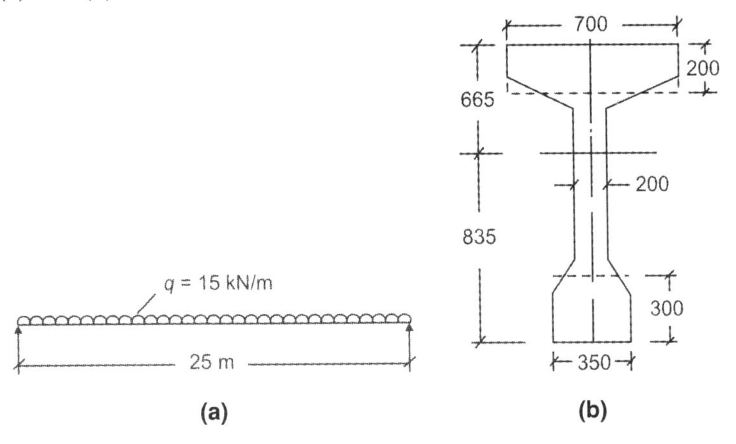

Fig. 8.5 (a) and (b)

Given data:

Dead load = 2 kN/m

Live load = 15 kN/m

f_{ck} = 40 N/mm^2 $\quad$ f_{ct} = 35 N/mm^2

η = 0.8

Force in each cable = 1750 kN

Based on preliminary calculations using the relation $\left(\frac{w_{min}}{w_{ud}}\right)$ and $M_u = 0.10 f_{ck} b d^2$, the cross-section shown in Fig. 8.5 is assumed.

Section properties

$A = 44.5 \times 10^4 \text{ mm}^2 \quad y_t = 665 \text{ mm}$

$y_b = 835 \text{ mm} \quad I_{xx} = 1117 \times 10^8 \text{ mm}^4$

$Z_t = 1.68 \times 10^8 \text{ mm}^3 \quad Z_b = 1.33 \times 10^8 \text{ mm}^3$

Self weight of girder = 10.68 kN/m

$M_{min} = (0.125 \times 10.68 \times 25^2) = 834 \text{ kN.m}$

Design working load = (10.68 + 2 + 15) = 27.68 kN/m

$\therefore$ Design moment $M_d = (0.125 \times 27.68 \times 25^2) = 2163 \text{ kN.m}$

$$f_t = \left(f_{tt} - \frac{M_{min}}{Z_t}\right) = \left(0 - \frac{834 \times 10^6}{1.68 \times 10^8}\right)$$

$$= -4.76 \text{ N/mm}^2$$

$$f_b = \left(\frac{f_{tw}}{\eta} + \frac{M_d}{\eta Z_b}\right) = \left(0 + \frac{2163 \times 10^6}{0.8 \times 1.33 \times 10^8}\right)$$

$$= 20.32 \text{ N/mm}^2$$

$$\therefore \quad P = \left[\frac{A(f_t Z_t + f_b Z_b)}{Z_t + Z_b}\right]$$

$$= \left[\frac{44.5 \times 10^4 [-4.76 \times 1.68 \times 10^8 + 20.32 \times 1.33 \times 10^8]}{(1.68 + 1.33)10^8}\right]$$

$$= 2810 \times 10^3 \text{ N} = 2810 \text{ kN}$$

$$\text{No. of cables} = \left(\frac{2810}{1750}\right) = 2$$

$$\therefore \quad e = \frac{Z_t Z_b (f_b - f_t)}{A(f_t Z_t + f_b Z_b)}$$

$$= \left\{\frac{1.68 \times 1.33 \times 10^8 \times 10^8 (20.32 + 4.76)}{44.5 \times 10^4 [-4.76 \times 1.68 \times 10^8 + 20.32 \times 1.33 \times 10^8]}\right\}$$

$$= 543 \text{ mm}$$

The reinforcements in the cross-section is shown in Fig. 8.6

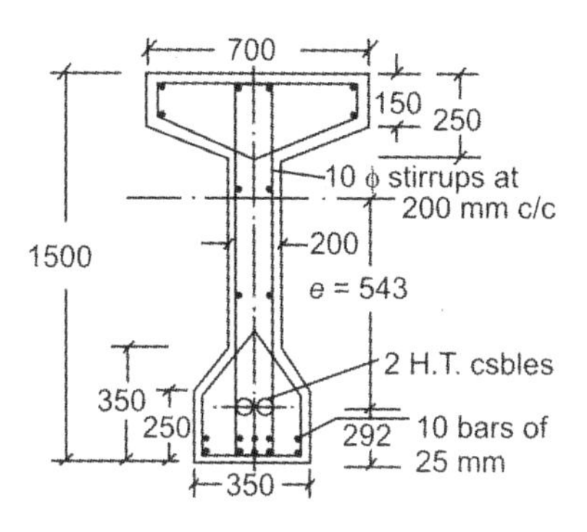

Fig. 8.6

Problem 8.5

Design a pre-tensioned roof purlin of effective span 6 m to support a service uniformly distributed load of 5 kN/m

Assume the following data:

$$f_{ct} = 50 \text{ N/mm}^2 \quad E_c = 5700\sqrt{f_{ck}} = 40.3 \text{ N/mm}^2$$

$$f_{ci} = 30 \text{ N/mm}^2 \quad \eta = 0.8 \text{ and}$$

7 mm high tensile wires with $f_p = 1600$ N/mm^2 are available for use. Design a suitable symmetrical I-section as class 1 type member and determine the number of wires required and the eccentricity. Sketch the cross-section showing details of reinforcements.

Solution

From IS: 1343-2012code, permissible stresses for M-50 grade concrete are computed as

$$f_{ct} = 0.475 f_{ci} = (0.475 \times 30) = 14.25 \text{ N/mm}^2$$

$$f_{cw} = 0.37 f_{ck} = (0.37 \times 50) = 18.5 \text{ N/mm}^2$$

For type-1 member, $f_{tt} = f_{tw} = 0$

Tensile strength of concrete $\quad f_t = 0.24\sqrt{f_{ck}} = 0.24\sqrt{50}$

$= 1.697$ N/m^2

Initial stress in high tensile steel $= (0.8 \times 1600) = 1280$ N/mm^2

Cross sectional dimensions

Approximate depth $= 50$ to 60 mm per metre span

For 6 mm span overall depth $= h = 300$ to 360 mm

Adopt a section as shown in Fig. 8.7

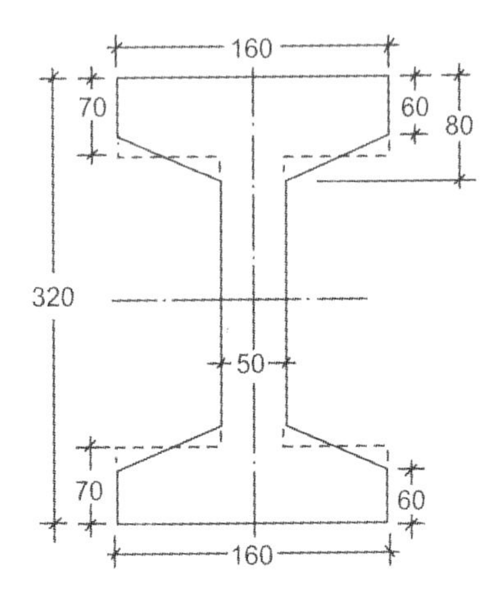

Fig. 8.7

Width of top flange $= b = \left(\frac{h}{2}\right) = \left(\frac{320}{2}\right) = 160$ mm

Flange depth $= (0.2\,h) = (0.2 \times 320) = 70$ mm

Thickness of web (minimum)= 50 mm

Properties of section

$$A = 31400 \text{ mm}^2 \qquad I = 3835 \times 10^5 \text{ mm}^4$$

$$M_g = \frac{(0.76 \times 6^2)}{8} = 3.42 \text{ kN.m}$$

$$Z = Z_t = Z_b = 240 \times 10^4 \text{ mm}^3$$

$$M_q = \frac{(5 \times 6^2)}{8} = 22.50 \text{ kN.m}$$

$$g = 0.76 \text{ kN/m} \quad q = 5 \text{ kN/m}$$

Check for minimum section modulus

$$Z_b \geq \left[\frac{M_q + (1-\eta)M_g}{\eta f_{ct} - f_{tw}}\right]$$

$$\geq \left[\frac{(22.50 \times 10^6) + (1-0.8)\,3.42 \times 10^6}{(0.8 \times 14.25) - 0}\right]$$

$$\geq 203.3 \times 10^4 \text{ mm}^3 \leq 240 \times 10^4 \text{ provided.}$$

Hence safe.

$$f_t = \left[f_{tt} - \left(\frac{M_g}{Z_t} \right) \right] = 0 - (3.42 \times 10^6)\,(230 \times 10^4)$$

$$= -1.486 \text{ N/mm}^2$$

$$f_b = \left[\left(\frac{f_{tw}}{\eta} \right) + \frac{(M_g + M_q)}{\eta Z_b} \right]$$

$$= \left[0 + \frac{(3.42 + 22.50)10^6}{0.8 \times 240 \times 10^6} \right] = 13.7 \text{ N/mm}^2$$

Prestressing force

$$P = \left[\frac{A(f_t Z_t + f_b Z_b)}{Z_t + Z_b} \right] = \left[\frac{A(f_t + f_b)}{2} \right]$$

$$= \left[\frac{31400(-1.486 + 13.7)}{2} \right]$$

$$= 191760 \text{ kN} = 191.76 \text{ kN}$$

Number of 7 mm diameter H.T. wires

Initially stressed to 1280 N/mm^2 $= \left(\dfrac{197726}{38.5 \times 1280} \right) \simeq 4$ wires

Eccentricity of the prestressing force is computed as

$$e = \left[\frac{Z_t Z_b (f_b - f_t)}{A(f_t Z_t + f_b Z_b)} \right] = \left[\frac{Z(f_b - f_t)}{A(f_t + f_b)} \right]$$

$$= \left[\frac{240 \times 10^4 (15.566)}{31400(12.594)} \right] = 91 \text{ mm}$$

Supplementary reinforcement

Longitudinal steel $\nless$ 0.15%

$$= (0.0015 \times 31400) = 47 \text{ mm}^2$$

Provide 4 bars of 6 mm diameter HYSD bars

Shear reinforcement $\nless$ 0.1%

$$= (0.001 \times 1000 \times 50) = 50 \text{ mm}^2$$

Provide 6 mm diameter single legged stirrups at 150 mm centres.

The details of reinforcement provided in the section is shown in Fig. 8.8.

Check for ultimate flexural strength

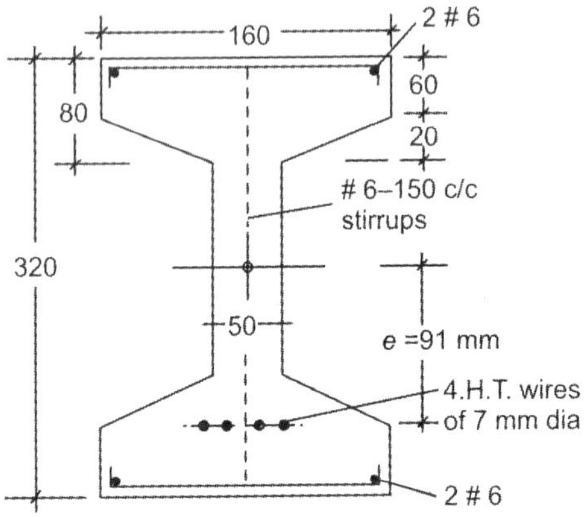

Fig. 8.8

$$M_u \text{ (required)} = 1.5\,(M_g + M_q) = 1.5(3.42 + 22.5) = 38.88 \text{ kN.m}$$

$$M_u = A_p f_{pu}(d - 0.42\,x_u)$$

$$\text{Ratio}\left[\frac{(A_p f_p)}{(bd\, f_{ck})}\right] = \left[\frac{(154\times1600)}{(160\times251\times50)}\right] = 0.123$$

From Table 11 of IS: 1343, or Table 4.1, read out,

$$\left(\frac{f_{pu}}{0.87\,f_p}\right) = 1.00 \qquad \text{and} \qquad \left(\frac{x_u}{d}\right) = 0.26$$

$\therefore$ $f_{pu} = (0.87 \times 1600) = 1392 \text{ N/mm}^2$

and $x_u = (0.26 \times 250.5) = 65 \text{ mm} < 70 \text{ mm}$, Hence safe.

$$M_u = 1392 \times 154\,(251 - 0.42 \times 65)$$
$$= 47.84 \times 10^6 \text{ N.mm}$$
$$= 47.84 \text{ kN.m} > 38.88 \text{ kNm, Hence safe.}$$

Check for ultimate shear strength at support

$$V_u \text{ (required)} = 1.5\,(V_g + V_q)$$
$$= 1.5(5.76 \times 0.5 \times 6) = 25.92 \text{ kN}$$

$$f_{cp} = \left(\frac{\eta P}{A}\right) = \left(\frac{(0.8\times197726)}{(31400)}\right) = 5.03 \text{ N/mm}^2$$

$$V_{co} = 0.67\, b_w\, h\sqrt{f_t^2 + 0.8\, f_{cp} f_t}$$

$$= 0.67 \times 50 \times 320\sqrt{1.697^2 + 0.8\times5.03\times1.697}$$
$$= 33399 \text{ N} = 33.399 \text{ kN} > 25.92 \text{ kN, Hence safe}$$

Check for deflection

$$a_p = \left(\frac{PeL^2}{8EI}\right) = \left[\frac{197726 \times 90.5 \times 6000^2}{8 \times 40.3 \times 10^3 \times 3700 \times 10^5}\right]$$

$$= -5.4 \text{ mm (upwards)}$$

$$a_{g+q} = \left[\frac{5(g+q)L^4}{384EI}\right]$$

$$= \left[\frac{5 \times 5.76 + 6000^4}{384 \times 40.3 \times 10^3 \times 3700 \times 10^5}\right]$$

$$= 6.5 \text{ mm (downwards)}$$

Assuming creep coefficient $\phi = 1.6$

Long term deflection is computed as

$$a_{RL} = (1 + \phi)\,[\eta a_p + a_{g+q}]$$

$$= (1 + 1.6)\,[-0.8(5.4) + 6.5]$$

$$= 5.668 \text{ mm (downwards)}$$

Maximum permissible deflection as per IS: 1343 code is

$$\left(\frac{\text{span}}{250}\right) = \left(\frac{6000}{250}\right) = 24 \text{ mm}$$

Hence, limit state of deflection is satisfied.

Problem 8.6

Design a post-tensioned prestressed concrete roof girder to suit the following data:

Effective span = 20 m

Live load = 12 kN/m

$f_{ck} = 50 \text{ N/mm}^2 \qquad f_{ci} = 41 \text{ N/mm}^2$

$f_t = 0.24\sqrt{50} = 1.7 \text{ N/mm}^2$

$E_c = 5700\sqrt{50} = 40.3 \times 10^3 \text{ N/mm}^2$

Loss ratio = $\eta = 0.85$

Cables containing 12 wires of 7 min diameter ($f_p = 1500 \text{ N/mm}^2$) are available for use.

Design the girder as type-1 member to conform to the specifications of the Indian standards.

Solution

Cross-sectional dimensions

An unsymmetrical I-section (tee-section) is selected.

Depth of girder = 50 mm per metre of span
$= (50 \times 20) = 1000$ mm

Adopt overall depth = h = 1000 mm

Width of top flange = b = 0.4 to 0.5 h
$= (0.4 \times 1000) = 400$ mm
$= (0.5 \times 1000) = 500$ mm

Adopt width of top flange = b = 500 mm which is nearly half the depth.

Thickness of web = b_w = Minimum of 150 mm to house cable ducts of 50 mm diameter with a cover of 50 mm on either side.

Figure 8.9 shows the cross-sectional dimensions of the tee-section selected based on empirical and practical considerations.

The dimensions of the bottom flange depends upon the number of cables and the rows. The dimensions should be such that a clear cover of 50 mm between the cables and their rows is available.

Assuming that three cables are provided in a horizontal row and the diameter of the cable is 50 mm, the minimum width required to house three cables in a row with a clear cover of 50 mm between them is 350 mm. A bottom flange 350 mm wide by 350 mm deep is provided as shown in Fig. 8.9.

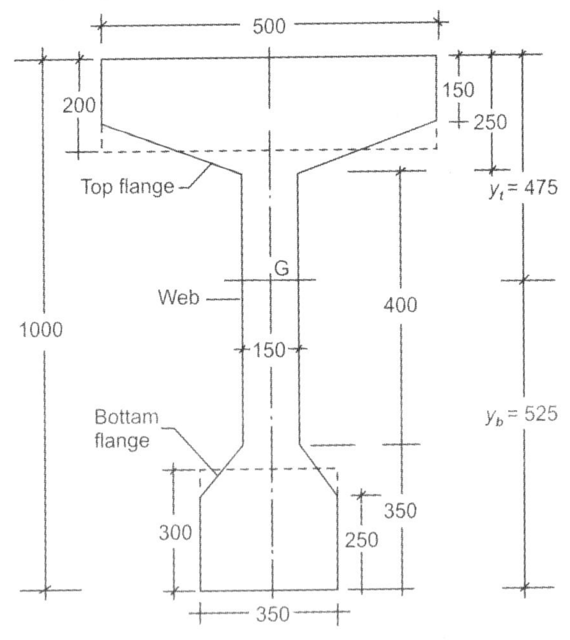

Fig. 8.9

Section properties

A = 305000 mm^2 $\qquad b$ = 500 mm

y_t = 475 mm $\qquad I$ = 31548 × 10^6 mm^4

y_b = 525 mm $\qquad Z_t$ = 66.4 × 10^6 mm^3

h = 1000 mm $\qquad Z_b$ = 60.1 × 10^6 mm^3

b_w = 150 mm

Design moments and shear forces

$$g = (0.305 \times 24) = 7.32 \text{ kN/m}$$

$$M_g = (0.125 \times 7.32 \times 20^2) = 366 \text{ kN.m}$$

$$M_q = (0.125 \times 12 \times 20^2) = 600 \text{ kN.m}$$

$$V = V_g + V_q = 0.5(7.32 + 10)\,20 = 193.2 \text{ kN}$$

Permissible stresses (IS: 1343-2012)

$$f_{ck} = 50 \text{ N/mm}^2 \quad \text{and} \quad f_{ci} = 41 \text{N/mm}^2$$

For post-tensioned work (referring to Fig. 8 of IS: 1343)

$$f_{ct} = 0.427 f_{ci} = (0.427 \times 41) = 17.5 \text{ N/mm}^2$$

For type-1 member, $f_{tt} = f_{tw} = 0$

From Fig. 7 of IS: 1343 corresponding to zone-I

$$f_t = 0.37 \quad f_{ck} = (0.37 \times 50) = 18.5 \text{ N/mm}^2$$

Check for minimum section modulus

$$Z_b \geq \left[\frac{M_q + (1-\eta)M_g}{\eta f_{ct} - f_{tw}}\right]$$

$$Z_b \geq \left[\frac{\left(600\times10^6 + (1-0.85)366\times10^6\right)}{(0.85\times17.5)-0}\right]$$

$$\geq 44 \times 10^6 \text{ mm}^3 < 60.1 \times 10^6 \text{ mm}^3 \text{ (provided).}$$

Hence safe.

Prestressing force and eccentricity

$$f_t = \left[f_{tt} - \left(\frac{M_g}{Z_t}\right)\right] = \left[0 - \frac{\left(366\times10^6\right)}{\left(66.4\times10^6\right)}\right]$$

$$= -5.51 \text{ N/mm}^2$$

$$f_b = \left[f_{tw} + \frac{\left(M_g + M_q\right)}{\eta Z_b}\right]$$

$$= \left[0 + \frac{\left(966\times10^6\right)}{\left(0.85\times60.1\times10^6\right)}\right] = 18.9 \text{ N/mm}^2$$

$$e = \left[\frac{Z_t Z_b(f_b - f_t)}{A(f_t Z_t + f_b Z_b)}\right]$$

$$= \left\{\frac{(60.1\times 66.4)10^{12}\,(18.9+5.51)}{305000\left[\left(-5.51\times 66.4\times 10^6\right)\right]+\left(18.9\times 60.1\times 10^6\right)}\right\}$$

$$= 416.28 \text{ mm}$$

Since several cables in two rows have to be arranged, a minimum effective cover of 150 mm is required.

Hence, maximum possible eccentricity

$$= e = (y_b - 150) = (525 - 150) = 375 \text{ mm}$$

Modified prestressing force is computed by the relation

$$P = \left[\frac{Af_b Z_b}{Z_b + A_e}\right]$$

$$= \left[\frac{\left(305000\times 18.9\times 60.1\times 10^6\right)}{\left(60.1\times 10^6\right)+(305000\times 375)}\right]$$

$$= 1986.5 \times 10^3 \text{ N} = 1986.5 \text{ kN}$$

Using Freyssinet cables containing 12 wires of 7 mm diameter initially stressed to 1100 N/m^2,

$$\text{Number of cables} = \left(\frac{1986.5\times 10^3}{12\times 38.5\times 1100}\right) \simeq 4 \text{ cables}$$

Supplementary reinforcement

According to IS: 1343 code, minimum longitudinal reinforcement should not be lass than 0.15% of cross-sectional area of concrete section.

$$A_{st} = (0.0015 \times 305000) = 457.5 \text{ mm}^2$$

Provide 12 mm diameter HYSD bars in the top and bottom flanges.

Minimum web reinforcement is 0.15% of the web area in plan.

$$A_{sv} = (0.0015 \times 150 \times 1000) = 225 \text{ mm}^2$$

Using 10 mm diameter two legged stirrups, spacing is given by

$$S_v = \left[\frac{(1000\times 2\times 28.5)}{225}\right] = 697 \text{ mm}$$

But the spacing should not be greater than the

a. Clear depth of web or 4 times the thickness of web whichever is smaller

b. 0.75 times the effective depth

Adpot 10 mm diameter two legged stirrups at 400 mm centres.

Fig. 8.10 shows the details of cables and supplementary reinforcement provided at the center of span section.

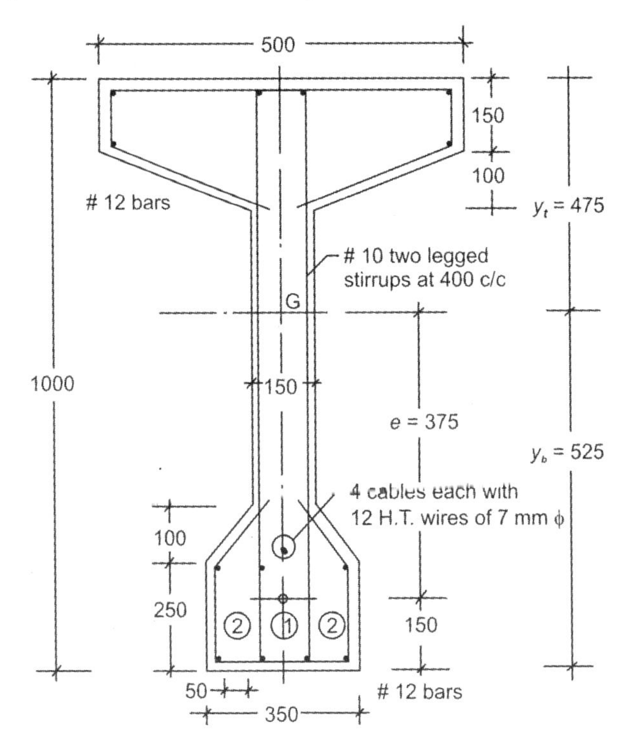

Fig. 8.10

Problem 8.7

A pre-tensioned electric pole has to be designed to suit the following data:

Maximum working moment due to wind pressure at the base of the pole is computed as 25.5 kN.m

Permissible compressive stress in concrete = f_{cw} = 18 N/mm^2

Permissible tensile stress is concrete = f_{tw} = 5 N/mm^2

Loss ratio = η = 0.7

High tensile wires of 5 mm diameter with f_p = 1600 N/mm^2 are available for use.

a. Design a suitable cross-section for the pole and the number of wires required.
b. Check for the limit states of collapse and cracking.

Solution

The section modulus required is given by the expression,

$$Z_t = Z_b \geq \left[\frac{2\,M_d}{(f_{cw} - f_{tw})}\right] = \left[\frac{2 \times 25.5 \times 10^6}{[8-(-5)]}\right]$$

$$= 223 \times 10^4 \text{ mm}^3$$

Using a rectangular section of width (b) = 150 mm, the overall depth is obtained as

$$h = \sqrt{\frac{6Z}{b}} = \sqrt{\frac{6 \times 223 \times 10^4}{150}} = 300 \text{ mm}$$

A pole with a cross-section of 150 mm by 320 mm is adopted for the base section.

Since the pole is subjected to equal bending in opposite directions, the member should be designed for uniform prestress.

Section modulus of the rectangular section

$$= Z = \left[\frac{150 \times 320^2}{6}\right] = 250 \times 10^4 \text{ mm}^3$$

Prestress in the member

$$= \left[\frac{f_{tw}}{\eta} + \frac{M_w}{\eta Z}\right] = \left[\frac{-5}{0.7} + \frac{25.5 \times 10^6}{0.7 \times 250 \times 10^4}\right]$$

$$= 7.5 \text{ N/mm}^2$$

$$\text{Initial prestressing force} = \left[\frac{(7.5 \times 150 \times 320)}{1000}\right] = 360 \text{ kN}$$

Permissible force in 5 mm diameter H.T. wires

$$= [(19.6 \times 0.8 \times 1600)/1000] = 25 \text{ kN}$$

$$\text{Number of wires required} = \left(\frac{360}{25}\right) = 14.5$$

Provide 16 wires at the base section with 8 wires distributed on the tension and compression sides arranged in two rows spaced 25 mm centres with an effective cover of 25 mm.

Problem 8.8

The post tensioned prestressed concrete slab of a National Highway culvert has an effective span of 10.4 m and a width of 8.261 m. M-40 grade concrete and 7 mm diameter high tensile wires with an ultimate tensile strength of 1500 N/mm^2 housed in cables with 12 wires and anchored by Freyssinet anchorages of 150 mm diameter are available for use. Compressive strength of concrete at transfer is 35 N/mm^2. Loss ratio may be taken as 0.8. The thickness of the concrete slab has been selected as 500 mm and wearing coat as 80 mm. Live load analysis indicates a maximum bending moment of 187 kN.m at the centre of span and a maximum shear force of 80.75 kN. Design the slab as class 1 type with no tensile stresses at any stage.

a. Check the adequacy of the section
b. Design the minimum prestressing force the eccentricity of cables at the centre of span
c. Check for the ultimate flexural strength and shear strength according to IS: 1343 code.

Solution

$f_{ct} = 15\ \text{N/mm}^2 \qquad f_{tt} = 0\ \text{N/mm}^2 \qquad \eta = 0.8$

$M_q = 187\ \text{kN.m} \qquad V_q = 80.75\ \text{kN} \qquad L = 10.4\ \text{m}$

Thickness of concrete deck slab = 500 mm

Thickness of wearing coat = 80 mm

Dead weight of wearing coat = $(0.08 \times 24) = 2.00\ \text{kN/m}^2$

Dead weight of deck slab = $(0.5 \times 24) = 12\ \text{kN/m}^2$

Total dead load = $(2 + 12) = 14\ \text{kN/m}^2$

Dead load bending moment

$$= M_g = (0.125 \times 14 \times 10.4^2) = 190\ \text{kN.m}$$

Dead load shear force at support section

$$= V_g = [0.5 \times 14 \times 10.4] = 72.8\ \text{kN}$$

a. Check for adequacy of the section

Section modulus $= Z_t = Z_b = Z = \left[\dfrac{1000 \times 500^2}{6}\right]$

$$= (41.66 \times 10^6)\ \text{mm}^3$$

Range of stress at bottom fibre

$$= f_{br} = (\eta f_{ct} - f_{tw}) = (0.8 \times 15 - 0)$$

$$= 12\ \text{N/mm}^2$$

The minimum section modulus is given by

$$Z_b \geq \left[\frac{M_q + (1-\eta)\, M_g}{f_{br}}\right]$$

$$= \left[\frac{(187 \times 10^6) + (1-0.8)(190 \times 10^6)}{12}\right]$$

$$= (18.75 \times 10^6)\ \text{mm}^3$$

Which is less than the provided value of $(41.66 \times 10^6)\ \text{mm}^3$ (hence, section is safe).

b. Minimum prestressing force and eccentricity of cables

$$f_t = \left[f_{tt} - \frac{M_g}{Z_t}\right] = \left[0 - \frac{(190 \times 10^6)}{(41.66 \times 10^6)}\right]$$

$$= -4.56 \text{ N/mm}^2$$

$$f_b = \left[\frac{f_{tw}}{\eta} + \frac{M_g + M_q}{\eta Z_b}\right] = \left[0 + \frac{(187 + 190) \times 10^6}{0.80\,(41.66 \times 10^6)}\right]$$

$$= 11.31 \text{ N/mm}^2$$

$$P = \frac{A(f_t Z_t + f_b Z_b)}{(Z_t + Z_b)}$$

$$= \frac{(1000 \times 500 \times 41.66 \times 10^6)\,(11.31 - 4.56)}{(2 \times 41.66 \times 10^6)}$$

$$= (1687.5 \times 10^3) \text{ N} = 1687.5 \text{ kN}$$

Using Freyssinet cables containing 12 wires of 7 mm diameter which are stressed to 1200 N/mm^2,

$$\text{force in each cable} = \left[\frac{(12 \times 38.5 \times 1200)}{1000}\right] = 554 \text{ kN}$$

$$\text{Spacing of cables} = \left(\frac{1000 \times 554}{1687.5}\right) = 328 \text{ mm}$$

$$e = \frac{Z_t Z_b (f_b - f_t)}{A(f_t Z_t + f_b Z_b)}$$

$$= \frac{(41.66^2)\,(10^{12})\,[11.31 + 4.56]}{[(1000 \times 500 \times (41.66 \times 10^6) + (-4.56 + 11.31)]}$$

$$= 195 \text{ mm.}$$

c. Check for ultimate flexural and shear strength

Flexural strength:

Considering 1 m width of the slab, $b = 100$ mm

$$A_p = \left[\frac{(12 \times 38.5 \times 1000)}{328}\right] = 1408 \text{ mm}^2$$

$$d = 445 \text{ mm}$$

$$f_p = 1500 \text{ N/mm}^2$$

i. Failure by yielding of steel

M_u = $[0.9\, d\, A_p fp] = [0.9 \times 445 \times 1408 \times 1500]$

= (845×10^6) N.mm = 846 kN.m

ii. Failure by crushing of concrete

M_u = $[0.176\, b\, d^2 f_{ck}] = [0.176 \times 1000 \times 445^2 \times 40]$

= (1394×106) N.mm = 1394 kN.m

The actual M_u is the lesser of the two values and is taken as 846 kN.m

Required ultimate moment

= $[1.5\, M_g + 2.5\, M_q] = [(1.5 \times 190) + (2.5 \times 187)]$

= 752.5 kN.m

Hence, the ultimate moment capacity of the section (M_u = 846 kN.m) is greater than the required ultimate moment (752.5 kN.m).

Shear strength:

Ultimate shear force

= $V_g = [1.5\, V_g + 2.5\, V_q]$

= $[(1.5 \times 72.8) + (2.5 \times 87.72)] = 328.5$ kN

According to IRC: 18-2000, the ultimate shear resistance of the support section uncracked in flexure is given by the expression,

$$V_{cw} = 0.67\, bh\, \sqrt{f_t^2 + 0.8\, f_{cp} f_t} + \eta P \sin \theta$$

where b = width of slab = 1000 mm

h = overall depth = 500 mm

f_t = principal tensile stress

$$= 0.24\sqrt{f_{ck}} = 0.24\sqrt{40} = 1.51\ \text{N/mm}^2$$

f_{cp} = compressive prestress at the centroidal axis

$$= \left(\frac{\eta P}{A}\right) = \left(\frac{0.8 \times 1687.5 \times 10^3}{1000 \times 500}\right) = 2.7\ \text{N/mm}^2$$

Eccentricity of cable at the centre of span = e = 195 mm

The cables are concentric at support section.

Slope of the cable at support

$$= \theta = \left(\frac{4e}{L}\right) = \left(\frac{4 \times 195}{10.4 \times 1000}\right) = 0.075$$

$\therefore$

$$V_{cw} = (0.67 \times 1000 \times 500)\, \sqrt{1.51^2 + 0.8 \times 2.7 \times 1.51} + (0.8 \times 1687.5 \times 10^3 \times 0.075)$$

= 889817 N = 890 kN

Since the ultimate shear force V_u is less than 50% of the ultimate shear resistance V_c, no shear reinforcement is required.

Problem 8.9

A continuous prestressed concrete box girder is to be designed for covering two spans of 50 m each to support the loadings from a National Highway. The I-section of the box girder at mid support section is subjected to dead and live load bending moments of magnitude 13438 and 1986 kN.m respectively. The web and flanges of the box girder are 300 mm thick and the webs are spaced at 2 m intervals. The overall depth of the girder is 2 m. M-60 grade concrete is used for casting the girder. The compressive stress in concrete at transfer is 20 N/mm^2.

The section has to be designed as class-1 type structure without permitting tensile stresses at any stage. Assume a loss ratio of 0.8.

a. Check the adequacy of the section provided for the girder to resist the loads.

b. The prestressing force required assuming an eccentricity of 700 mm.

Solution

Effective width of flange = 2000 mm $\quad f_{ct} = 20$ N/mm^2

Thickness of flange and web = 300 mm $\quad f_{tw} = 0$ N/mm^2

Overall depth of girder = 2000 mm $\quad \eta = 0.8$

Sectional properties of the symmetrical I-girder are as follows:

$$\text{Cross-sectional area} = A = (1.62 \times 10^6)\ \text{mm}^2$$

$$\text{Second moment of area} = I = (94 \times 10^{10})\ \text{mm}^4$$

$$\text{Section modulus} = Z = Z_t = Z_b$$

$$= (940 \times 10^6)\ \text{mm}^3$$

a. Check for adequacy of section

$$M_g = 13438\ \text{kN.m}$$

$$M_q = 1986\ \text{kN.m}$$

$$f_{br} = (\eta f_{ct} - f_{tw}) = (0.8 \times 20 - 0) = 16\ \text{N/mm}^2$$

$$f_b = \left[\frac{f_{tw}}{\eta} + \frac{M_g}{\eta Z_b}\right] = \left[0 + \frac{13438 \times 10^6}{0.80\,(41.66 \times 10^6)}\right]$$

$$= 17.86\ \text{N/mm}^2$$

$$Z_b \geq \left[\frac{M_q + (1-\eta)\,M_g}{f_{br}}\right]$$

$$= \left[\frac{(1986 \times 10^6) + (1 - 0.8)(13438 \times 10^6)}{16}\right]$$

$= (292 \times 10^6)\ \text{mm}^3 < (940 \times 10^6)\ \text{mm}^3$ (provided)

Hence, section provided is safe to resist the imposed loads.

b. Prestressing force

Eccentricity provided at support section = e = 700 mm

$$P = \left[\frac{Af_b Z_b}{Z_b + A_e}\right] = \left[\frac{(1.62 \times 10^6)(17.86)(940 \times 10^6)}{(940 \times 10^6) + (1.62 \times 10^6) 700}\right]$$

= 13109932 N = 13110 kN.

Problem 8.10

A prestressed concrete slab of a cable stayed bridge deck is 400 mm thick, spans over 10 m between the stiffening girders. The analysis of IRC Class-AA loads on the bridge deck indicates the negative moment at support due to dead loads as 100 kN.m and live load moment as 256 kN.m. The positive bending moment at the centre of span of slab due to dead load is computed as 50 kN.m and live load bending moment as 169 kN.m. Assuming the allowable compressive stress in concrete at transfer and working loads as 20 N/mm^2, check the adequacy of the section and determine the required prestressing force and eccentricity at support and centre of span sections. Assume 20% loss of prestress and no tensile stresses are permitted at any stage.

Solution

Thickness of slab = 400 mm

Cross-sectional area per metre width

$= A = (400 \times 1000) = (4 \times 10^5)\ \text{mm}^2$

$$\text{Section modulus} = Z = Z_t = Z_b = \left[\frac{1000 \times 400^2}{6}\right]$$

$= (26.66 \times 10^6)\ \text{mm}^3$

At support section

Dead load moment = M_g = 100 kN.m

Live load moment = M_q = 256 kN.m

At the centre of span section

Dead load moment = M_g = 50 kN.m

Live load moment = $M_q = 169$ kN.m

$f_{ct} =$ 20 N/mm² and $f_{tw} = 0$

Loss ratio = $\eta = 0.8$

$f_{br} = (\eta f_{ct} - f_{tw}) = (0.8 \times 20 - 0) = 16 \text{ N/mm}^2$

$$Z_b \geq \left[\frac{M_q + (1-\eta)M_g}{f_{br}}\right]$$

$$= \left[\frac{(256 \times 10^6) + (1-0.8)(100 \times 10^6)}{16}\right]$$

$= (17.25 \times 10^6) \text{ mm}^3 < (26.66 \times 10^6) \text{ mm}^3$ (provided)

Hence, section provided is safe to resist the imposed loads.

The minimum prestressing force required at the support section is computed using the relation

$$P = \frac{A(f_t Z_t + f_b Z_b)}{(Z_t + Z_b)}$$

$$f_t = \left[f_{tt} - \frac{M_g}{Z_t}\right] = \left[0 - \frac{100 \times 10^6}{(26.66 \times 10^6)}\right] = -3.75 \text{ N/mm}^2$$

$$f_b = \left[\frac{f_{tw}}{\eta} + \frac{M_g + M_q}{\eta Z_b}\right] = \left[0 + \frac{(100+256) \times 10^6}{0.80(26.66 \times 10^6)}\right]$$

$= 16.69 \text{ N/mm}^2$

$$P = \left[\frac{(4 \times 10^5)(26.66 \times 10^5)(16.69 - 3.75)}{(2 \times 26.66 \times 10^6)}\right]$$

$= (2588 \times 10^3) \text{ N} = 2588 \text{ kN}$

The eccentricity of the cables at support section is given by the expression

$$e = \left[\frac{Z_t Z_b (f_b - f_t)}{A(f_t Z_t + f_b Z_b)}\right]$$

$$= \left[\frac{(26.66 \times 10^6)^2 (16.69 + 3.75)}{(4 \times 10^5)(26.66 \times 10^6)(16.69 - 3.75)}\right]$$

$= 105$ mm (towards the top of the slab)

The positive bending moment at the centre of span is of smaller magnitude and hence the eccentricity required at the centre of span is computed by limiting the tensile stress at top fibre to zero under the loading condition of prestress together with the self weight stress.

At the centre of span section, we have P = 2588 kN and M_g = 50 kN.m. Using the stress relation,

$$\left[\frac{P}{A} - \frac{Pe}{Z_t} + \frac{M_g}{Z_t}\right] = 0$$

$$\left[\frac{(2588 \times 10^3)}{(4 \times 10^5)} - \frac{(2588 \times 10^3)\,e}{(26.66 \times 10^6)} + \frac{(50 \times 10^6)}{(26.66 \times 10^6)}\right] = 0$$

Solving, eccentricity at the centre of span = e = 86 mm (towards the soffit of slab).

9

Composite Construction of Prestressed and *in situ* Concrete

Problem 9.1

A rectangular pretensioned concrete beam has a breadth of 100 mm and depth of 230 mm, and the prestress after all losses have occurred is 12 N/mm^2 at the soffit and zero at the top. The beam is incorporated in a composite I beam by casting a top flange of breadth 300 mm and depth 50 mm.

Calculate the maximum uniformly distributed live load that can be supported on a simply supported span of 4.5 m, without any tensile stresses occurring,

a. if the slab is externally supported while casting, and
b. if the pretensioned beam supports the weight of the slab while casting.

Solution

The given section is shown in Fig. 9.1.

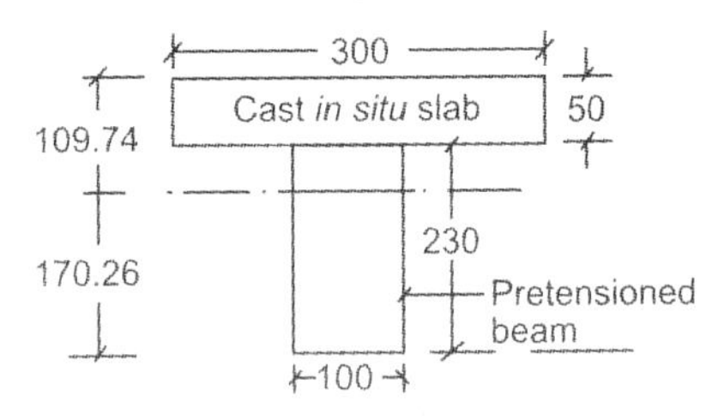

Fig. 9.1

Section properties of pretensioned beam

$$A = 23000 \text{ mm}^2$$

$$Z = \left(\frac{100 \times 230^2}{6}\right) = 8.81 \times 10^5 \text{ mm}^3$$

$$\text{Self weight} = (0.1 \times 0.23 \times 24) = 0.552 \text{ kN/m}$$

$$\text{Self weight moment} = (0.125 \times 0.552 \times 4.5^2) = 1.397 \text{ kN.m}$$

$$\text{Stresses at top and bottom} = \pm\left(\frac{1.397 \times 10^6}{8.81 \times 10^5}\right) = \pm 1.585 \text{ N/mm}^2$$

$$\text{Self weight of cast } \textit{in situ} \text{ slab} = (0.05 \times 0.3 \times 24) = 0.36 \text{ kN/m}$$

$$\text{Moment due to slab weight} = \left(\frac{0.36 \times 4.5^2}{8}\right) = 0.91 \text{ kN.m}$$

Stresses due to slab weight in precast section

$$= \left(\frac{0.91 \times 10^6}{8.81 \times 10^5}\right) = \pm 1.034 \text{ N/mm}^2$$

The section properties of composite section are as follows:

$$I_{xx} = 2.83 \times 10^8 \text{ mm}^4$$

$$Z_t = 2.57 \times 10^6 \text{ mm}^3$$

$$Z_b = 1.66 \times 10^6 \text{ mm}^3$$

Let w = Live load on composite section.

Hence we have live load on composite section,

$$= (0.3 \times 1 \times w) = 0.3\,w \text{ kN/m}$$

$$\text{Live load moment} = \frac{0.3\,w(4.5)^2}{8} = 0.76\,w \text{ kN.m}$$

Live load stresses in composite section:

$$\text{At top} = \left[\frac{0.76\,w \times 10^6}{2.57 \times 10^6}\right] = 0.295\,w \text{ N/mm}^2$$

$$\text{At bottom} = \left[\frac{0.76\,w \times 10^6}{1.66 \times 10^6}\right] = 0.45\,w \text{ N/mm}^2$$

For unpropped condition, the stresses are as shown in Fig. 9.2.

a. For the condition of zero tensile stress at soffit,

   $12 - 1.585 - 1.034 - 0.457\,w = 0$

   Solving $w = 20.5 \text{ kN/m}^2$

b. For propped condition,

   Moment due to slab weight = 0.91 kN.m

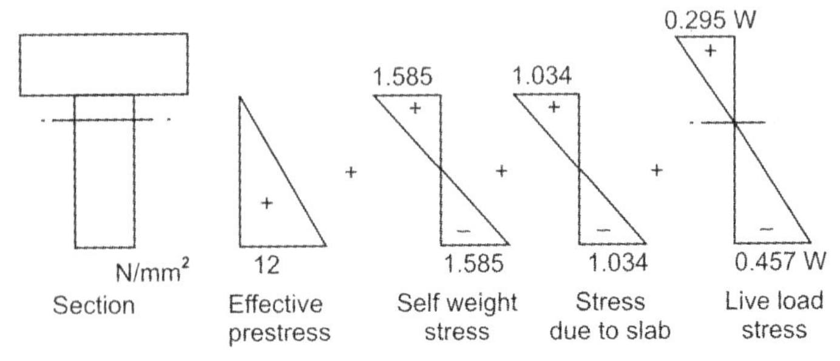

Fig. 9.2

$$\text{Stress at top} = \left(\frac{0.91\times 10^6}{2.57\times 10^6}\right) = 0.35\ \text{N/mm}^2$$

$$\text{At bottom} = \left(\frac{0.91\times 10^6}{1.66\times 10^6}\right) = 0.548\ \text{N/mm}^2$$

Hence, the condition is

$12 - 1.585 - 0.548 - 0.457\ w = 0$

Solving: $w = 21.6\ \text{kN/m}^2$

Problem 9.2

A composite bridge deck of span 12 m is made up of a precast prestressed symmetrical I-section and an *in situ* cast slab. The precast I beams are spaced at 750 mm centres and the top slab of the *in situ* concrete is 120 mm thick. The cross-sectional details of the precast I beam are as follows:

Thickness of top and bottom flanges = 110 mm
Width of top and bottom flanges = 200 mm
Thickness of web = 75 mm
Depth of precast I beam = 500 mm
Self weight of precast concrete = 24 kN/m^3
Self weight of cast *in situ* concrete = 23.5 kN/m^3

The prestressed beam is supported during the placing of *in situ* concrete. The form work load is estimated as 0.2 kN/m of the span.

If the compressive prestress in the beam is 15 N/mm^2 at the bottom and zero at the top, calculate the maximum stresses developed in the precast and *in situ* cast concrete under an imposed load of 5 kN/m^2 (assuming).

a. the modular ratio of cast *in situ* to precast concrete as 1.0.

b. the modular ratio of cast *in situ* to precast concrete as 0.8.

Solution

The precast section is shown in Fig. 9.3.

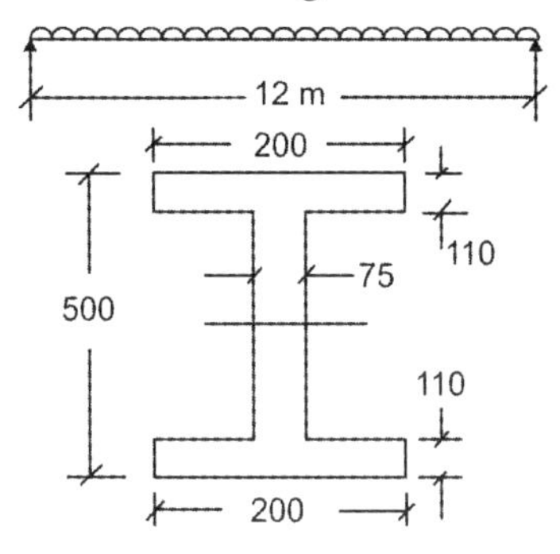

Fig. 9.3

Properties of precast P.S.C. section

$$I = \left[\frac{200 \times 500^3}{12} - \frac{125 \times 280^3}{12}\right]$$

$$= 1.855 \times 10^9 \text{ mm}^4$$

$$Z_b = Z_t = Z = \left(\frac{1.855 \times 10^9}{250}\right) = 7.42 \times 10^6 \text{ mm}^3$$

$$A = [2(200 \times 110) + (280 \times 75)]$$

$$= 65 \times 10^3 \text{ mm}^2 = 0.065 \text{ m}^2$$

Self weight of precast beam $= (0.065 \times 24) = 1.56$ kN/m

Weight of *in situ* concrete slab 120 mm thick over a length of 750 mm $= (0.12 \times 0.75 \times 23.5) = 2.115$ kN/m

Weight of form work load $= 0.2$ kN/m

$\therefore$ Total load of cast *in situ* slab $= (2.115 \times 12) = 25.38$ kN

Moments

Moment due to self weight of precast section

$$= \left(\frac{1.56 \times 12^2}{8}\right) = 28.08 \text{ kN.m}$$

Moment due to cast *in situ* slab

$$= \left(\frac{25.38 \times 12}{8}\right) = 38.07 \text{ kN.m}$$

Moment due to weight of form work

$$= \left(\frac{0.2 \times 12^2}{8}\right) = 3.6 \text{ kN.m}$$

Stresses in precast section

1. Stress due to self weight of precast section

$$= \left(\frac{28.08 \times 10^6}{7.42 \times 10^6}\right) = 3.78 \text{ N/mm}^2$$

2. Stress due to weight of cast *in situ* slab

$$= \left(\frac{38.07 \times 10^6}{7.42 \times 10^6}\right) = 5.13 \text{ N/mm}^2$$

Stress due to weight of form work

$$= \left(\frac{3.6 \times 10^6}{7.42 \times 10^6}\right) = 0.48 \text{ N/mm}^2$$

Moment due to live load

$$= \frac{(5 \times 0.75) \times 12^2}{8} = 67.5 \text{ kN.m}$$

The centroidal axis of the composite section is shown in Fig. 9.4.

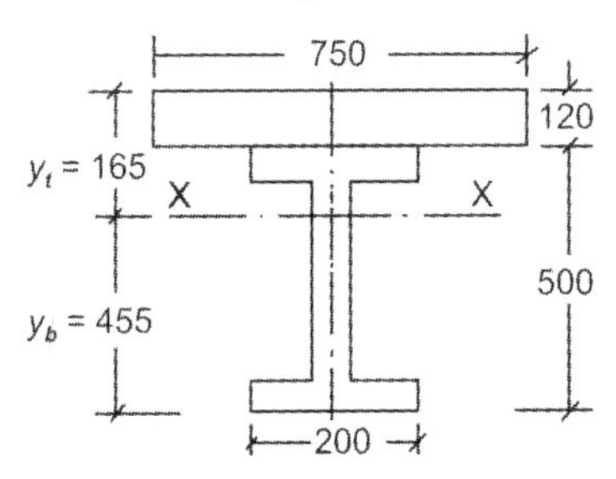

Fig.9.4

Properties of composite section

$$\bar{y} = y_t = 165 \text{ mm}$$
$$I_{xx} = 5.68 \times 10^9 \text{ mm}^4$$
$$Z_t = 3.44 \times 10^7 \text{ mm}^3$$
$$Z_b = 1.24 \times 10^7 \text{ mm}^3$$

Case (a)

Modular ratio of cast *in situ* concrete to precast concrete is 1.0.

L.L. Stresses

Stress at top $= \left(\frac{67.5 \times 10^6}{3.44 \times 10^7}\right) = 1.96 \text{ N/mm}^2$

$$\text{Stress at bottom} = \left(\frac{67.5\times10^6}{1.24\times10^7}\right) = 5.44\ \text{N/mm}^2$$

The resultant stresses are shown in Fig. 9.5

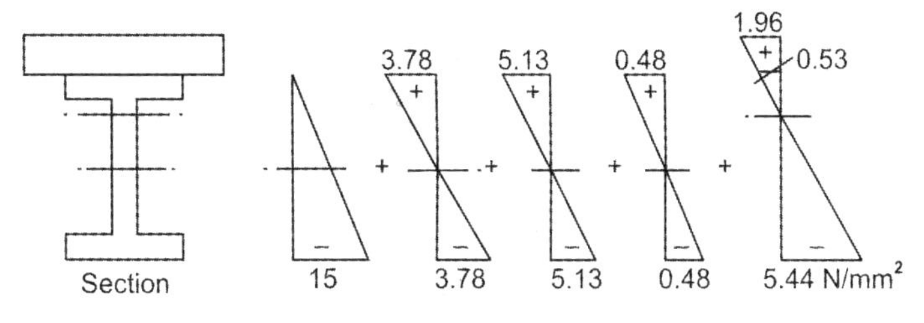

Fig. 9.5

Max. stress at top = 1.96 N/mm^2 (compression)

At junction of precast concrete

$$= (3.78 + 5.13 + 0.48 + 0.53)$$
$$= 9.92\ \text{N/mm}^2\ \text{(compression)}$$

Case (b)

If the modular ratio of cast *in situ* to precast concrete is 0.8 then ratio of modulus of elasticity of precast to cast *in situ* slab

$$= \left(\frac{1}{0.8}\right) = 1.25$$

Properties of equivalent composite section

Area of *in situ* slab = (750 × 120) = 9 × 10^4 mm^2

Area of P.S.C. beam = (65 × 10^3) mm^2

If $\bar{y}$ = Distance of centroid from top

$$\bar{y} = \left[\frac{(750\times120\times60)+(1.25\times65\times10^3)370}{(9\times10^4)+(1.25\times65\times10^3)}\right]$$

$$= 207\ \text{mm}$$

$$I = 6.51\times10^9\ \text{mm}^4$$

$$Z_t = 3.14\times10^7\ \text{mm}^3$$

$$Z_b = 1.57\times10^7\ \text{mm}^3$$

Live load stresses

$$\text{At the top} = \left(\frac{67.5\times10^6}{3.14\times10^7}\right) = 2.14\ \text{N/mm}^2$$

$$\text{At the bottom} = \left(\frac{67.5\times10^6}{1.57\times10^7}\right) = 4.29\ \text{N/mm}^2$$

Resultant stresses

At the top = 2.14 N/mm^2 (compression)

At the junction (precast section) = 10.28 N/mm^2 (compression)

Problem 9.3

The mid span section of a composite *T* beam comprises a pretensioned beam, 300 mm wide and 900 mm deep and an *in situ* cast slab 900 mm wide and 150 mm deep. The effective prestressing force located 200 mm from the soffit of the beam is 2180 kN. The moment due to the weight of the precast section is 273 kN.m at mid span. After this is erected in place, the top slab is cast producing a moment of 136.5 kN.m at mid span.

After the slab concrete is hardened, the composite section is to carry a maximum live load moment of 750 kN.m. Compute the resultant final stresses at

a. the top of slab, and
b. the top and bottom of the precast section.

Also estimate the ultimate moment capacity of the composite cross section using the Indian standard code provisions:

Area of steel	= 2340 mm^2
Cube strength of slab concrete	= 35 N/mm^2
Tensile strength of steel	= 1680 N/mm^2

Solution

The given prestressed section is shown in Fig. 9.6.

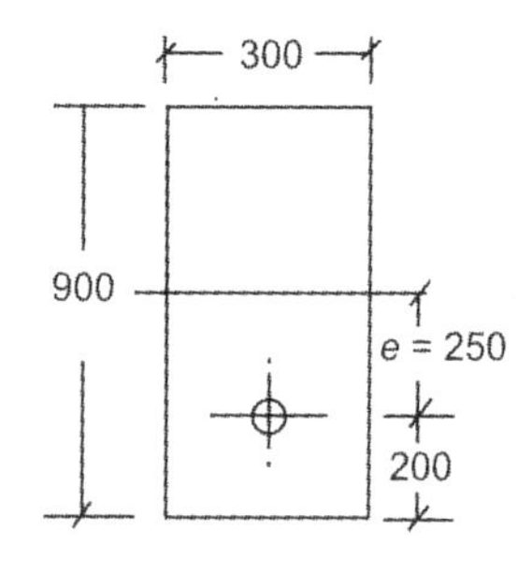

Fig. 9.6

P = 2180 kN $\qquad e$ = 250 mm

$A = 27 \times 10^4$ mm^2

$$Z = \left(\frac{300 \times 900^2}{6}\right) = 4.05 \times 10^7 \text{ mm}^3$$

Prestress

At top $= \left[\dfrac{2180\times10^3}{27\times10^4} - \dfrac{2180\times10^3\times250}{4.05\times10^7}\right]$

$= -4.38\ \text{N/mm}^2$

At bottom $= \left[\dfrac{2180\times10^3}{27\times10^4} + \dfrac{2180\times10^3\times250}{4.05\times10^7}\right]$

$= +21.52\ \text{N/mm}^2$

Moment due to self weight of precast section = 273 kN.m

$\therefore$ Stresses at top and bottom $= \left(\dfrac{273\times10^6}{4.05\times10^7}\right)$

$= \pm6.74\ \text{N/mm}^2$

Stresses due to cast *in situ* slab $= \left(\dfrac{136.5\times10^6}{4.05\times10^7}\right)$

$= \pm3.37\ \text{N/mm}^2$

The composite section is shown in Fig. 9.7.

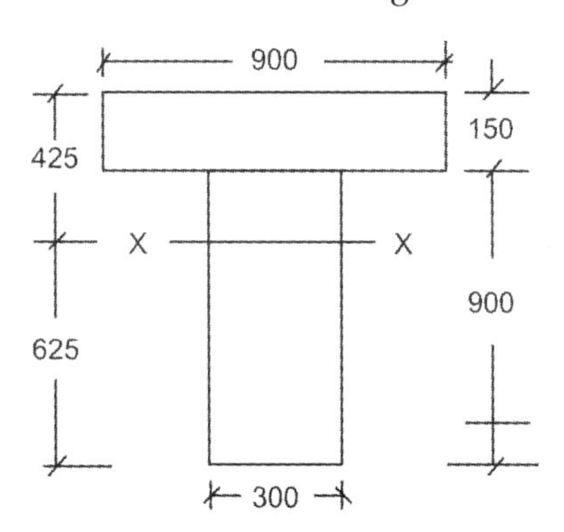

Fig. 9.7

Properties of composite section

$I_{xx} = 432 \times 10^8\ \text{mm}^4$

$Z_t = 1.02 \times 10^8\ \text{mm}^3$

$Z_b = 0.7 \times 10^8\ \text{mm}^3$

Live load moment = 750 kN.m

$$\text{Stress at top} = \left(\frac{750\times 10^6}{1.02\times 10^8}\right)$$

$$= 7.35\ \text{N/mm}^2\ \text{(compression)}$$

$$\text{Stress at bottom} = \left(\frac{750\times 10^6}{0.7\times 10^8}\right) = -10.71\ \text{N/mm}^2\ \text{(tension)}$$

The stresses due to various moments are shown in Fig. 9.8

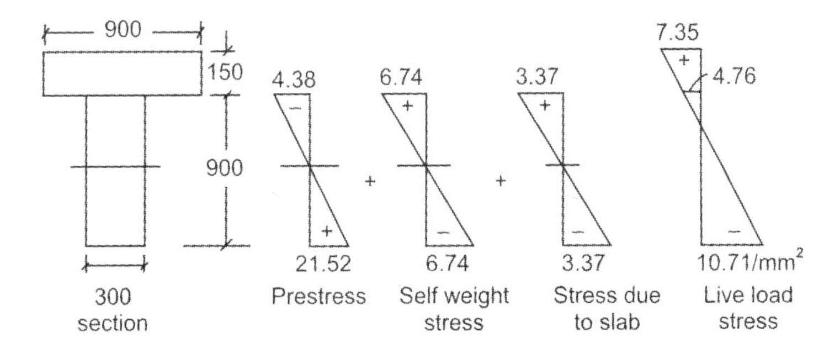

Fig. 9.8

Resultant stresses

a. At the top of slab = 7.35 N/mm^2 (compression)
b. At the top of precast section

$$= 6.74 + 3.37 + 4.76 - 4.38$$
$$= 10.49\ \text{N/mm}^2\ \text{(compression)}$$

At bottom of the precast section

$$= 21.52 - 6.74 - 3.37 - 10.71$$
$$= 0.7\ \text{N/mm}^2\ \text{(compression)}$$

Problem 9.4

The deck of a prestressed concrete bridge with an overall depth of 300 mm is made up of an inverted *T* section, with *in situ* concrete laid over it. The precast prestressed *T* section has the following dimensions and properties:

Width and depth of slab	= 300 and 80 mm
Width and depth of stem	= 70 and 160 mm
Height of centroid from soffit	= 80 mm
Prestress at bottom	= 11 N/mm^2 (compression)

Second moment of area = 1472×10^5 mm^4
Prestress at top = 1 N/mm^2 (tension)
Modulus of elasticity of concrete = 35 kN/mm^2

The bridge has a span of 6 m and the precast beams are required to support the weight of the web concrete infill without any propping. When the infill, which may be assumed to have a modulus of elasticity of 28 kN/mm^2, has hardened, a uniformly distributed live load of 13 kN/m^2 is applied.

Calculate the resultant final stresses at

a. the top and bottom of the precast beams, and
b. the highest and lowest points in the concrete infill

Solution

The composite section is shown in Fig. 9.9

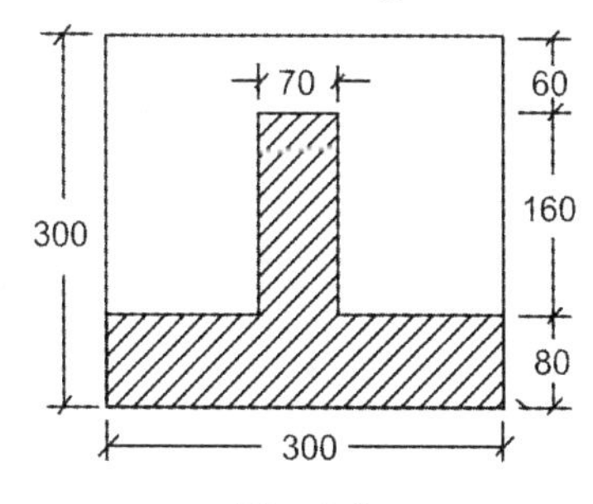

Fig. 9.9

The properties of P.S.C. section are shown in Fig. 9.10.

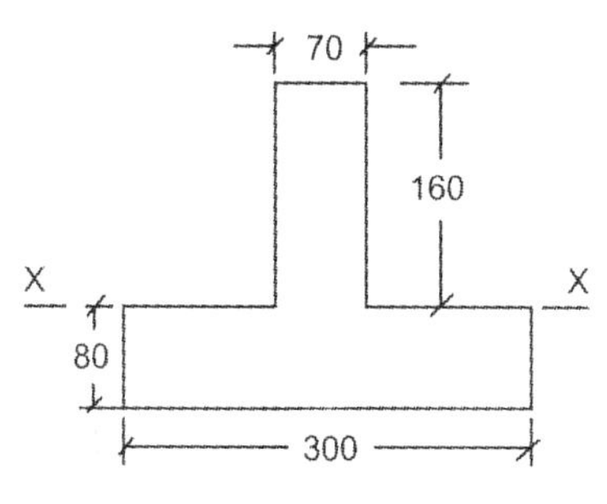

Fig. 9.10

Area of cast *in situ* slab = [90000 – 35200] = 54800 mm^2

$I_{xx} = 1472 \times 10^5$ mm^4 $\quad A = 35200$ mm^2

$E_c = 35$ kN/mm^2 $\quad Z_t = 9.2 \times 10^5$ mm^3

$Z_b = 18.4 \times 10^8$ mm^3

Prestress at top $= -1\ \text{N/mm}^4$ (tension)
Prestress at bottom $= 11\ \text{N/mm}^4$ (compression)
Self weight of P.S.C. beam
$= (0.0352 \times 24) = 0.844\ \text{kN/m}$
Moment due to self weight

$$= \left(\frac{0.844 \times 6^2}{8}\right) = 3.8\ \text{kN.m}$$

Stress due to self weight

At top $= \left(\dfrac{3.8 \times 10^6}{9.2 \times 10^5}\right) = 4.13\ \text{N/mm}^2$ (compression)

At bottom $= \left(\dfrac{3.8 \times 10^6}{18.4 \times 10^5}\right) = -2.06\ \text{N/mm}^2$ (tension)

Weight of cast *in situ* slab
$= (0.0548 \times 24) = 1.31\ \text{kN/m}$
Moment due to weight of slab

$$= \left(\frac{1.31 \times 6^2}{8}\right) = 5.91\ \text{kN.m}$$

Stress in P.S.C. unit due to weight of cast *in situ* slab:

At top $= \left(\dfrac{5.91 \times 10^6}{9.2 \times 10^5}\right) = 6.42\ \text{N/mm}^2$ (compression)

At bottom $= \left(\dfrac{5.91 \times 10^6}{18.4 \times 10^5}\right) = -3.21\ \text{N/mm}^2$ (tension)

Composite section properties

Ratio os moduli $= \left(\dfrac{35}{28}\right) = 1.28$

Area of in situ slab $= 54800\ \text{mm}^2$
Area of P.S.C. section $= 35200\ \text{mm}^2$
If $\bar{y}$ = Distance of centroid from bottom

$$\bar{y} = \frac{(300 \times 60 \times 270) + (230 \times 160 \times 160) + (1.25 \times 35200 \times 80)}{(54800) + (1.25 \times 35200)}$$

$= 100\ \text{mm}$

$$I = \left[1.25\,[1472\times10^5 + 35200\times20^2] + \left(\frac{230\times160^3}{12}\right)\right.$$

$$\left. + (36800\times60^2) + \left(\frac{300\times60^3}{12}\right) + (18000\times170^2)\right]$$

$= 9381 \times 10^5$ mm^4

$\therefore$ $Z_t = 46.9 \times 10^5$ mm^3, $Z_b = 93.81 \times 10^5$ mm^3

Live load $= 0.3 \times 13 = 3.9$ kN/m

Moment due to live load

$= (0.125 \times 3.9 \times 6^2) = 17.55$ kN.m

Stress at top $= \left(\dfrac{17.55\times10^6}{46.9\times10^5}\right)$

$= 3.74$ N/mm^2 (compression)

Stress at bottom $= \left(\dfrac{17.55\times10^6}{93.81\times10^5}\right) = -1.87$ N/mm^2 (tension)

The stress distribution diagrams are shown in Fig. 9.11.

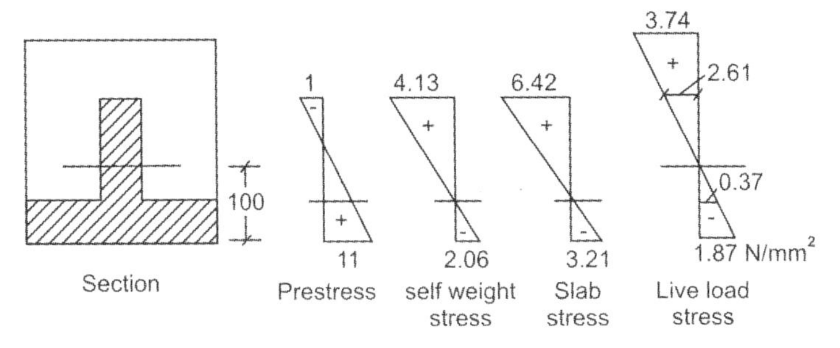

Fig. 9.11

Resultant stresses

Precast section

At top $= (4.13 + 6.42 + 2.61 - 1.0)$

$= 13.61$ N/mm^2 (compression)

At bottom $= -(2.06 + 3.21 + 1.87) + 11.0$

$= 3.86$ N/mm^2 (compression)

Cast in situ slab

At top $= 3.74$ N/mm^2 (compression)

At bottom $= -0.37$ N/mm^2 (tension)

Problem 9.5

A composite T section girder consists of a pretensioned rectangular beam, 120 mm wide and 240 mm deep, with an *in situ* cast slab 360 mm wide and 60 mm deep, laid over the beam. The pretensioned beam contains eight wires of 5 mm diameter, located 30 mm from the soffit. The tensile strength of high tensile steel is 1600 N/mm^2 and the cube strength of concrete in the top slab is 20 N/mm^2.

a. Estimate the flexural strength of the composite section.
b. Calculate the ultimate shear which will cause separation of the two parts of the girder if,
   i. the surface of contact is roughened to withstand a shear stress of 1 N/mm^2, and
   ii. 10 mm mild steel strirrups (two legged) are placed at 100 mm centres. Ultimate shear stress across stirrups = 190 N/mm^2.

Solution

The composite section is shown in Fig. 9.12

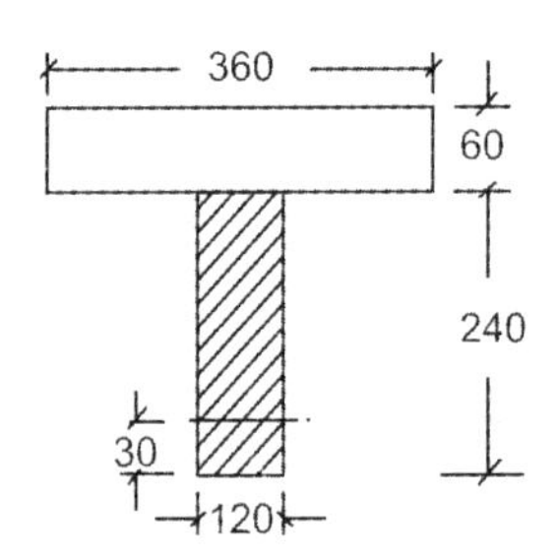

Fig. 9.12

Given:
$$A_p = 8 \times \frac{\pi}{4}(5)^2 = 157 \text{ mm}^2$$
$$d = (60 + 240 - 30) = 270 \text{ mm}$$
$$f_{ck} = 20 \text{ N/mm}^2$$
$$f_p = 1600 \text{ N/mm}^2$$
$$\left(\frac{A_p f_p}{bd\, f_{ck}}\right) = \left(\frac{157 \times 1600}{360 \times 270 \times 20}\right) = 0.13$$

From Table 4.1 or (Table 11 of IS: 1343-2012)

$$\left(\frac{f_{pu}}{0.87 f_p}\right) = 1.0$$

$$\therefore \quad f_{pu} = 1392 \text{ N/mm}^2$$

$$\left(\frac{x_u}{d}\right) = 0.28$$

$$\therefore \quad x_u = 270 \times 0.28 = 76.74 > 60 \text{ mm}$$

(flange thickness)

$\therefore$ Neutral axis lies outside flange

So,
$$A_{pf} = \left[\frac{0.44 f_{ck} (b - b_w) t}{0.87 f_p}\right]$$

$$= \left[\frac{0.44 \times 20 \times (360 - 120) 60}{0.87 \times 1600}\right] = 91.03 \text{ mm}^2$$

$$\therefore \quad A_{pw} = (A_p - A_{pf}) = (157 - 91.03) = 65.97 \text{ mm}^2$$

$$\therefore \quad \left(\frac{A_{pw}}{b_w d} \cdot \frac{f_p}{f_{ck}}\right) = \left(\frac{65.97 \times 1600}{120 \times 270 \times 20}\right) = 0.16$$

From, Table 4.1

$$f_{pu} = 0.87 f_p = 1392 \text{ N/mm}^2$$

$$\left(\frac{x_u}{d}\right) = 0.3478$$

$$\therefore \quad x_u = (0.3478 \times 270) = 93.90 \text{ mm}$$

$$\therefore \quad M_u = f_{pu} A_{pw} (d - 0.42 x_u) + 0.44 f_{ck} (b - b_w) t \left(d - \frac{t}{2}\right)$$

$$= 1392 \times 65.97 (270 - 0.42 \times 93.9) + 0.44$$

$$\times 20 (240) 60 \left(270 - \frac{60}{2}\right)$$

$$= 52 \times 10^6 \text{ N.mm} = 52 \text{ kN.m}$$

Ultimate shear

The centroidal axis of the section is shown in Fig. 9.13

$$\bar{y} = 115.71 \text{ mm}$$

$$I_{xx} = 4.22 \times 10^8 \text{ mm}^4$$

Given: $\tau = 1\ \text{N/mm}^2$

$b = 120$ mm

$$S = 360 \times 60\left(115.71 - \frac{60}{2}\right) = 360 \times 60 \times 85.71$$

V_u = Ultimate shearing force

$$= \left(\frac{\tau \cdot I \cdot b}{S}\right) = \left(\frac{1 \times 4.22 \times 10^8 \times 120}{360 \times 60 \times 85.71}\right)$$

$$= 27.4 \times 10^3\ \text{N} = 27.4\ \text{kN}$$

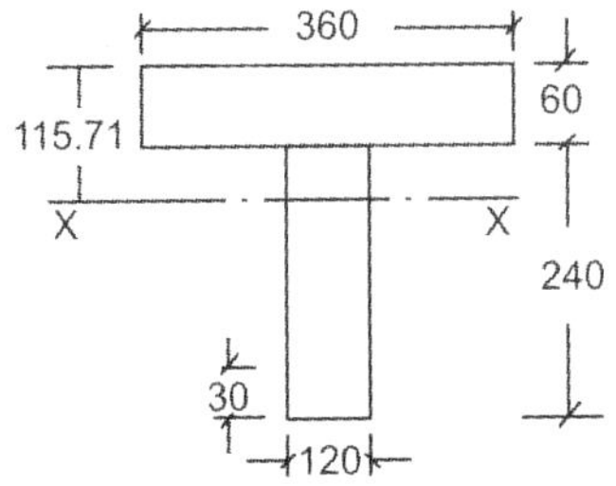

Fig. 9.13

ii. When 10 mm ϕ steel stirrups are placed at 100 mm centres, ultimate shear across stirrups = 190 N/mm^2

Shear resistance of one pair of stirrups

$$= \left[2 \times \frac{\pi}{4}(10)^2 \times 190\right] = 30{,}000\ \text{N}$$

Unit shear stress $= \left(\dfrac{30{,}000}{120 \times 100}\right) = 2.5\ \text{N/mm}^2$

Ultimate shear $V_u = \left(\dfrac{27.4 \times 2.5}{1}\right) = 68.5\ \text{kN}$

Problem 9.6

A composite bridge deck is made up of an *in situ* cast slab 120 mm thick and symmetrical I-section of precast pretensioned beams having the flange width and thickness of 200 mm and 110 mm respectively. Thickness of web = 75 mm. Overall depth of I-section = 500 mm, spacings of I beams = 750 mm (centres). The modulus of elasticity of in situ slab concrete is 30 kN/mm^2. Estimate the stresses developed in the composite member due to a differential shrinkage of 100×10^{-6} between the precast and cast *in situ* elements.

Solution

The composite section is shown in Fig. 9.14.

$$I_{xx} = 60.37 \times 10^8 \text{ mm}^4$$
$$Z_t = 0.31 \times 10^8 \text{ mm}^3$$
$$Z_b = 0.14 \times 10^8 \text{ mm}^3$$
$$Z_j = 0.86 \times 10^8 \text{ mm}^3$$

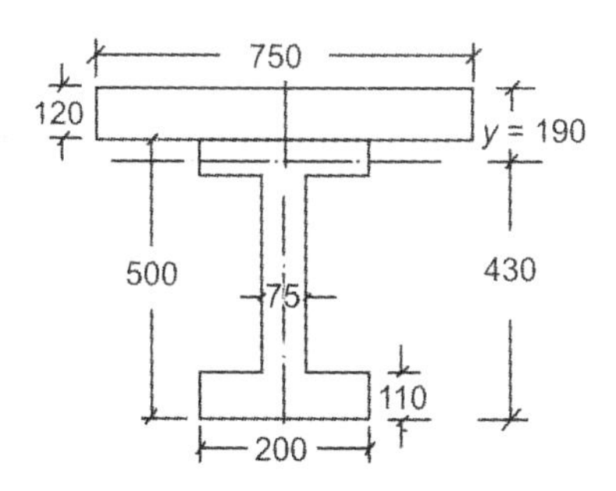

Fig. 9.14

Differential shrinkage $= \varepsilon_{cs} = (100 \times 10^{-6})$

Area of *in situ* concrete $= 90{,}000 \text{ mm}^2$

Uniform tensile stress induced in cast *in situ* slab

$$\varepsilon_{cs} E_c = (100 \times 10^{-6})(30 \times 10^3) = 3 \text{ N/mm}^2$$

Force $N_{sh} = (\varepsilon_s \,.\, E_c \,.\, A_i)$

$$= (100 \times 10^{-6} \times 30 \times 10^3 \times 9 \times 10^4)$$
$$= (270 \times 10^3) \text{ N} = 270 \text{ kN}$$

Eccentricity of force $N_{sh} = (190 - 60) = 130$ mm

$\therefore$ Moment $= (270 \times 10^3 \times 130) = 3.51 \times 10^7$ N.mm

Total area of composite section $= (1.55 \times 10^5) \text{ mm}^2$

$$\text{Direct comp. stress} = \left(\frac{270 \times 10^3}{1.55 \times 10^5}\right) = 1.74 \text{ N/mm}^2$$

Bending stresses

$$\text{Top fibre} = \left(\frac{3.51 \times 10^7}{0.31 \times 10^8}\right) = 1.13 \text{ N/mm}^2$$

$$\text{Bottom fibre} = \left(\frac{3.51\times10^7}{0.41\times10^8}\right) = 2.50\ \text{N/mm}^2$$

$$\text{Junction} = \left(\frac{3.51\times10^7}{0.86\times10^8}\right) = 0.36\ \text{N/mm}^2$$

Differential shrinkage stresses

a. In P.S.C. beam [+ compression, – tension]

At top of beam = (1.74 + 0.36) = 2.1 N/mm^2

At bottom of beam = (1.74 – 2.50) = –0.76 N/mm^2

b. In *in situ* slab

At top of slab = (1.74 + 1.13 – 3) = –0.13 N/mm^2

At bottom of slab = (1.74 + 0.36 – 3) = –0.9 N/mm^2 (junction)

Problem 9.7

Design the required depth of a composite deck slab of a bridge using the standard inverted *T* beam M_1 of C and CA (Table 9.1) to support an imposed load of 15 kN/m^2 on an effective span of 15 m. Determine the minimum prestressing force required and the corresponding eccentricity. Assume grade 40 concrete in the precast beam with a compressive strength at transfer of 35 N/mm^2. The compressive strength of concrete in the *in situ* cast slab is 30 N/mm^2.

$$f_{ct} = 17.5\ \text{N/mm}^2, \qquad f_{tw} = -2.9\ \text{N/mm}^2$$

Solution

The standard *T* beam of C and CA is obtained from Table 9.1 and is shown in Fig. 9.15.

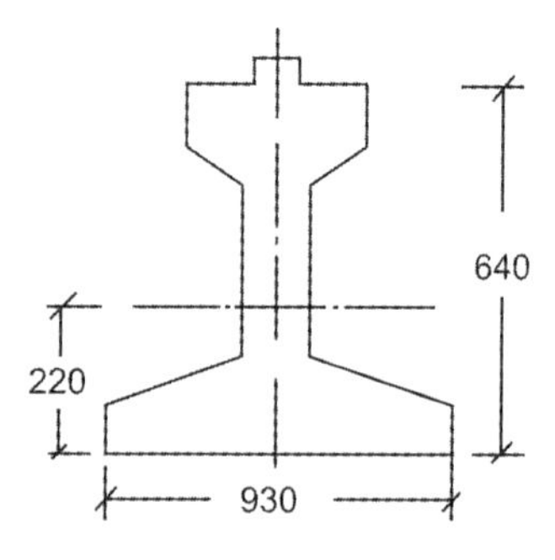

Fig. 9.15

Table 9.1 Section properties of MOT/C and CA standards beams

Number	Overall depth (mm)	Area (mm^2)	Height of centroid above bottom fibre (mm)	Section moduli $mm^3 \times 10^6$		Dead load (kN/m)
				Top fibre	Bottom fibre	
M1	640	284,650	220	24.72	47.17	6.71
M2	720	316,650	265	35.64	61.04	7.46
M3	800	348,650	310	46.96	74.31	8.21
M4	880	323,050	302	43.41	85.95	7.61
M5	960	355,050	357	59.39	100.33	8.37
M6	1040	387,050	409	75.39	116.23	9.12
M7	1200	393,450	454	66.46	123.16	8.52
M8	1200	393,450	454	87.39	143.57	9.27
M9	1280	425,450	512	108.09	161.96	10.02
M10	1360	457,450	568	128.65	179.36	10.78

$$\text{Overall depth} = 640 \text{ mm}$$
$$\text{Area} = 284650 \text{ mm}^2$$
$$Z_t = 24.72 \times 10^6 \text{ mm}^3$$
$$Z_b = 47.17 \times 10^6 \text{ mm}^3$$
$$\text{Dead load} = 6.71 \text{ kN/m}$$
$$\text{Imposed load} = q = 15 \text{ kN/m}^2$$
$$\text{Span} = 15 \text{ m}$$
$$\text{Width} = 930 \text{ mm}$$

$M - 40$ Grade and

$$f_{ci} = 35 \text{ N/mm}^2 \qquad f_{ct} = 17.5 \text{ N/mm}^2$$
$$f_w = -2.9 \text{ N/mm}^2 \qquad \eta = 0.8$$

$$\text{Self weight moment } (M) = (0.125 \times 6.71 \times 15^2) = 188.7 \text{ kN.m}$$

$$\text{Live load moment } M' = \left(\frac{0.93 \times 15 \times 15^2}{8}\right) = 392.3 \text{ kN.m}$$

The required depth of cast *in situ* slab is assumed as

$$(640 + 110) = 750 \text{ mm}$$

$\therefore$ Load due to self weight of precast beam and cast *in situ* slab

$$= (0.75 \times 0.93 \times 24) = 16.74 \text{ kN/m}$$

$$\text{Corresponding moment } M = (0.125 \times 16.74 \times 15^2)$$
$$= 471 \text{ kN.m}$$

The minimum section modulus required for the composite section is given by

$$Z'_b \geq \left[\frac{Z_b M'}{Z_b(\eta f_{ct} - f_{tw}) - (M - \eta M_{\min})}\right]$$

Substituting the values

$$Z'_b \geq \left[\frac{47.17 \times 10^6 \times 392.3 \times 10^6}{47.17 \times 10^6(0.8 \times 17.5 + 2.9) - (471 \times 10^6 - 0)}\right]$$

$$\geq 56.7 \times 10^6 \text{ mm}^3$$

If h = overall depth of section

$$\left(\frac{bh^2}{6}\right) = Z'_b$$

$$\therefore \quad \left(\frac{930 \times h^2}{6}\right) = 56.7 \times 10^6$$

$\therefore$ h = 605 mm < 750 mm (provided). Hence safe.

$$f_b = \left[\frac{f_{tw}}{\eta} + \frac{M}{\eta Z_b} + \frac{M'}{Z'_b}\right]$$

$$= \left[\frac{-2.9}{0.8} + \frac{471 \times 10^6}{0.8 \times 47.17 \times 10^6} + \frac{392.3 \times 10^6}{0.8 \times 56.7 \times 10^6}\right]$$

$$= 17.50 \text{ N/mm}^2$$

$$f_t = \left(f_{tt} - \frac{M_{\min}}{Z_t}\right) = (-2.9 - 0) = -2.9 \text{ N/mm}^2$$

$$\therefore \quad P = \frac{284650\,[-2.9 \times 24.72 \times 10^6 + 17.50 \times 47.17 \times 10^6]}{(47.17 + 24.72)\,10^6}$$

$$= 2984 \times 10^3 \text{ N} = 2984 \text{ kN}$$

$$e = \frac{Z_t Z_b (f_b - f_t)}{A(f_t Z_t + f_b Z_b)}$$

$$= \frac{24.72 \times 47.17 \times 10^2\,(17.50 + 2.9)}{284650[-2.9 \times 24.72 \times 10^6 + 17.50 \times 47.17 \times 10^6]}$$

$$= 111 \text{ mm}$$

Problem 9.8

A composite tee-beam is made up of a pretensioned rib 300 mm wide by 600 mm deep and a cast *in situ* slab 500 mm wide by 150 mm thick having a modulus of elasticity of 28 kN/mm^2. If the differential shrinkage is 150 × 10^{-6} units, determine the srinkage stresses developed in the precast and cast *in situ* units.

Solution

Area of *in situ* concrete $= A_i = (500 \times 150) = 75000$ mm^2

Differential shrinkage $= \in_{cs} = 150 \times 10^{-6}$ units

Uniform tensile stress induced in the cast *in situ* slab

$$\in_{cs} E_c = (100 \times 10^{-6})(28 \times 10^3) = 4.2 \text{ N/mm}^2$$

Force $N_{sh} = \in_{cs} E_c A_i = (4.2 \times 75000) = 315 \times 10^3$ N

The centroid of the composite section is located 340 mm from the top fibre.

Eccentricity of the compressive force N_{sh} from the centroid of the composite section $= (340 - 75) = 265$ mm

∴ Moment $= (315 \times 10^3)(265) = 83475 \times 10^3$ N.mm

Second moment of area of composite section $= 12985 \times 10^6$ mm^4

Section modulus of the various fibres are computed as

$$\text{Top fibre} \quad = Z_t = \frac{(12985 \times 10^6)}{340} = 3819 \times 10^4 \text{ mm}^3$$

$$\text{Bottom fibre} \quad = Z_b = \frac{(12985 \times 10^6)}{410} = 3167 \times 10^4 \text{ mm}^3$$

$$\text{Junction} \quad = Z_j = \frac{(12985 \times 10^6)}{190} = 6834 \times 10^4 \text{ mm}^3$$

$$\text{Direct compressive stress} = \left(\frac{315 \times 10^3}{255 \times 10^3}\right) = 1.23 \text{ N/mm}^2$$

$$\text{Bending stress top fibre} = \frac{(83475 \times 10^3)}{(3819 \times 10^4)} = 2.18 \text{ N/mm}^2$$

$$\text{Bottom fibre} = \frac{(83475 \times 10^3)}{(3167 \times 10^4)} = 2.63 \text{ N/mm}^2$$

$$\text{Junction} = \frac{(83475 \times 10^3)}{(6834 \times 10^4)} = 1.22 \text{ N/mm}^2$$

The differential shrinkage stresses in the precast pretensioned rib and cast *in situ* slab are computed by the addition of direct stresses and bending stresses.

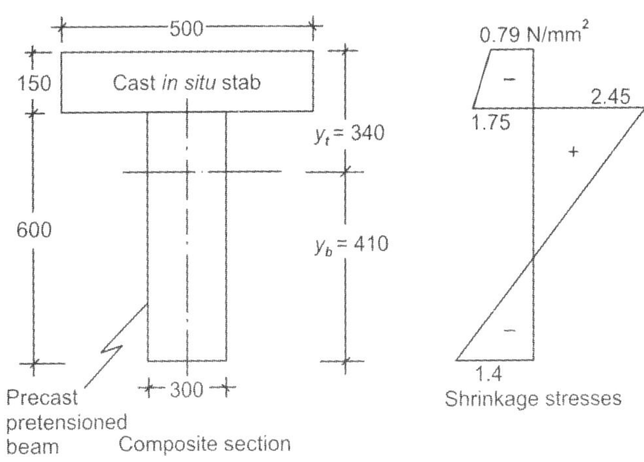

Fig. 9.16

Differential shrinkage stresses

a. In precast pretensioned beam

At top of beam = (1.23 + 1.22) = 2.45 N/mm²

At bottom of beam = (1.23 – 2.63) = –1.4 N/mm² (tension)

b. In cast *in situ* slab

At top of slab = (1.23 + 2.18 – 4.2) = 0.79 N/mm² (tension)

At bottom of slab (junction) = (1.23 + 1.22 – 4.2) = –1.75 N/mm² (tension)

The resultant shrinkage stress distribution is shown in Fig. 9.16.

Problem 9.9

A composite section is made up of a precast pretensioned rib 100 mm wide by 200 mm deep with a cast *in situ* slab 400 mm wide by 40 mm thick. The beam with an effective span of 5 m is prestressed by an effective force of 150 kN at an eccentricity of 33.33 mm. The live load on the composite beam is 8 kN/m. The modulus of elasticity of precast beam concrete is 35 kN/mm² while that of concrete in the cast *in situ* slab is 28 kN/mm². Loss ratio is 0.85. Assuming propped type of construction, determine the long term deflection of the composite beam if creep coefficient is 1.6.

Solution

Properties of precast prestressed beam

$$A = (200 \times 100) = 2 \times 10^4\ \text{mm}^2$$

$$g = (0.1 \times 0.2 \times 24) = 0.48\ \text{kN/m}$$

$$I = \frac{(100 \times 200^3)}{12} = 66.66 \times 10^6\ \text{mm}^4$$

$$P = 150\ \text{kN} \qquad e = 33.33\ \text{mm}$$

$$E_c = 35\ \text{kN/mm}^2 \qquad L = 5\ \text{m}$$

Properties of composite section

$$A = (400 \times 40) = 16 \times 10^3\ \text{mm}^2$$

$$g = (0.04 \times 0.4 \times 24) = 0.384\ \text{kN/m}$$

$$E_c = 28\ \text{kN/mm}^2$$

The centroid of the composite section consisting of two different units is determined by taking moments about an axis passing through the soffit of the beam.

Let y_b = distance of centroid from soffit and the

$$\text{Ratio of modulus of elasticity} = \left(\frac{35}{28}\right) = 1.25$$

$$(16 + 1.25 \times 20)10^3 \times y_b = (16 \times 10^3 \times 220) + (1.25 \times 20 \times 10^3 \times 100)$$

Solving $y_b = 146$ mm

$$I_e = \frac{(400 \times 40^3)}{12} + (16 \times 10^3 \times 74^2)$$

$$+ 1.25\left[\frac{(100 \times 200^3)}{12} + 20 \times 10^3 \times 46^2\right]$$

$$= 226 \times 10^6\ \text{mm}^4$$

Deflection of the prestressed beam is computed as

$$a_p = \left(\frac{PeL^2}{8EI}\right) = \left[\frac{150 \times 33.33 \times 5000^2}{8 \times 35 \times 66.66 \times 10^6}\right]$$

$$= -6.69\ \text{mm (upwards)}$$

Deflection of the composite beam due to self weight of prestressed beam, self weight of cast *in situ* slab and live load is computed as

$$a_{g+q} = \left(\frac{5wL^4}{384EI}\right)$$

where, $g = (0.48 + 0.384) = 0.864$ kN/m

$q = 8$ kN/m

$w = (g + q) = (0.864 + 8) = 8.864$ kN/m

$= 0.00864$ kN/mm

$$a_{g+q} = \left[\frac{5 \times 0.00864 \times 5000^4}{384 \times 35 \times 226 \times 10^6}\right] = 9.12 \text{ mm}$$

Given $\phi = 1.6$ and $\eta = 0.85$

Long term deflection is computed as

$$a_{RL} = (1 + \phi)[\eta a_p + a_{g+q}]$$
$$= (1 + 1.6)[0.85\,(-6.69) + 9.12] = 8.944 \text{ mm}$$

According to IS: 1343 code the maximum permissible long term deflection is limited to $\left(\dfrac{\text{span}}{250}\right)$

$$\left(\frac{\text{span}}{250}\right) = \left(\frac{500}{250}\right) = 20 \text{ mm}$$

Hence, the actual deflection is well within the maximum permissible deflection.

Problem 9.10

A composite tee-beam is made up of a pretensioned rib 200 mm wide by 600 mm deep and a cast *in situ* slab 500 mm wide by 150 mm thick. The compressive strength of concrete in the pretensioned rib and cast *in situ* slab are 40 and 20 N/mm^2 respectively. The pretensioned rib is prestressed by pretensioned wires of area 400 mm^2 located at 100 mm from the soffit and are initially stressed to 1200 N/mm^2. Estimate the ultimate flexural strength of the composite section using IS: 1343 code recommendations. Adopt $f_p = 1600$ N/mm^2.

Solution

Data:

$b = 500$ mm $\qquad f_p = 1600$ N/mm^2

$b_w = 200$ mm $\qquad f_{ck}$ (slab) $= 20$ N/mm^2

$t = 150$ mm $\qquad f_{ck}$ (rib) $= 40$ N/mm^2

$d = 650$ mm $\qquad A_p = 400$ mm^2

The effective reinforcement ratio is computed as

$$\left(\frac{A_p f_p}{f_{ck}\, bd}\right) = \left(\frac{400 \times 1600}{20 \times 500 \times 650}\right) = 0.098$$

Referring to Table 11 of IS: 1343 code, read out the values as

$$\left(\frac{f_{pu}}{0.87 f_p}\right) = 1.0 \text{ and } \left(\frac{x_u}{d}\right) = 0.217$$

$\therefore$ $f_{pu} = (1.0 \times 0.87 \times 1600) = 1392 \text{ N/mm}^2$ and

$x_u = (0.217 \times 650) = 141 \text{ mm} < 150 \text{ mm}$

The ultimate flexural strength of the composite section is

$$M_u = A_p f_{pu} [d - 0.42\, x_u]$$
$$= (400 \times 1392)[650 - 0.42 \times 141]$$
$$= 329 \times 10^6 \text{ N.mm} = 329 \text{ kN.m}$$

Problem 9.11

A precast pretensioned rib 100 mm wide by 200 mm deep is to be connected to a M-25 grade cast *in situ* concrete slab, 400 mm wide and 40 mm thick. Estimate the ultimate shearing force which will cause separation of the two elements for the following two cases conforming to BS: 8110 code specifications.

a. If the surface is rough tamped and without links to withstand a horizontal shear stress of 0.6 N/mm², and
b. With nominal links and the contact surface are as cast to withstand a horizontal shear stress of 1.2 N/mm².

Assume the moduli of elasticity of precast and cast *in situ* concrete are equal.

Solution

Area of cast *in situ* slab $A = (400 \times 40) = 16000 \text{ mm}^2$

Width of precast rib $= b = 100$ mm

Centroid of the composite T-section located at 87 mm from top

Second moment of area of composite section

$$I = (1948 \times 10^5) \text{ mm}^4$$

Distance of centroid of cast *in situ* slab from centroid of composite section $y = (87 - 20) = 67$ mm

Case a:

If V_u = Ultimate shearing force

τ = Shear stress = 0.6 N/mm²

$$\tau = \left[\frac{V_u A y}{I b}\right]$$

$$V_u = \left[\frac{\tau I b}{A y}\right] = \left[\frac{(0.6 \times 1948 \times 10^5 \times 100)}{(16000 \times 67)}\right]$$

$$= 10902 \text{ N} = 10.902 \text{ kN}$$

Case b:

When nominal links are used, the design ultimate shear stress,

$$\tau = 1.2 \text{ N/mm}^2$$

Ultimate shear resistance is expressed as

$$V_u = \left[\frac{\tau I b}{A y}\right] = \left[\frac{(1.2 \times 1948 \times 10^5 \times 100)}{(16000 \times 67)}\right]$$

$$= 21804 \text{ N} = 21.804 \text{ kN}.$$

Problem 9.12

The cross-section of a composite bridge deck is made up of a prestressed symmetrical I-beam having flanges 200 mm wide and thickness 100 mm, the overall depth of I-beam being 500 mm. The I-beams spaced at 750 mm centres are connected by a cast *in situ* slab 120 mm thick made of M-25 grade concrete. The centroid of composite section is located at a distance of 180 mm above the centroid of the prestressed I-section. If the design ultimate shear force at the support is 160 kN and the second moment of area of the composite section is (5584×10^6) mm^4, evaluate the shear stress at the interface and design the shear connection with nominal link reinforcements.

Solution

Design ultimate shear force = V_u = 160 kN

Second moment of area of composite section

$$I = (5584 \times 10^6) \text{ mm}^4$$

Width of flange at interface of cast *in situ* slab b = 200 mm

First moment of the area of slab about the centroid of the composite section is given by

$$Ay = \left[(750 \times 120\left(620 - 250 - 180 - \frac{120}{2}\right)\right]$$

$$= (11.70 \times 10^6) \text{ mm}^3$$

Shear stress at the junction

$$\tau = \left[\frac{V_u A y}{I b}\right] = \left[\frac{160 \times 10^3 \times 11.70 \times 10^6}{5584 \times 10^6 \times 200}\right]$$

$$= 1.67 \text{ N/mm}^2$$

Using nominal link reinforcements of 0.15 percent of the cross-sectional area,

$$A_{sv} = (0.0015 \times 200 \times 1000) = 300 \text{ mm}^2/\text{m}$$

Using Fe-250 grade steel of 6 mm diameter two legged links,

$$\text{Spacing of links} = \left(\frac{1000 \times 2 \times 28}{300}\right) = 186 \text{ mm}$$

Adopt a spacing of 180 mm for the vertical links.

Problem 9.13

A composite bridge deck is made up of a pretensioned rectangular beam having a width of 300 mm and depth of 600 mm. The cast *in situ* slab is 500 mm wide by 150 mm thick. The ultimate shear force at the support section is 392 kN.

a. Estimate the horizontal shear stress at the junction of precast and *in situ* slab
b. Neglecting the shear resistance between the surfaces, design suitable vertical reinforcements to resist the shear force at support section using Fe-415 HYSD bars.

Solution

Design ultimate shear force $V_u = 392$ kN

Area of cast *in situ* slab $A = (500 \times 150) = 75000 \text{ mm}^2$

Centroid of composite section is located at 340 mm from the top fibre.

Second moment of area of composite section

$$I = (12985 \times 10^4) \text{ mm}^4$$

Fe-415 grade HYSD bars for link reinforcement.

First moment of area of cast *in situ* slab about the centroid is calculated as,

$$Ay = [(500 \times 150)(340 - 75)] = (19.875 \times 10^6) \text{ mm}^3$$

Shear stress at the junction

$$\tau = \left[\frac{V_u Ay}{Ib}\right] = \left[\frac{392 \times 10^3 \times 19.875 \times 10^6}{12985 \times 10^6 \times 300}\right]$$

$$= 2 \text{ N/mm}^2$$

Effective depth of the composite section

$$d = (750 - 50) = 700 \text{ mm}$$

Using 10 mm diameter Fe-415 HYSD bars as two legged vertical links, spacing is computed as,

$$S_V = \left[\frac{A_{sv} 0.87 f_y d}{V_u}\right] = \left[\frac{2 \times 79 \times 0.87 \times 415 \times 700}{392 \times 10^3}\right]$$

$$= 101 \text{ mm}$$

Adopt 10 mm diameter vertical links at a spacing of 100 mm near the supports.

Problem 9.14

A composite beam of rectangular section is made up of a precast prestressed inverted T-beam having a rib, 100 mm wide by 800 mm deep and a slab 400 mm wide and 200 mm thick. The *in situ* concrete has a thickness of 800 mm and a width of 150 mm on either side of the prestressed rib to form a rectangular section. The precast beam T-beam is reinforced with high tensile wires (f_{pu} = 1600 N/mm^2) having an area of 800 mm^2 and located 100 mm from the soffit of the beam. If the cube strength of concrete in the *in situ* cast slab and prestressed beam is 20 and 40 N/mm^2 respectively, estimate the flexural strength of the composite section.

Solution

Width of section b = 400 mm $\quad f_p$ = 1600 N/mm^2

Effective depth d = 900 mm $\quad A_p$ = 800 mm^2

Cube strength of concrete in prestressed beam

$$f_{ck} = 40 \text{ N/mm}^2$$

Cube strength of concrete in cast *in situ* slab

$$f_{ck} = 20 \text{ N/mm}^2$$

The compression zone of composite beam comprises the precast and the *in situ* cast elements, of which the former is 25 percent and the latter is 75 percent. Hence, the average compressive strength of concrete in the stress block is computed as,

$$f_{ck} = [(40 \times 0.25) + (20 \times 0.75)] = 25 \text{ N/mm}^2$$

The effective reinforcement ratio is given by

$$\left[\frac{A_p f_p}{bd\, f_{ck}}\right] = \left[\frac{800 \times 1600}{400 \times 900 \times 25}\right] = 0.142$$

Referring to Table 4.1, interpolate the stress in steel by the ratio

$$\left[\frac{f_{pu}}{0.87\, f_p}\right] = 1.00$$

$\therefore$ $$f_{pu} = (0.87 \times 1600) = 1392 \text{ N/mm}^2$$

and the ratio $\left(\dfrac{x_u}{d}\right) = 0.29$ or $x_u = (0.29 \times 900) = 261$ mm

$\therefore$ $$M_u = f_{pu} A_p (d - 0.42\, x_u)$$

$$= \left[\frac{(1392 \times 800)\{900 - (0.42 \times 261)\}}{10^6}\right] = 880 \text{ kN.m}$$

Problem 9.15

A composite slab deck comprising a prestressed inverted T-section with cast *in situ* slab is to be designed. The composite deck is rectangular in section with a width of 300 mm and overall depth of 600 mm. The composite slab is required to support an imposed load of 16 kN/m^2 over a span of 14 m. The compressive stress in concrete and the tensile stress may be assumed to be 20 and 1 N/mm^2 respectively. Assume 15 percent loss in prestress. Determine the prestressing force required for the section and the corresponding eccentricity.

Solution

Self weight of composite slab

$$g = (0.3 \times 0.6 \times 24) = 4.32 \text{ kN/m}$$

Self weight moment

$$M = (0.125 \times 4.32 \times 14^2) = 105 \text{ kN.m}$$

Live load moment

$$M' = (0.125 \times 0.3 \times 1 \times 16 \times 14^2) = 118 \text{ kN.m}$$

Loss ratio $\eta = 0.85$

$$f_{ct} = 20 \text{ N/mm}^2$$

$$f_{tw} = -1 \text{ N/mm}^2$$

Section modulus of composite slab

$$Z_b' = \left[\frac{300 \times 600^2}{6}\right] = (18 \times 10^6) \text{ mm}^3$$

The required section modulus of the precast pretensioned inverted T-section is given by

$$Z_b \geq \left[\frac{Z_b'(M - \eta M_{\min})}{Z_b'(\eta f_{ct} - f_{tw}) - M'}\right]$$

$$= \left[\frac{18 \times 10^6 \ (105 \times 10^6 - 0)}{18 \times 10^6 \ (0.85 \times 20 + 1) - (118 \times 10^6)}\right]$$

$$\geq (9.2 \times 10^6) \text{ mm}^3$$

An inverted T-section with the following dimensions will provide the required section modulus.

Thickness and width of top flange = 100 mm and 150 mm respectively.

Thickness and width of bottom flange = 100 mm and 300 mm respectively.

Thickness of web = 60 mm
Overall depth = 500 mm
The section properties are as follows:

$$A = (63 \times 10^3)\ \text{mm}^2$$
$$Z_t = (6 \times 10^6)\ \text{mm}^3$$
$$Z_b = (9.5 \times 10^6)\ \text{mm}^3$$

Prestress at bottom fibre is evaluated using the expression,

$$f_b = \left[\frac{f_{tw}}{\eta} + \frac{M}{\eta Z_b} + \frac{M'}{\eta Z_b'}\right]$$
$$= \left[\frac{-1}{0.85} + \frac{(105 \times 10^6)}{(0.85 \times 9.5 \times 10^6)} + \frac{(118 \times 10^6)}{(0.85 \times 18 \times 10^6)}\right]$$
$$= 19.54\ \text{N/mm}^2$$
$$f_t = \left[f_{tt} - \frac{M_{min}}{Z_t}\right] = [-1 - 0] = -1\ \text{N/mm}^2$$

The minimum prestressing force required is obtained by using the expression,

$$P = \frac{A(f_t Z_t + f_b Z_b)}{(Z_t + Z_b)}$$
$$= \left[\frac{63 \times 10^3 \times 10^6 \{(-1 \times 6) + (19.54 \times 9.5)\}}{(15.5 \times 10^6)}\right]$$
$$= 730\ \text{kN}$$

The corresponding eccentricity is computed as,

$$e = \left[\frac{Z_t Z_b (f_b - f_t)}{A(f_t Z_t + f_b Z_b)}\right]$$
$$= \left[\frac{(9.5 \times 6 \times 10^{12})(19.62 + 1)}{(63 \times 10^3)\,10^6\,\{(19.54 \times 9.5) - (1 \times 6)\}}\right]$$
$$= 102\ \text{mm.}$$

10

Statically Indeterminate Structures

Problem 10.1

A two-span continuous prestressed concrete beam *ABC* (*AB* = *BC* = 15 m) has a uniform cross-section with a width of 250 mm and a depth of 600 mm. The cable carrying an effective prestressing force of 500 kN is parallel to the axis of the beam and located at an eccentricity of 200 mm.

a. Determine the secondary and resultant moment developed at mid support section *B*.
b. If the beam supports an imposed load of 2.4 kN/m, calculate the resultant stresses developed at the top and bottom of the beam at *B*. Also calculate the resultant line of thrust through the beam *AB*.

Solution

The continuous beam with its cross-section is shown in Fig. 10.1.

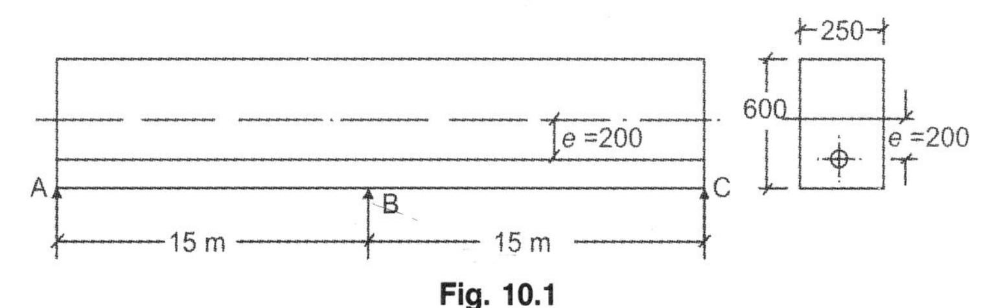

Fig. 10.1

$$P = 500 \text{ kN}$$

$$e = 200 \text{ mm}$$

Self weight of beam

$$g = (0.25 \times 0.6 \times 24) = 3.6 \text{ kN/m}$$

$$\left(\frac{gL^2}{8}\right) = \left(\frac{3.6 \times 15^2}{8}\right) = 101.25 \text{ kN.m}$$

a. Using the tendon reaction method, we have the moments as shown below:

A	B	C
$-Pe \longrightarrow -\dfrac{Pe}{2}$		$+\dfrac{Pe}{2} \longleftarrow +Pe$

$$\therefore \quad \left(\frac{Pe}{2}\right) = \left(\frac{500\times 200}{2}\right) = 50 \times 10^3 \text{ kN.mm}$$

∴ Resultant moment at B = RM at B = 50 kN.m

∴ SM at B = (RM – PM) = [50 – (–500 × 0.2)] = +150 kN.m

b. Due to UDL (LL) of q = 2.4 kN/m

Dead load of g = 3.6 kN/m

Total load w = 6.0 kN/m

$$\text{Moment at } B \text{ due to loads} = \left(\frac{w\cdot L^2}{8}\right) = \frac{(6\times 15^2)}{8}$$

$$= -168.75 \text{ kN.m}$$

$$\therefore \text{ Resultant moment at } B = \left[-168.75 + \left(\frac{Pe}{2}\right)\right]$$

$$= [-168.75 + 50]$$

$$= -118.75 \text{ kN.m (hogging)}$$

$$\text{Section modulus} = Z_t = Z_b = Z = \left(\frac{bd^2}{2}\right)$$

$$= (250 \times 600^2)/6 = 15 \times 10^6 \text{ mm}^3$$

$$\therefore \quad \text{Stress at top} = \left(\frac{M}{Z}\right) = \left[\frac{(-118.75\times 10^6)}{(15\times 10^6)}\right]$$

$$= -7.91 \text{ N/mm}^2 \text{ (tension)}$$

$$\text{Stress at bottom} = \left[\left(\frac{2P}{A}\right) + \left(\frac{M}{Z}\right)\right]$$

$$= \left[\frac{(2\times 500\times 10^3)}{(250\times 600)} + 7.91\right]$$

$$= +14.57 \text{ N/mm}^2 \text{ (compression)}$$

Line of thrust

At A, eccentricity e = –200 mm

$$\text{Shift at } B = \left(\frac{R.M.}{P}\right) = \left[\frac{(-118.75\times 10^3)}{(500)}\right]$$
$$= -237.5 \text{ mm}$$

Shift at centre of AB is computed as follows:

$$\text{RM at centre of } AB = \left[\left(-\frac{Pe}{4}\right) + \left(\frac{wL^2}{16}\right)\right]$$
$$= [(-500 \times 0.2)/4 + (6 \times 15^2)/8]$$
$$= +59.37 \text{ kN.m (sagging)}$$

$\therefore$ Shift of pressure line at centre at AB is given by

$$\text{Shift} = \left[\frac{R.M.}{P}\right] = \left[\frac{(59.37\times 10^6)}{(360\times 10^3)}\right]$$
$$= 164.9 \text{ mm}$$

The resultant thrust line in span AB and BC is shown in Fig. 10.2.

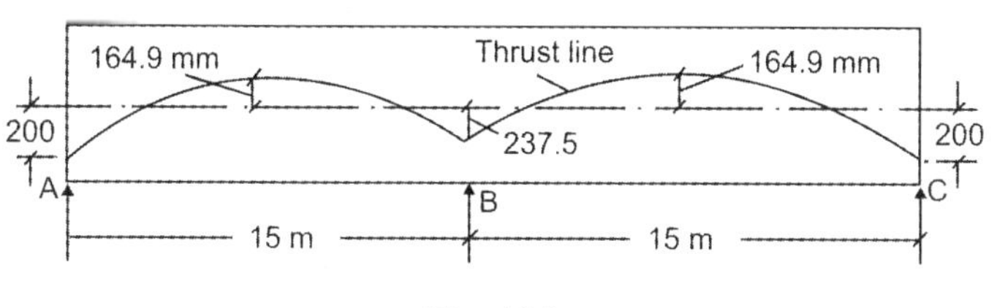

Fig. 10.2

Problem 10.2

A two-span continuous beam ABC ($AB = BC = 10$ m) is of rectangular section, 200 mm wide and 500 mm deep. The beam is prestressed by a parabolic cable, concentric at end supports and having an eccentricity of 100 mm towards the soffit of the beam at centre of spans and 200 mm towards the top of beam at mid support B. The effective force in the cable is 500 kN.

a. Show that the cable is concordant.

b. Locate the pressure line in the beam when it supports an imposed load of 5.6 kN/m, in addition to its self weight.

Solution

The two span continuous beam is shown in Fig. 10.3.

$$\text{Equivalent UDL} = w_e = \left(\frac{8Pe}{L^2}\right) = \left(\frac{8\times 500\times 10^3 \times 200\times 10^{-3}}{10^2}\right)$$
$$= 8 \text{ kN/m (upwards)}$$

Fixed end moments due to loading is

$$M_A = 0 \qquad M_C = 0$$

$$M_B = +\left(\frac{w_e L^2}{8}\right) = \left(\frac{8 \times 10^2}{8}\right) = 100 \text{ kN.m}$$

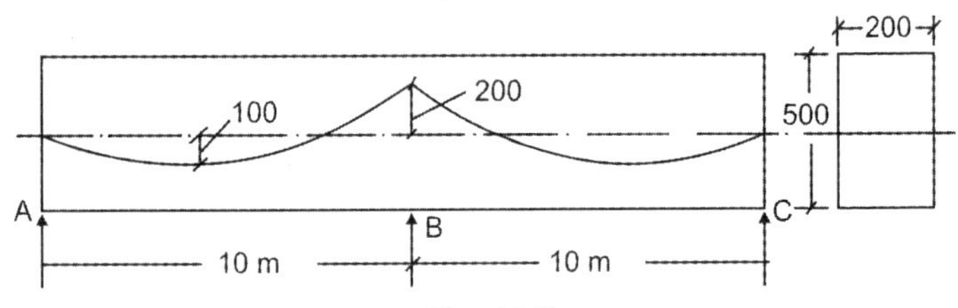

Fig. 10.3

$$\text{SM at } B = [\text{RM} - \text{PM}] = [100 - 500 \times 0.2] = 0$$

a. Hence the cable is concordant

b. Self weight of beam $= (0.2 \times 0.5 \times 24) = 2.4$ kN.m

Live load on the beam $= 5.6$

Total load $= 8.0$ kN/m (downwards)

Resultant loading on beam $= (8 - 8) = 0$

Due to the axial prestress of 500 kN. The pressure line coincides with the centroidal axis.

Problem 10.3

A continuous concrete beam *ABC* (*AB* = *BC* = 10 m) has a uniform rectangular cross-section, 100 mm wide and 300 mm deep. A cable carrying an effective prestressing force of 360 kN varies linearly with an eccentricity of 50 mm towards the soffit at the end support to 50 mm towards the top of the beam at mid support *B*.

a. Determine the resultant moment at B due to prestressing only.
b. If the eccentricity of the cable at *B* is +25 mm, show that the cable is concordant.

Solution

The two span continuous beam with the cable is shown in Fig. 10.4.

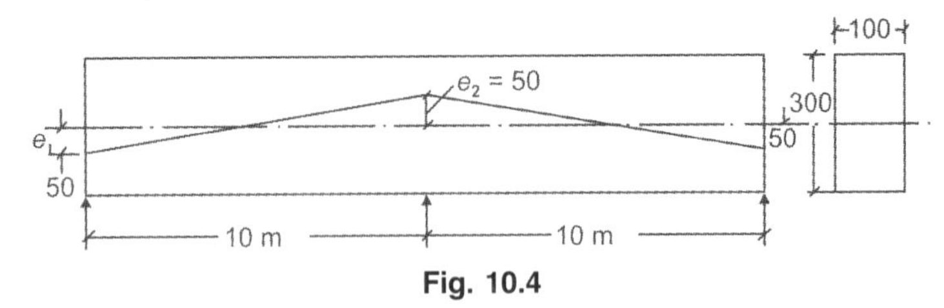

Fig. 10.4

$$P = 360 \text{ kN} \qquad e = 50 \text{ mm}$$

a.

A	B	C
$-Pe \longrightarrow -\dfrac{Pe}{2}$	$+\dfrac{Pe}{2} \longleftarrow +Pe$	

$$\therefore \quad \text{R.M. at } B = \left(\frac{Pe}{2}\right) = \left(\frac{360 \times 0.05}{2}\right) = +9 \text{ kN.mm}$$

b. If $\quad e_1 = 50$ mm at A, $\quad e_2 = -25$ mm at B

$$\text{Then R.M. at } B = +\frac{Pe}{2} = +9 \text{ kN.m}$$

R.M. = (S.M. + P.M.) at B = $(9 - 360 \times 0.025) = 0$

Since secondary moment is zero at B. The cable is concordant.

Problem 10.4

A three-span continuous prestressed concrete prismatic beam of depth 2 m is prestressed by a design prestressing force of 3300 kN. The spans are $AB = CD = 30$ m and $BC = 40$ mm. The eccentricities of the prestressing force at different locations along the beam are given in the following table:

x, measured from A	0	15	30	50	70	85	100
Eccentricity e, mm	0	–260	+450	–410	+450	–260	0

Note: – below the centroid, + above the centroid

a. If the cable profile is parabolic between the supports, evaluate the equivalent load and hence, compute the moments produced at support of mid span points due to prestressing.
b. Also locate the position of pressure line
c. Suggest a suitable concordant profile.

Solution

The three-span continuous beam with the parabolic cable is shown in Fig. 10.5.

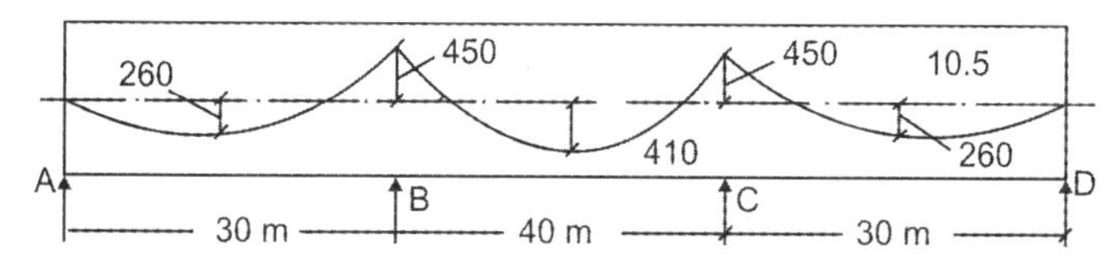

Fig. 10.5

$$P = 3300 \text{ kN}$$

Equivalent load on AB is (and CD)

$$w_e = \left(\frac{8\,Pe}{L^2}\right) = \left(\frac{8 \times 3300 \times 0.485}{30^2}\right) = 14.2 \text{ kN/m}$$

Equivalent load on BC is

$$w_e = \left(\frac{8\,Pe}{L}\right) = \left(\frac{8 \times 3300 \times 0.86}{40^2}\right) = 14.2 \text{ kN/m}$$

Fixed end moments

$$\text{Span} \quad AB = \left(\frac{wL^2}{12}\right) = \left(\frac{14.2 \times 30^2}{12^2}\right) = 1065 \text{ kN.m}$$

$$\text{Span} \quad BC = \left(\frac{w\,L^2}{12}\right) = \left(\frac{14.2 \times 40^2}{12}\right) = 1893 \text{ kN.m}$$

Moment distribution

Distribution factors at B and C is given by

$$d_{BA} = d_{BC} = 0.5$$

and $\quad d_{CB} = d_{CD} = 0.5$

Final moments are obtained by moment distribution as shown below:

A	B		C		D
	0.5	0.5	0.5	0.5	
+1055	−1065	+1893	−1893	+1065	−1065
−1065 ⟶	−532			+532 ⟵	+1065
	−148	−148	+148	+148	
		+74	−74		
	−37	−37	+37	+37	
		18.5	18.5		
	−9.25	−9.25	+9.25	+9.25	
		4.12	−4.12		
	−2.06	−2.06	+2.06	+2.06	
		1.03	−1.03		
	−0.51	−0.51	+0.51	+0.51	
0	−1794	+1794	−1794	+1794	0 kN.m

The bending moment is shown in Fig. 10.6,

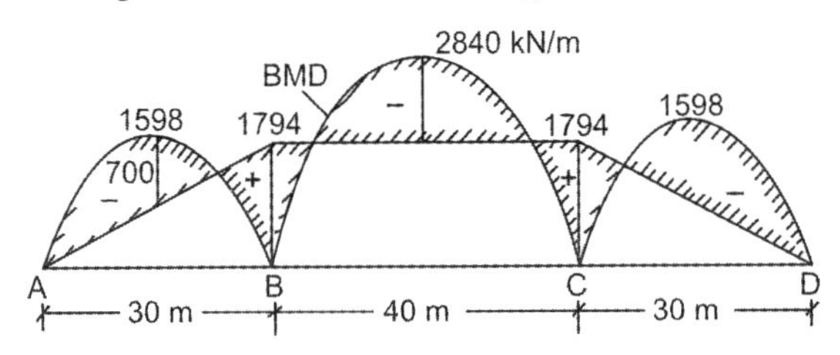

Fig. 10.6

Positive moments at centre of spans of

$$AB = \left(\frac{wL^2}{8}\right) = \left(\frac{14.2 \times 30^2}{8}\right) = 1598 \text{ kN.m}$$

$$BC = \left(\frac{wL^2}{8}\right) = \left(\frac{14.2 \times 40^2}{8}\right) = 2840 \text{ kN.m}$$

Resultant moments

Mid span of AB and CD = –700 kN.m

Mid span of BC = –1050 kN.m

Location of pressure line (shift)

At A = 0

$$\text{At } B \text{ and } C = \left(\frac{R.M.}{P}\right) = \left(\frac{+1794 \times 10^3}{3300}\right) = +543 \text{ mm}$$

$$\text{At mid span of } AB = \left(\frac{-700 \times 10^3}{3300}\right) = -212 \text{ mm}$$

$$\text{At mid span of } BC = \left(\frac{-1050 \times 10^3}{3300}\right) = -318 \text{ mm}$$

These values are the shift in the pressure line from the centroidal axis as shown in Fig. 10.7.

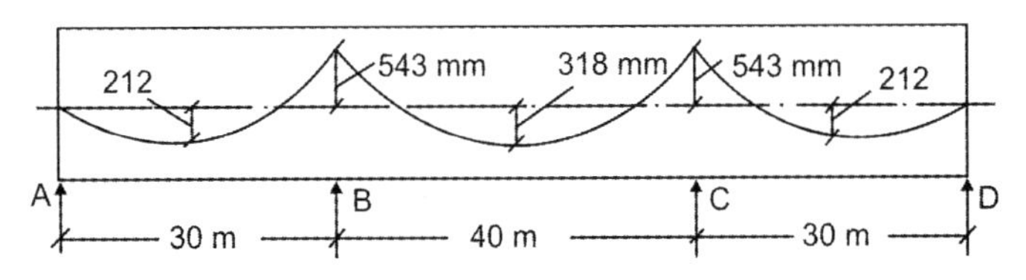

Fig. 10.7

Problem 10.5

A two-span continuous concrete beam *ABC* (*AB* = *BC* = 12 m) has a rectangular section 300 mm wide and 800 mm deep. The beam is prestressed by a cable carrying an effective force of 700 kN. The cable has a linear profile in the span *AB* and parabolic profile in span *BC*. The eccentricities of the cable are +50 mm at *A*, –100 mm at a distance of 7 m from *A* and +200 mm at support *B* and –200 mm at and mid span of *BC* (– below and + above centroid axis)

a. Evaluate the resultant moment developed at *B* due to the prestressing force.
b. Sketch the line of thrust in the beam if it supports a uniformly distributed load of 5 kN/m which includes the self weight of the beam.
c. Find the resultant stress distributed at the mid support section for condition (b).

Solution

The two-span continuous beam is shown in Fig. 10.8

$$\tan\theta_1 = \left(\frac{150}{7000}\right) = 0.021$$

$$\tan\theta_2 = \left(\frac{300}{5000}\right) = 0.06$$

For small angles $\tan\theta = \theta$

$P = 700$ kN

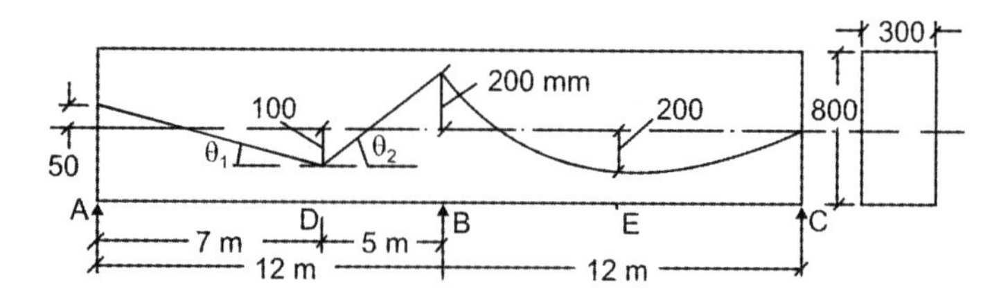

Fig. 10.8

Concentrated force at $D = 700(0.021 + 0.06) = 57$ kN (upwards)

$$\text{Equivalent U.D.L. on span } BC = \left(\frac{8Pe}{L^2}\right) = \left(\frac{8 \times 700 \times 0.3}{12^2}\right)$$

$= 11.67$ kN/m

The equivalent loads on the beam are shown in Fig. 10.9.

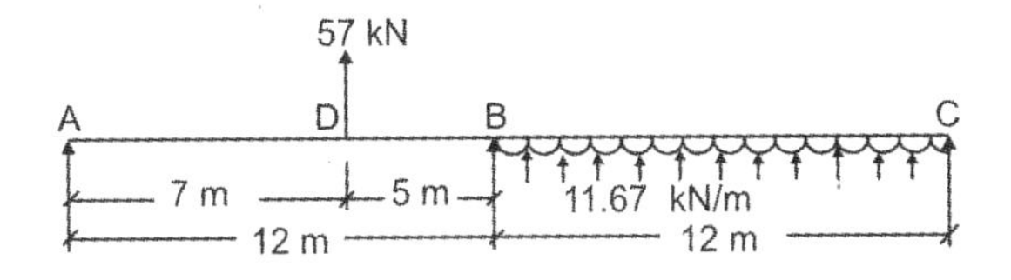

Fig. 10.9

Fixed end moments

$$M_{FAB} = \left(\frac{57 \times 7 \times 5^2}{12^2}\right) = +69.3 \text{ kN.m}$$

$$M_{FBA} = \left(\frac{57 \times 5 \times 7^2}{12^2}\right) = -97.0 \text{ kN.m}$$

$$M_{FBC} = \left(\frac{11.67 \times 12^2}{12}\right) = +140 \text{ kN.m}$$

$$M_{FCB} = \left(\frac{11.67 \times 12^2}{12}\right) = -140 \text{ kN.m}$$

At A, moment due to prestressing force $= P.e$
$= (700 \times 0.05)$
$= 35$ kN.m

Moment distribution

A	B: 0.5	B: 0.5	C
+35.0	–	–	–
+69.3	–97	+140	–140
–69.3	–21.5	–21.5	+140
	–34.65	+70	
	–17.615	–17.615	
+35	–170	+170	0 kN.m

a. The resultant moment developed at B = +170 kN.m

b. For this case the equivalent loads on the beam is shown in Fig. 10.10.

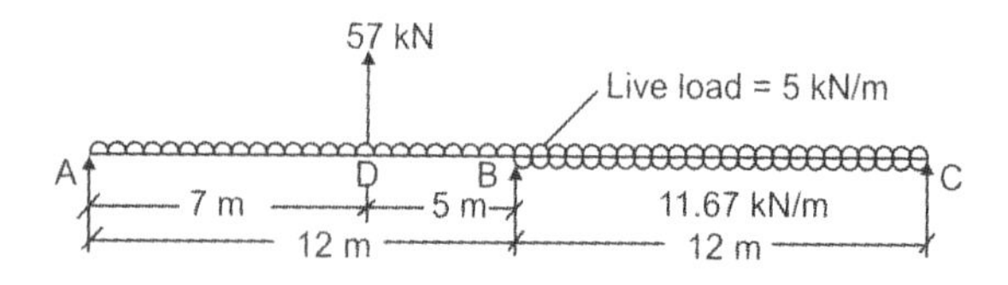

Fig. 10.10

Fixed end moments

F.E.M. at A due to prestressing force

$$P = P.e = (700 \times 0.05) = +35 \text{ kN.m}$$

$$\text{Due to U.D.L.} = \left(\frac{5 \times 12^2}{12}\right) = -60 \text{ kN.m}$$

Due to concentrated load

$$= \left(\frac{56 \times 7 \times 5^2}{12^2}\right) = +69 \text{ kN.m}$$

$\therefore$ Resultant F.E.M. at A = +9 kN.m

F.E.M. at B = (+60 – 97) = –37 kN.m

For span BC,

Resultant U.D.L. = (11.67 – 5) = 6.67 kN/m

$$\therefore \quad \text{F.E.M. at } B = \left(\frac{6.67 \times 12^2}{12}\right) = +80 \text{ kN.m}$$

F.E.M. at C = –80 kN.m

Moment distribution

A	B (0.5)	B (0.5)	C
+35	–	–	–
+9	–37	+80	–80
–9	–21.5	–21.5	+80
	–4.5	+40	
	–17.75	–17.75	
+35	–80	+80	0 kN.m

The bending moment diagram is shown in Fig. 10.11.

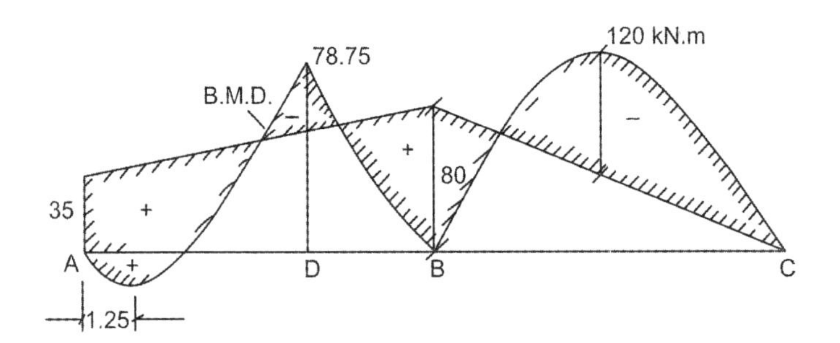

Fig. 10.11

Location of pressure line (shift = R.M./P).

At $A = +50$ mm

1.25 m from $A = \left(\dfrac{43.7 \times 10^3}{700}\right) = +62.4$ mm

7 m from $A = \left(\dfrac{-17.1 \times 10^3}{700}\right) = -24.4$ mm

At $B = \left(\dfrac{+80 \times 10^3}{700}\right) = +115.4$ mm

At centre of $BC = \left(\dfrac{-79.6 \times 10^3}{700}\right) = -113.7$ mm

Resultant stress at B

$$Z = \frac{300 \times 800^2}{6} = 32 \times 10^6 \text{ mm}^3$$

$$\text{Stress at top fibre} = \left(\frac{700 \times 10^3}{24 \times 10^4} + \frac{80 \times 10^6}{32 \times 10^6}\right)$$

$$= 5.44 \text{ N/mm}^2 \text{ (compression)}$$

$$\text{Bottom fibre} = \left(\frac{700 \times 10^3}{24 \times 10^4} - \frac{80 \times 10^6}{32 \times 10^6}\right)$$

$$= 0.39 \text{ N/mm}^2 \text{ (compression)}$$

Problem 10.6

A prestressed portal frame *ABCD* fixed at *A* and *D* has the following section properties:

Member	Length, m	Sectional area, cm^2	Second moment of area, cm^4	Section modulus, cm^3
AB and *CD*	4.5	190	16,000	1280
BC	6.0	190	24,000	1600

The columns *AB* and *CD* are prestressed by a straight cable (effective force in cable = 160 kN) with an eccentricity of 50 mm towards the inside of frame at *A* and *D* and 50 mm towards the outside of frame at *B* and *C*. The horizontal member *BC* is prestressed by a cable (effective force = 140 kN) with a parabolic profile having an eccentricity of 50 mm above the centroid at *B* and *C* and 100 mm below the centroid at the centre of *BC*. The overall depth of *AB* and *CD* is 250 mm and that of *BC* is 300 mm.

a. Calculate the secondary moments developed at *A* and *B*

b. Find the displacement of the line of thrust in the members *AB* and *BC*.

Solution

The portal frame being symmetrical, half the frame is analysed by introducing two redundant reactions R_1 and R_2 at the mid span section of transom *BC* as shown in Fig. 10.12. The moments m_0 developed due to the prestressing and the moments m_1 and m_2 induced by applying unit reactions R_1 and R_2. The flexibility coefficients are evaluated by using the product integrals, as given below:

$$u_1 = \int \frac{(m_0 m_1\, ds)}{EI}$$

$$= \frac{160}{EI}\left[\left(\frac{1}{6}\times 4.5\times 4.5\times 0.1\right)-\left(\frac{1}{2}\times 4.5\times 4.5\times 0.05\right)\right]$$

$$= -\left(\frac{27}{EI}\right)$$

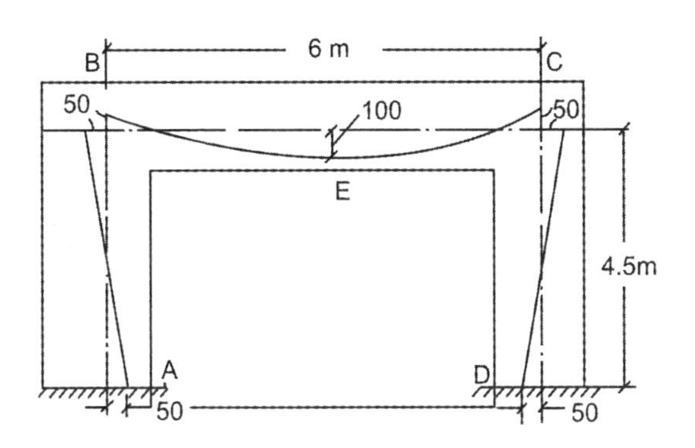

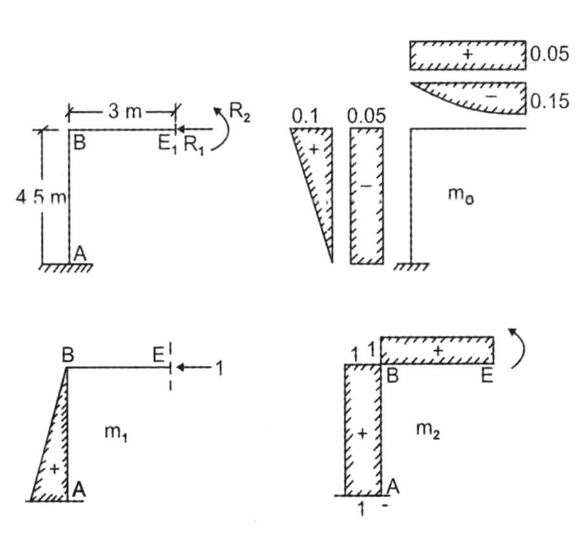

Fig. 10.12

$$u_2 = \int \frac{(m_0 m_2\, ds)}{EI}$$

$$= \frac{160}{EI}\left[\left(\frac{1}{2}\times 4.5\times 1\times 0.1\right)-(4.5\times 0.05\times 1)\right]$$

$$+\frac{140}{EI}\left[(3\times 0.05\times 1)-\left(\frac{2}{3}\times 3\times 0.15\times 1\right)\right] = -\left(\frac{21}{EI}\right)$$

$$f_{11} = \int \frac{(m_1\cdot m_1\cdot ds)}{EI} = \frac{1}{EI}\left(\frac{1}{3}\times 4.5\times 4.5\times 4.5\right) = \left(\frac{30.375}{EI}\right)$$

$$f_{12} = f_{21} = \int \frac{(m_1 m_2\, ds)}{EI} = \frac{1}{EI}\left(\frac{1}{2}\times 4.5\times 4.5\times 1\right) = \left(\frac{10.125}{EI}\right)$$

$$f_{11} R_1 + f_{12} R_2 + u_1 = 0$$
$$f_{21} R_1 + f_{22} R_2 + u_2 = 0$$
$$\therefore \quad 30.375 R_1 + 10.125 R_2 - 27 = 0$$
$$10.125 R_1 + 7.5 R_2 - 21 = 0$$

Solving $\quad R_1 = -0.074$ kN

$\quad R_2 = +2.9$ kN.m

Secondary moments

$\therefore \quad M_A = (2.9 - 0.074 \times 4.5) = 2.57$ kN.m

$\quad M_B = 2.9$ kN.m

Problem 10.7

A prestressed two-hinged portal frame has columns *AB* and *CD* 4.8 m and transom *BC* = 9 m. The members have a cross-section, 125 mm wide and 300 mm deep throughout. The columns are prestressed by a straight cable with an eccentricity of 75 mm towards outside of frame at *B* and *C* with zero eccentricity at the hinges *A* and *D*. The transom is prestressed by a straight cable having an eccentricity of 100 mm below the centroid at the centre of *CB* and 75 mm above the centoid at *B* and *C*.

If all cables are tensioned to a force of 225 kN (neglecting self weight), calculate:

a. The secondary moment developed at *B*,
b. The resultant stress distribution at *B* and centre of *BC*,
c. The displacement of the line of thrust at *B* and centre of *BC*.

Solution

The prestressed portal frams *ABCD* is shown in Fig. 10.13.

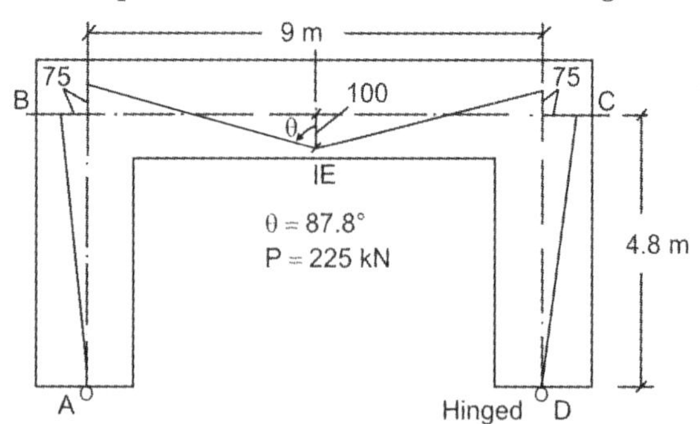

Fig. 10.13

Using tendon reaction method,
the combined effect of the web anchorages at B is
$(225 \times 0.075 - 225 \times 0.075) = 0$

In span BC, the tendon reaction at E gives rise to a concentrated load given by $= 2P \cos \theta$

$= (2 \times 225 \times 0.038) = 17.1$ kN

Fixed end moment at B and C are

$$\text{F.E.M.} = \left(\frac{WL}{8}\right) = \left(\frac{17.1 \times 9}{8}\right) = 19.23 \text{ kN.m}$$

Moment distribution

A	B		C		D
	0.5	0.5	0.5	0.5	
		+19.23	–19.23		
	–9.62	–9.62	+9.62	+9.62	

The resultant moment at $B = 9.62$ kN.m

Primary moment at $B = P.e$

$= (225 \times 0.075) = 16.8$ kN.m

$\therefore$ Secondary moment S.M. = (R.M. – P.M.)

$= (9.62 - 16.8) = -7.18$ kN.m

The section at B is 125 by 300. Hence

$$Z = \left(\frac{125 \times 300^2}{6}\right) = 187.5 \times 10^4 \text{ mm}^3$$

Stresses at B (due to moment)

$$\text{At top} = \left(\frac{9.62 \times 10^6}{187.5 \times 10^4}\right) = 5.13 \text{ N/mm}^2$$

(compression outside face)

At bottom $= 5.13 \text{ N/mm}^2$ (tension inside face)

$$\text{Direct stress} = \left(\frac{P}{A}\right) = \left(\frac{225 \times 10^3}{125 \times 300}\right) = 6 \text{ N/mm}^2$$

Resultant stress at B

At top $= (6 + 5.13) = 11.13 \text{ N/mm}^2$ (compression)

At bottom $= (6 - 5.13) = 0.87 \text{ N/mm}^2$ (tension)

Problem 10.8

A continuous beam ABC ($AB = BC = 20$ m) has a rectangular section 400 mm wide and 600 mm deep throughout the two spans and is prestressed by a concordant cable having a cross-sectional area of 1700 mm^2 located 60 mm from the soffit of the beam at mid span points and 60 mm from the top of the beam at B. If the beam supports uniformly distributed superimposed service load of 4.24 kN/m throughout the spans, estimate the load factor against failure assuming,

a. elastic distribution of moments,

b. complete redistribution of moments

Adopt $f_{pu} = 1600$ N/mm^2 and $f_{cu} = 40$ N/mm^2, $D_c = 24$ kN/m^3

Solution

The two span continuous beam is shown in Fig. 10.14.

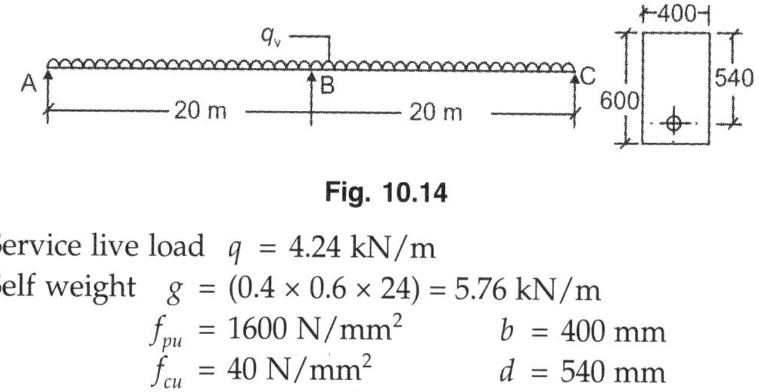

Fig. 10.14

Service live load $q = 4.24$ kN/m

Self weight $g = (0.4 \times 0.6 \times 24) = 5.76$ kN/m

$f_{pu} = 1600$ N/mm^2 $\quad b = 400$ mm

$f_{cu} = 40$ N/mm^2 $\quad d = 540$ mm

$A_{ps} = 1700$ mm^2

$$\left(\frac{A_{ps} \cdot f_{pu}}{bd\, f_{cu}}\right) = \left(\frac{1700 \times 1600}{400 \times 540 \times 40}\right) = 0.31$$

Refer Table 4.1 to interpolate the values of

$$\left(\frac{f_{pb}}{f_{pu}}\right) = 0.85 \quad \text{and} \quad \left(\frac{x}{d}\right) = 0.558$$

$$\therefore \quad M_u = f_{pb} \cdot A_{ps}\,(d - 0.5x)$$
$$= 0.85 \times 1600 \times 1700\,(540 - 0.5 \times 0.558 \times 540)$$
$$= 901 \times 10^6 \text{ N.mm} = 901 \text{ kN.m}$$

a. *Assuming elastic distribution of moments*

$$M_u = 0.125\,(g + q_u)\,L^2$$
$$901 = 0.125\,(5.76 + q_u)\,20^2$$
$$\therefore \quad q_u = 1226 \text{ kN/m}$$

$$\therefore \quad L{\cdot}F = \left(\frac{12.26}{4.24}\right) = 2.89$$

b. *Assuming complete redistribution of moments*

$$q_u = \left(\frac{12\,M_u}{L^2} - g\right) = \left(\frac{12 \times 901}{400} - 5.76\right) = 21.27 \text{ kN/m}$$

$$\therefore \quad L{\cdot}F = (21.27/4.24) = 5.01$$

Problem 10.9

Design a prestressed concrete beam continuous over two equal spans of 9 m to support live loads of 30 kN each at the centre of spans. The loads may be applied independently or jointly. Permissible stresses being zero in tension and 15 N/mm^2 in compression. Loss ratio = 0.85. Determine a concordant profile and show it on an elevation of the beam. Allowing for a minimum cover of 100 mm, sketch a suitable transformed profile to reduce the slope of the tendons at the central support to a minimum. Check for limit states of serviceability and collapse.

Solution

The two span continuous beam with live load is shown in Fig. 10.15. Live load moments at various points $A, 1, 2, 3, 4, 5$ and B are computed for loads applied jointly and independently.

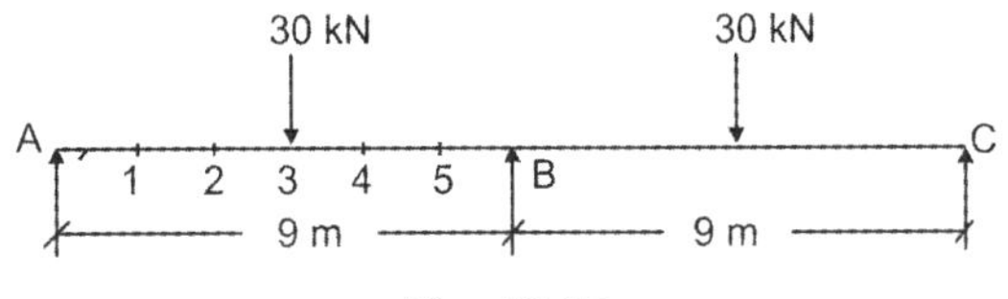

Fig. 10.15

Case a: *Both Loads applied jointly*

By conducting moment distribution the moment diagram shown in Fig. 10.16 is obtained.

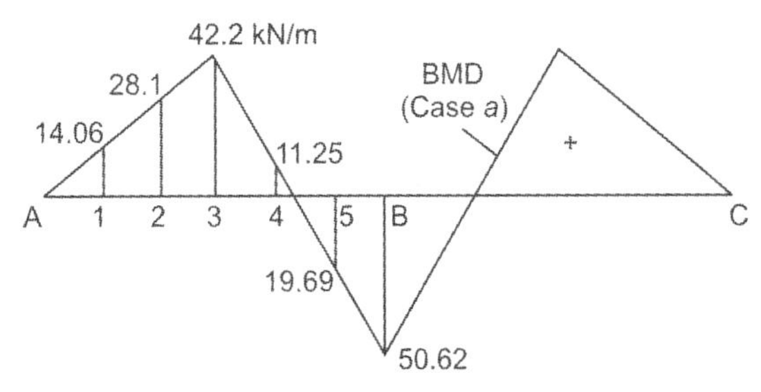

Fig. 10.16

Case b:

For load acting at point 3 only the B.M. diagram shown in Fig. 10.17 is obtained.

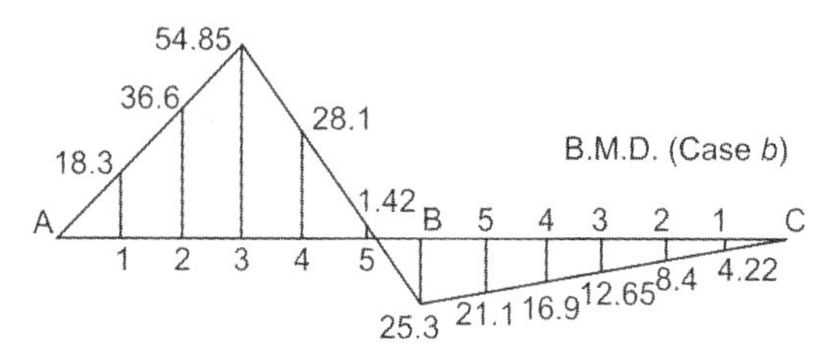

Fig. 10.17

Dead load moments

$$M_{max} = \left(\frac{WL}{4}\right) = \left(\frac{30 \times 9}{4}\right) = 67.5 \text{ kN.m}$$

Assuming a rectangular section with $b = 200$ mm

$$Z = \left(\frac{bh^2}{f_c}\right) = \left(\frac{M_r}{f_c}\right)$$

$$\therefore \quad \left(\frac{200h^2}{6}\right) = \left(\frac{67.5 \times 10^6}{15}\right)$$

$\therefore \quad h = 367$ mm

Adopt $\quad h = 400$ mm

$\quad b = 200$ mm

Self weight of beam = (0.4 × 0.2 × 240) = 1.92 kN/m

By moment distribution, the moment diagram shown in Fig. 10.18 is obtained.

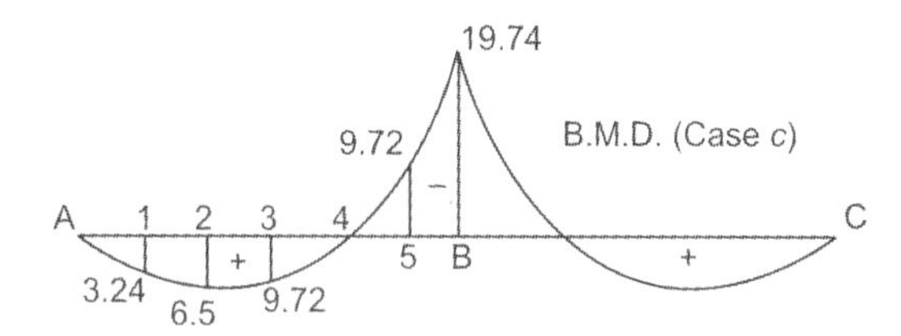

Fig. 10.18

The dead and live load moments for different cases are compiled in Table 10.1.

Table 10.1: Bending moments

Section	Loading pattern			Max. L.L. moment, kN.m		D.L. moment, kN.m
	Case *a*	Case *b*	Case *c*	ML_p	ML_n	M_s
A	0	0	0	0	0	0
1	+14.06	+18.30	–4.22	+18.30	–4.22	+3.24
2	+28.10	+36.6	–8.40	+36.60	--8.40	+6.50
3	+42.2	54.95	–12.65	+54.85	–12.65	+9.75
4	+11.25	+28.10	–16.90	+28.10	–16.90	0
5	–19.69	+1.42	–21.10	+1.42	–21.10	–9.75
B	–50.625	–25.3	–25.3	0	–50.625	–19.44

∴ Minimum prestressing force

$$P = \left(\frac{M_r}{h/3}\right) = \left[\frac{(M_{max} - M_{min})}{h/3}\right]$$

$$= \left[\frac{67.5 \times 10^3}{(400/3)}\right] = 506.25 \text{ kN}$$

The limiting zone for thrust line is calculated as shown in Table 10.2. The profile of concordant cable is determined as shown in Table 10.3.

Table 10.3: Determination of concordant profile

Section	*I*	Simpson's rule coefficient (*q*)	(*dx*/3)	$K = \left(\frac{qdx}{31}\right)$	Unit moment diagram, m	*K.m.*	First trial *e*, cm	*K.m.e.*
A	1	1	2/3	0.667	0	0	0	0
1	1	4	2/3	2.667	0.167	0.440	6.5	2.86
2	1	2	2/3	1.333	0.333	0.440	6.3	2.77
3	1	4	2/3	2.667	0.500	1.330	6.1	8.12
4	1	2	2/3	1.333	0.667	0.880	3.3	2.90
5	1	4	2/3	0.667	0.833	2.230	4.45	–9.92
B	1	1	2/3	0.667	1.000	0.667	–10.0	–6.70

$\Sigma K.m.e = 0$

Table 10.2: Determination of limiting zone for thrust line

Location point	ML_p (kN.m)	ML_n (kN.m)	M_g (kN.m)	Maximum moment M_2 $(ML_p + M_g)$	Minimum moment M_1 $(ML_n + M_g)$	Range of moment $(M_2 - M_1)$	(M_2/P) from upper kern (mm)	(M_2/P) from lower kern (mm)	Eccentricity $e_2 = \left(\frac{M_2}{P} - 133.3\right)$ (mm)	Eccentricity $e_1 = \left(\frac{M_1}{P} + 133.3\right)$ (mm)
A	0	0	0	0	0	0	0	0	–133.33	+133.33
1	+18.30	–4.22	3.24	+21.54	–0.98	+22.52	42.55	–1.94	–90.80	+131.40
2	+36.60	–8.40	6.50	+43.10	–1.90	+45.00	85.14	–3.75	–48.20	+129.60
3	+54.85	–12.65	9.72	+64.57	–2.93	67.50	127.55	–5.80	–5.80	+127.50
4	+28.10	–16.90	0	28.10	–16.90	45.00	55.51	–33.38	–77.80	+99.95
5	+1.42	–21.10	–9.72	–8.30	–30.82	22.52	–16.40	–60.88	–149.70	+72.45
B	0	–50.62	–19.44	–19.44	–70.06	50.62	–38.40	–138.40	–171.70	–5.10
	Prestressing force			P = 506 kN						

The limiting zone and concordant cable profile is shown in Fig. 10.19.

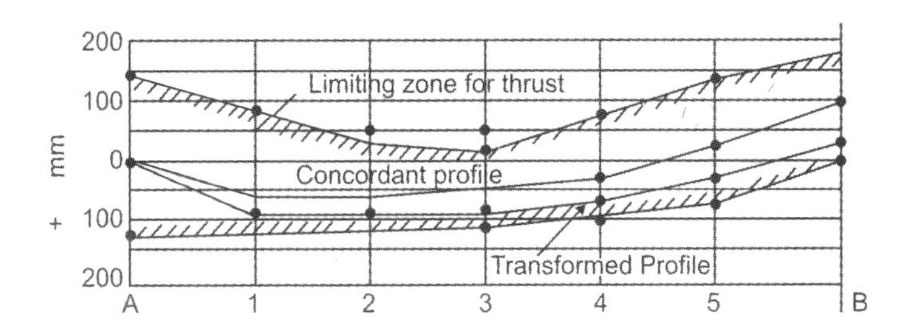

Fig. 10.19

Linear transformation of cable profile

If minimum cover of 100 mm is sufficient shift of 39 mm can be given at point–3, as shown in Fig. 10.20. The shift at B is 78 mm.

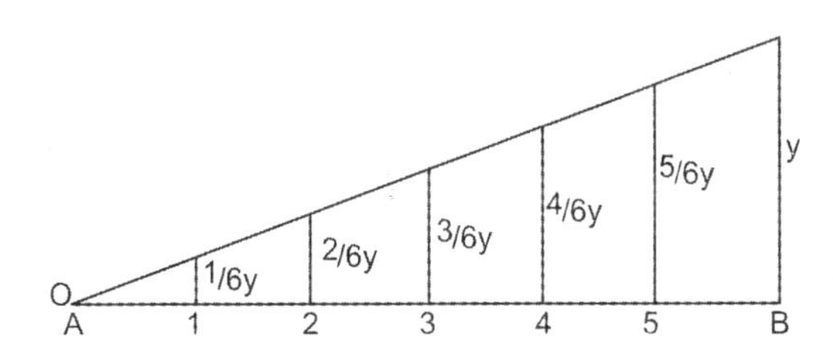

Fig. 10.20

Pont	*A*	1	2	3	4	5	*B*
Shift (mm)	0	+13	+26	+39	+52	+65	+78
Eccentricity (mm)	0	78	89	100	85	+20.5	–22

Check for stresses

$$\left(\frac{P}{A}\right) = 6.30 \text{ N/mm}^2$$

$$Z = 533.33 \times 10^4 \text{ mm}^3$$

$$e = 61 \text{ mm}$$

$$\left(\frac{Pe}{Z}\right) = 5.8 \text{ N/mm}^2$$

At mid span of AB (point–3)

Fiber	Prestress	Maximum load stress	Minimum load stress	Maximum stress	Minimum stress
Top	0.5	+12.1	–0.54	12.6	–0.04
Bottom	12.1	–12.1	+0.54	0	+12.64
At section *B*		$e = 100$ mm	$(Pe/Z) = 9.5\ \text{N/mm}^2$		
Top	15.80	–3.60	–13.14	+12.14	+2.65
Bottom	–3.20	+3.65	+13.14	+0.46	+9.95

Problem 10.10

Design a two-pinned portal frame 7.5 m high with a span of 12 m to support a uniformly distributed live load of 15 kN/m on the transom with stress limits of 14 N/mm^2 compression and zero tension. Assume that the transom and legs have the same section.

Solution

The two pinned portal frame is shown in Fig. 10.21.

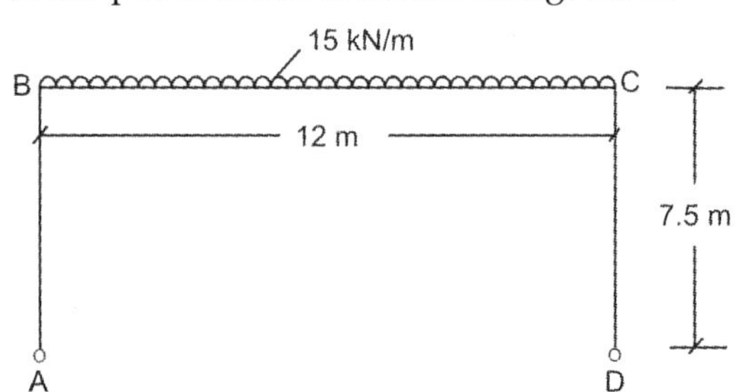

Fig. 10.21

Assuming the transom *BC* and the legs *AB* and *CD* have the same section, the moments are obtained by moment distribution.

Stiffness factors *Relative stiffness*

$AB \quad (3/4)\ (I/7.5) = 6$

$BC \quad (I/12) = 5$

$CD \quad (3/4)\ (I/7.5) = 6$

Distribution factors

At *B*, $d_{BA} = (6/11) \qquad d_{BC} = (5/11)$

At *C*, $d_{CB} = (5/11) \qquad d_{CD} = (6/11)$

Fixed end moments

At *B* and *C*, span *BC*, $M_F = \left(\dfrac{15 \times 12^2}{12}\right) = 180$ kN.m

Simply supported beam bending moment

$$= \left(\frac{15 \times 12^2}{8}\right) = 270 \text{ kN.m}$$

Determination of moments

A	B (6/11)	B (5/11)	C (5/11)	C (6/11)	D
0	0	−180	+180	0	0
	+98.2	+81.8	−81.8	−98.2	
		−40.9	+40.9		
	+22.3	+18.6	−18.6	−22.3	
		−9.3	+9.3		
	+5.07	+4.23	−4.23	−5.07	
		−2.12	+2.12		
	+0.26	+0.22	−0.22	−0.26	
		−0.11	+0.11		
	+0.06	+0.05	−0.05	−0.06	
0	+127	−127	+127	−127	0 kN.m

The bending moment diagram is shown in Fig. 10.22.

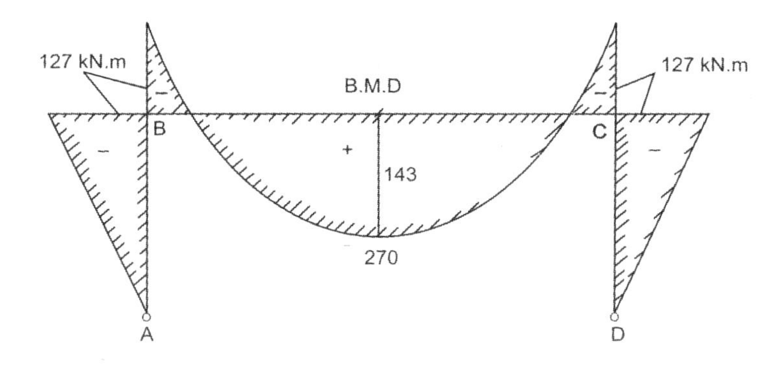

Fig. 10.22

Moment at knee	= −127 kN.m
Transom mid span moment	= +143 kN.m
Vertical reaction in legs	= 90 kN
Horizontal thrust	= 16.95 kN

For the transom:

$$M_L = 143 \text{ kN.m}$$

$$(N_1 - N_2) = 16.95 \text{ kN}$$

Assuming a breadth of $B = 300$ mm

$$M_L = (1/6)\ f_c bh^2 + (1/6)\ (N - N_2)\, h$$

$$143 \times 10^6 = \{(1/6) \times 14 \times 300 \times h^2\} + \{(1/6) \times 16.95 \times h \times 10^3\}$$

Solving $h = 450$ mm

Similarly for the legs $M_L = 127$ kN.m

$$(N_2 - N_1) = -90 \text{ kN}$$

$$\therefore \quad 127 \times 10^6 = \left[\frac{1}{6} \times 14 \times 300 h^2 + \frac{1}{6} \times 90 \times 10^3 \times h\right]$$

$\therefore$ Solving $h = 436$ mm

Adopt a uniform section (rectangular) with $b = 300$ mm
$h = 450$ mm

for both transom and legs.

Dead load moments

Dead load = $(0.3 \times 0.45 \times 24) = 3.24$ kN/m

$$\text{Fixed end moment} = \left(\frac{3.24 \times 12^2}{12}\right) = 38.88 \text{ kN.m}$$

The dead load moments can be obtained by proportion comparing with live load moments.

$$\text{Dead load moment at } B \text{ and } C = \left(\frac{127}{180} \times 38.88\right) = 27.5 \text{ kN.m}$$

The dead load B.M. diagram is shown in Fig. 10.23.

Knee moment = 27.5 kN.m

Transom mis span moment = 30.82 kN.m

Vertical reaction in legs = 19.44 kN

Horizontal thrust = 3.37 kN

The prestressing force required in the transom and legs is obtained

from

$$N_p = \left(\frac{1}{2} f_c bh - N_g - N_1\right)$$

or

$$N_p = \left(\frac{1}{2} f_c bh - N_g - N_2\right)$$

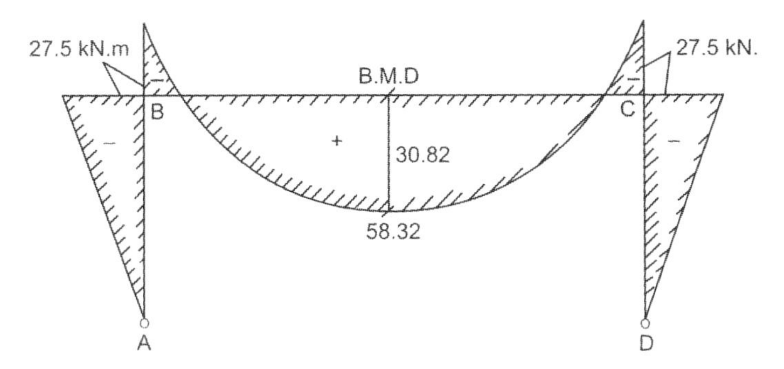

Fig. 10.23

For the transom

$$N_p = \left[\frac{1}{2}\times\frac{14\times300\times450}{1000} - 3.37 - 16.95\right] = 925 \text{ kN}$$

For the legs

$$N_p = \left[\frac{1}{2}\times\frac{14\times300\times450}{1000} - 19.44 - 90\right] = 836 \text{ kN}$$

Determination of the cable zone

a. Transom (under dead load only)

$$\left(\frac{N}{N_p}\right) = \left(\frac{925+3.37}{925}\right) = 1.0036$$

b. Transom (L.L. + D.L.)

$$\left(\frac{N}{N_p}\right) = \left(\frac{925+3.37+16.95}{925}\right) = 1.022$$

c. Top of legs (D.L. only)

$$\left(\frac{N}{N_p}\right) = \left(\frac{836+19.44}{836}\right) = 1.023$$

d. Top of legs

$$\left(\frac{N}{N_p}\right) = \left(\frac{836+19.44+90}{836}\right) = 1.13$$

Determination of limiting zone

1. **Leg foot zone limits**

   Since there are no moments due to the loads at the hinged support

   $$\left(\frac{h}{6}\right) = \left(\frac{450}{6}\right) = 75 \text{ mm}$$

   Distance measured from the centroidal axis is as follows;

   a. Upper limit (towards outside of frame)

   $$= (-75 \times 1.03) = -77 \text{ mm}$$

   b. Lower limit (towards inside of frame)

   $$= (+75 \times 1.13) = 95 \text{ mm}$$

2. **Leg top zone limits**

   a. Upper limit (outer)

   $$\left[\frac{M_2 + M_g}{N_p} - \frac{h}{6}\frac{N}{N_p}\right] = \left[\frac{-27.5 \times 10^3}{836} - 75(1.023)\right]$$

   $$= -110 \text{ mm}$$

   b. Lower (inner limit)

   $$\left[\frac{M_1 + M_g}{N_p} + \frac{h}{6}\frac{N}{N_p}\right] = \left[\frac{-154.6 \times 10^3}{836} + 75(1.13)\right]$$

   $$= -100 \text{ mm}$$

3. **Transom midspan zone limits**

   a. Upper limit

   $$\left[\frac{M_2 + M_g}{N_p} - \frac{h}{6}\frac{N}{N_p}\right] = \left[\frac{173.82 \times 10^3}{925} - 75(1.022)\right]$$

   $$= 111 \text{ mm}$$

   b. Lower limit

   $$\left[\frac{M_1 + M_g}{N_p} + \frac{h}{6}\frac{N}{N_p}\right] = \left[\frac{30.82 \times 10^3}{925} + 75(1.0036)\right]$$

   $$= 109 \text{ mm}$$

4. **Transom end zone limits**

   a. Upper limit

   $$\left[\frac{M_2 + M_g}{N_p} - \frac{h}{6}\frac{N}{N_p}\right] = \left[\frac{-27.5 \times 10^3}{925} - 75 \times 1.0036\right]$$

   $$= -105 \text{ mm}$$

b. Lower limit

$$\left[\frac{M_1 + M_g}{N_p} + \frac{h}{6}\frac{N}{N_p}\right] = \left[\frac{-154.6 \times 10}{925} + 75 \times 1.022\right]$$

$$= -91 \text{ mm}$$

These limits are plotted in the diagram of the portal frame is shown in Fig. 10.24.

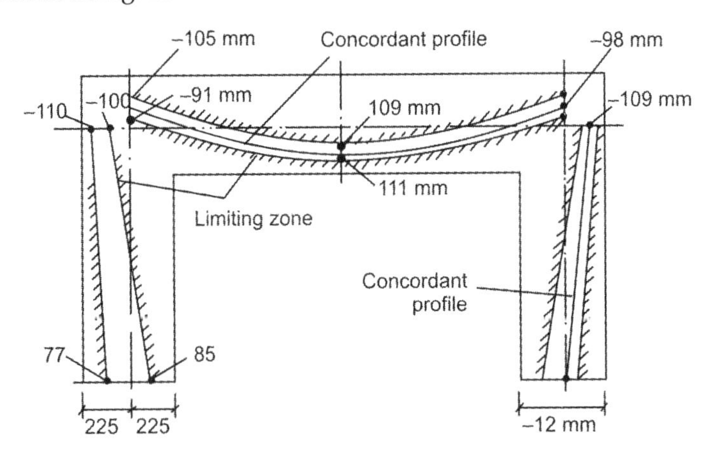

Fig. 10.24

Determination of bending concordant profile

A suitable cable within the limiting zone can be determined by considering the moments developed for the two cases of loading

$$e = \left[-\frac{M_g}{P} + \frac{M_1 + M_2}{2P}\right]$$

At transom ends

$$e = \left[\frac{-27.5 \times 10^3}{925} - \frac{127.1 \times 10^3}{2 \times 925}\right] = -98 \text{ mm}$$

At transom mid span

$$e = \left[\frac{30.82 \times 10^3}{925} + \frac{143 \times 10^3}{2 \times 925}\right] = +111 \text{ mm}$$

Eccentricity at leg top

$$e = \left[\frac{-27.5 \times 10^3}{836} - \frac{127.1 \times 10^3}{2 \times 836}\right] = -109 \text{ mm}$$

Additional eccentricity required for feet seperation is given by

$$e_2 = \left(\frac{-3N_{p1}I_2L}{N_{p2}A_1h^2}\right)$$

Here $A_1 = (300 \times 450)$ $\quad h = 7.5$ m

$$I_2 = \frac{200 \times 450^3}{12} \qquad N_{p1} = 925 \text{ kN}$$

$L = 12$ m

$$\therefore \quad e_1 = \left[\frac{-3 \times 925 \times (1/12) \times 300 \times 450^3 \times 12 \times 10^3}{836 \times (300 \times 450)(7500)^2}\right]$$

$$= -12 \text{ mm}$$

The eccentricities of the concordant profile is shown in Fig. 10.24.

Problem 10.11

A continuous concrete beam *ABCD* having spans *AB* = *CD* = 10 m and *BC* = 15 m, is prestressed by a parabolic cable concentric at supports *A* and *D* and eccentricities of 100 mm towards the soffit at mid spans *AB* and *CD* and 250 mm at mid span of *BC*. The eccentricities at supports *B* and *C* being 200 mm towards the top. The effective force in the cable is 500 kN. If the beam is of rectangular section 500 mm wide by 1000 mm deep and supports a live load of 4 kN/m, determine the pressure line in the beam.

Solution

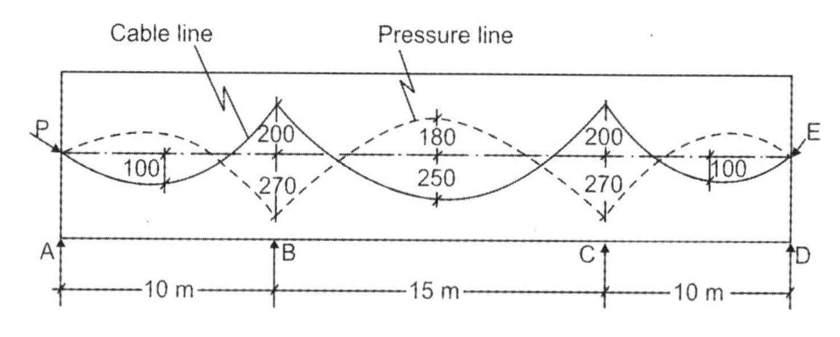

Fig. 10.25

Equivalent loads (Referring to Fig. 10.25)

$$w_1 = \left(\frac{8Pe}{L^2}\right) = \frac{(8 \times 500 \times 0.2)}{10^2} = 8 \text{ kN/m}$$

$$w_2 = \left(\frac{8Pe}{L^2}\right) = \frac{(8 \times 500 \times 0.45)}{15^2} = 8 \text{ kN/m}$$

$$w_3 = \left(\frac{8Pe}{L^2}\right) = \frac{(8 \times 500 \times 0.2)}{10^2} = 8 \text{ kN/m}$$

Self weight of beam = (0.5 × 1 × 24) = 12 kN/m

Live load on the beam = 4 kN/m

Total load on the beam = 16 kN/m

Referring to Fig. 10.26, the fixed end moments are computed as

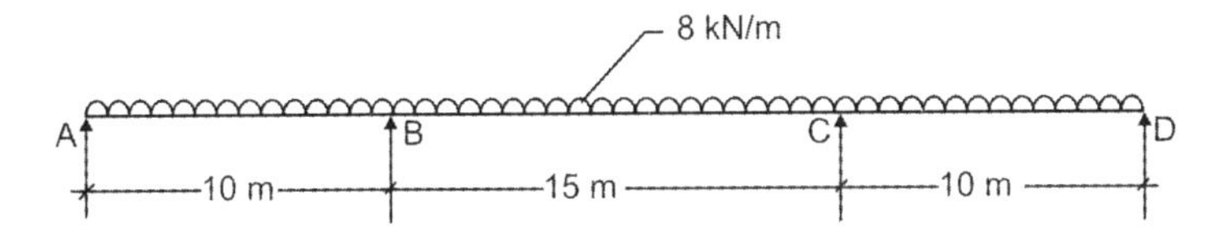

Fig. 10.26

$$M_{FAB} = M_{FCD} = \left(\frac{wL^2}{12}\right) = \frac{(8 \times 10^2)}{12} = 66.66 \text{ kN.m}$$

$$M_{FBC} = \left(\frac{wL^2}{12}\right) = \frac{(8 \times 15^2)}{12} = 150 \text{ kN.m}$$

Distribution factors

$$k_{AB} = k_{CD} = \left(\frac{3}{4}\right)\left(\frac{1}{10}\right) = 0.075$$

$$k_{BC} = k_{CB} = \left(\frac{1}{15}\right) = 0.066$$

$$d_{BA} = d_{CD} = \left[\frac{0.075}{0.075 + 0.066}\right] = 0.53$$

$$d_{BC} = d_{CD} = \left[\frac{0.066}{0.066 + 0.075}\right] = 0.47$$

Moment distribution is carried out to determine the moments at the continuous supports B and C as shown below:

A	B		C		D
	0.53	0.47	0.47	0.53	
–66.66	+66.66	–150	+150	–66.66	+66.66 kN.m
+66.66	33.33			–33.33	–66.66
	+26.50	+23.50	–23.50	–26.50	
		–12.70	+12.70		
	+6.70	+6.00	–6.00	–6.70	
		–3.00	+3.00		
	+1.59	+1.41	–1.41	–1.59	
		–0.70	±0.70		
	+0.37	+0.33	+0.33	–0.37	
0	+135	–135	+135	–135	0 kN.m

Resultant moments

Free bending moment at centre of *AB* and *CD*

$$= \frac{(8 \times 10^2)}{8} = 100 \text{ kN.m}$$

Free bending moment at centre of *BC*

$$= \frac{(8 \times 15^2)}{8} = 225 \text{ kN.m}$$

R.M. at *A* and *D* = 0

R.M. at centre of *AB* and *CD* = (100 – 67.5) = 32.5 kN.m

R.M. at support *B* = 135 kN.m

R.M. at support *C* = 135 kN.m

R.M. at centre of *BC* = (225 – 135) = 90 kN.m

Shift of pressure line $= \left(\frac{\text{R.M.}}{P}\right)$

Shift at *A* = 0

Shift at centre of *AB* and *CD* $= \frac{(32.5 \times 10^6)}{(500 \times 10^3)}$

= 65 mm (above C.G.)

$$\text{Shift at supports } B \text{ and } C = \frac{(-135 \times 10^6)}{(500 \times 10^3)}$$

$$= -270 \text{ mm (below C.G.)}$$

$$\text{Shift at centre } BC = \frac{(90 \times 10^6)}{(500 \times 10^3)}$$

$$= 180 \text{ mm (above C.G.)}$$

The resultant moment diagram is shown in Fig. 10.27 and the pressure line along the length of the beam is shown in Fig. 10.25.

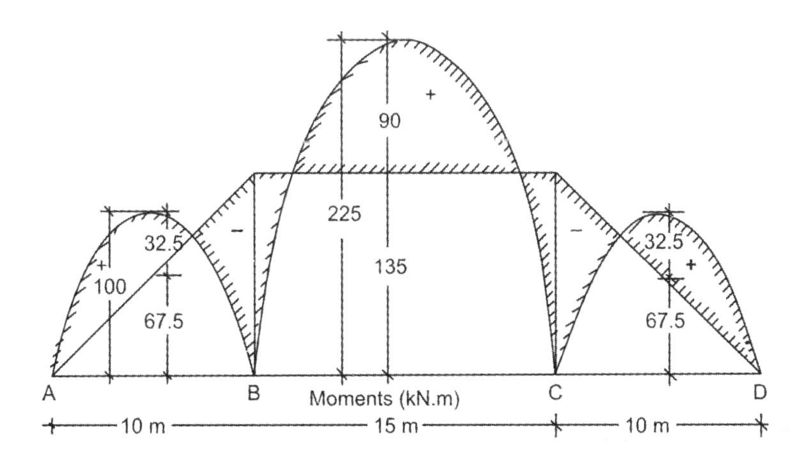

Fig. 10.27

Problem 10.12

A continuous beam ABC ($AB = BC = 20$ m) with an overall depth of 1 m is prestressed by a cable carrying an effective force of 300 kN. The cable profile is parabolic between the supports with zero eccentricity at the end supports A and C. The cable has an eccentricity of 100 mm towards the soffit at centre of spans and 200 mm towards the top at mid support B

a. Calculate the secondary moment at B and show that the cable is concordant.

b. Locate the pressure line when the beam supports a total U.D.L. of 3 kN/m which includes the self weight.

Solution

The two span continous beam is shown in Fig. 10.28.

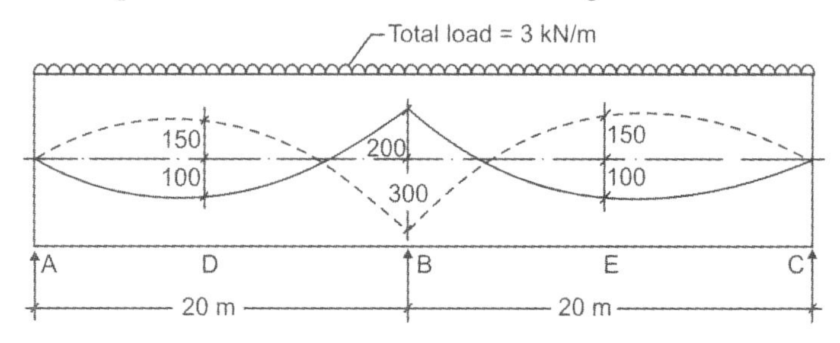

Fig. 10.28

Equivalent U.D.L. $w_e = \left(\frac{8Pe}{L}\right) = \left[\frac{(8 \times 300 \times 0.2)}{20^2}\right]$

$= 1.2$ kN/m

Moment at $B = M_B = \left(\frac{w_e L^2}{8}\right) = \frac{(1.2 \times 20^2)}{8}$

$= 60$ kN/m

S.M. at B = [R.M. – P.M.] = [60 – (300 × 0.2)] = 0

Hence, the cable is concordant.

Total load on the beam = 3 kN/m (acting downwards)

Equivalent U.D.L. = 1.2 kN/m (acting upwards)

Resultant load = [3 – 1.2] = 1.8 kN/m

Moment at B due to this load $= M_B = \frac{(1.8 \times 20^2)}{8} = -90$ kN.m

Moment at centre of AB and $BC = M_d = \frac{(1.8 \times 20^2)}{16} = 45$ kN.m

Shift of pressure line at $B = \left(\frac{M_B}{P}\right) = \frac{-(90 \times 10^6)}{(300 \times 10^3)}$

$= -300$ mm (below C.G.)

Shift of pressure line at $D = \left(\frac{M_D}{P}\right) = \frac{-(45 \times 10^6)}{(300 \times 10^3)}$

$= +150$ mm (above C.G.)

The pressure line distribution along the length of the beam is shown in Fig. 10.28.

Problem 10.13

A three span continuous prestressed concrete beam *ABCD* has a rectangular section 500 mm wide by 1000 mm deep. The span *AB* = *CD* = 10 m and *BC* = 20 m. The eccentricities of the parabolic cable is 50 mm towards the soffit at mid spans of *AB* and *CD* and 300 mm towards the soffit at mid span of *BC*. At supports *B* and *C* the eccentricity towards the top of the beam is 100 mm. The cable is concentric at *A* and *D* and carries at an effective force of 2000 kN. Calculate the equivalent load due to prestressing force and determine the location of the pressure line in the beam due to presstressing force and dead and live loads. Assume live load as 4 kN/m.

Solution

The prestressed continuous beam is shown in Fig. 10.29.

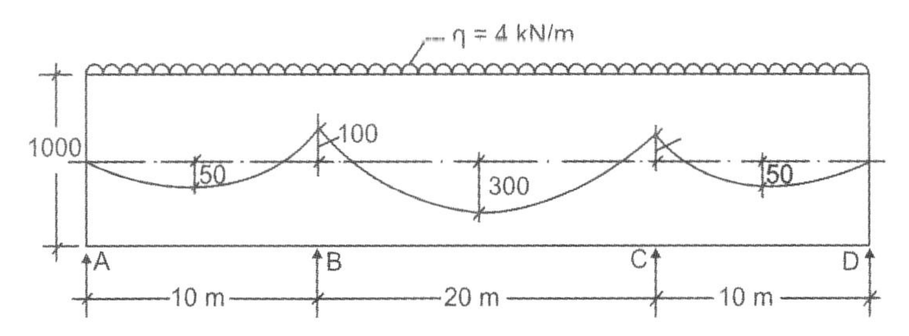

Fig. 10.29

Data:
$$P = 2000 \text{ kN}$$
$$q = 4 \text{ kN/m}$$
$$g = (0.5 \times 1 \times 24) = 12 \text{ kN/m}$$
$$\text{Total load} = (g + q) = (12 + 4) = 16 \text{ kN/m}$$

For spans *AB* and *CD*,

$$\text{equivalent load} = \left(\frac{8Pe}{L^2}\right) = \frac{(8 \times 2000 \times 0.1)}{10^2} = 16 \text{ kN/m}$$

For central span *BC*,

$$\text{equivalent load} = \left(\frac{8Pe}{L^2}\right) = \frac{(8 \times 2000 \times 0.4)}{20^2} = 16 \text{ kN/m}$$

Hence, the equivalent load on the beam acting upwards is computed as 16 kN/m. This upward load will counteract the downward load of 16 kN/m.

Hence, the pressure line coincides with the centre line of the beam and the stress distribution at every section is given by

$$\left(\frac{P}{A}\right) = \frac{(2000 \times 10^3)}{(500 \times 1000)} = 4 \text{ N/mm}^2.$$

Problem 10.14

A concrete beam of rectangular section 500 mm wide by 1000 mm deep is continuous over two spans $AB = BC = 10$ m. The beam is prestressed by a cable carrying an effective force of 2000 kN. The cable is parabolic and concentric at supports A, B and C and has an eccentricity of 100 mm at the centre of spans. Determine the secondary moment developed at mid support section due to prestressing. If the beam supports a live load of 8 kN/m throughout the spans AB and BC, estimate the fibre stresses developed at mid support section due to the combined effect of prestressing, dead and live loads.

Solution

Data: $P = 2000$ kN $\quad q = 8$ kN/m

$g = (0.5 \times 1 \times 24) = 12$ kN/m

Due to prestressing, the equivalent load is computed as

$$w_e = \left(\frac{8Pe}{L^2}\right) = \frac{(8 \times 2000 \times 0.1)}{10^2}$$

$$= 16 \text{ kN/m (upwards)}$$

a. Secondary moment at B due to prestressing

$$\text{S.M. at } B = \left(\frac{w_e L^2}{8}\right)$$

$$= \frac{(16 \times 10^2)}{8} = 200 \text{ kN.m}$$

b. Stress distribution at support section B

Net downward load = $w = (20 - 16) = 4$ kN/m

Resultant moment at B due to this load is given by

$$\left(\frac{wL^2}{8}\right) = \frac{(4 \times 10^2)}{8} = 50 \text{ kN.m (hogging)}$$

$$\text{Section modulus} = Z = \left(\frac{bD^2}{6}\right) = \frac{(500 \times 1000^2)}{6}$$

$$= 83.33 \times 10^6 \text{ mm}^3$$

$$\text{Flexural stress at} \quad B = \left[\frac{P}{A} \pm \frac{M}{Z}\right]$$

$$= \left[\frac{(2000 \times 10^3)}{(500 \times 10^6)} \pm \frac{(50 \times 10^6)}{(83.33 \times 10^6)}\right]$$

$$= (4 \pm 0.6)$$

Stress at top fibre $= 3.4 \text{ N/mm}^2$

Stress at bottom fibre $= 4.6 \text{ N/mm}^2$ (compression)

Problem 10.15

A two span continuous beam of concrete having a rectangular section 200 mm wide by 800 mm deep is prestressed by a parabolic cable in span $AB = 10$ m and linearly varying cable in span $BC = 10$ m. The effective force in the cable is 500 kN. The cable is concentric at supports A, B and C and has an eccentricity of 200 mm at centre of span AB and BC. If the beam supports a uniformly distributed load of 8 kN/m over span AB and a concentrated load of 40 kN at centre of span BC. Compute the equivalent loads on the beam due to prestressing. Determine the resultant stress distribution at centre support B due to prestressing, dead and live loads.

Solution

Data: $P = 500$ kN

Self weight $= g = (0.2 \times 0.8 \times 24) = 3.84$ kN.m

$e = 200$ mm $\qquad L = 10$ m

$$M_g = \left(\frac{gL^2}{8}\right) = \left(\frac{3.84 \times 10^2}{8}\right) = 48 \text{ kN.m}$$

Equivalent load on span AB is computed as

$$w_e = \frac{(8Pe)}{L^2} = \frac{(8 \times 500 \times 0.2)}{10^2} = 8 \text{ kN/m}$$

Equivalent load on span BC is computed as

$$W = 2P \sin\theta = 2P \tan\theta$$

$$= \frac{(2 \times 500 \times 0.2)}{5} = 40 \text{ kN}$$

Since, the live loads on the beam is counteracted by the equivalent loads on both spans acting upwards, there will be no moments developed. Hence, the only moments developed are due to the dead load of the beam given by $M_B = M_q = 48$ kN.m

$$Z = \text{Section modulus} = \left(\frac{bd^2}{6}\right) = \frac{(200 \times 800^2)}{6}$$

$$= 21.33 \times 10^6 \text{ mm}^3$$

Resultant stressess at support section B

$$\text{Stress at top} = \left[\frac{P}{A} - \frac{M_B}{Z}\right] = \left[\left(\frac{500 \times 10^3}{200 \times 800}\right) - \left(\frac{48 \times 10^6}{21.33 \times 10^6}\right)\right]$$

$$= 0.875 \text{ N/mm}^2 \text{ (compression)}$$

$$\text{Stress at soffit} = \left[\frac{P}{A} + \frac{M_B}{Z}\right] = \left[\left(\frac{200 \times 10^3}{200 \times 800}\right) + \left(\frac{48 \times 10^6}{21.33 \times 10^3}\right)\right]$$

$$= 5.375 \text{ N/mm}^2 \text{ (compression)}$$

Problem 10.16

A continuous beam of two equal spans of 30 m each has a rectangular section, 500 mm wide by 1000 mm deep throughout the spans. The beam is prestressed by a concordant cable having high tensile strands of cross-sectional area 3000 mm^2, 100 mm from the top of the beam at mid support section. If the beam supports an uniformly distributed service load of 8 kN/m throughout the span lengths, estimate the load factor against failure.

Given $f_{pu} = 1700$ N/mm^2, $f_{ck} = 50$ N/mm^2 and density of concrete as 24 kN/m^3, estimate the load factor against failure, assuming

a. Elastic distribution of moments

b. Complete redistribution of moments.

Solution

Continuous beam of two equal spans of 30 m each

Cross-section of the beam:

b = 500 mm and D = 1000 mm

Effective depth d = 900 mm with cover of 100 mm

A_p = 3000 mm^2

f_p = 1700 mm^2

f_{ck} = 50 N/mm^2

Dead weight of the beam

g = (0.5 × 1 × 24) = 12 kN/m

Live load on the beam $q = 8$ kN/m
Span length $L = 30$ mm

Referring to Table 4.1, compute the effective reinforcement ratio given by

$$\left[\frac{f_p A_p}{f_{ck} b d}\right] = \left[\frac{1700 \times 3000}{50 \times 500 \times 900}\right] = 0.22$$

Interpolating from the table, we have the ratio

$$\left[\frac{f_{pu}}{0.87 f_p}\right] = 0.93 \text{ and } \left(\frac{x_u}{d}\right) = 0.44$$

$$\therefore \quad M_u = f_{pu} A_p (d - 0.42\, x_u)$$
$$= (0.93 \times 0.87 \times 1700)\,[900 - (0.42 \times 0.44 \times 900)]$$
$$= (3027 \times 10^6) \text{ N.mm} = 3027 \text{ kN.m}$$

a. Assuming elastic distribution of moments

$$M_u = 0.125\,(g + q_u)\,L^2$$
$$3027 = 0.125\,(12 + q_u) \times 30^2$$

Solving, $q_u = 14.9$ kN/m

$$\text{Load factor} = \left[\frac{14.9}{6}\right] = 2.48$$

b. Assuming complete redistribution of moments

$$q_u = \left[\frac{12\, M_u}{L^2} - g\right] = \left[\frac{(12 \times 3027)}{30^2} - 12\right]$$
$$= 28.36 \text{ kN/m}$$

$$\text{Load factor against collapse} = \left(\frac{28.36}{8}\right) = 3.54.$$

Problem 10.17

A two span continuous beam ABC ($AB = BC = 10$ m) is of rectangular section 200 mm wide by 500 mm deep. The beam is prestressed by a parabolic cable, concentric at end supports and having an eccentricity of 100 mm towards the soffit of the beam at centre of spans and 200 mm towards the top at mid supports. The effective force in the cable is 500 kN.

a. Show that the cable is concordant.
b. Locate the pressure line in the beam when it supports a live load of 5.6 kN/m in addition to its self weight.

Solution

Cross-section of beam (rectangular):

$$b = 200 \text{ mm and } D = 500 \text{ mm}$$

Eccentricity of cable:

At end supports,

$$e = 0$$

At mid support,

$$e = 200 \text{ mm towards top fibre}$$

At centre of spans,

$$e = 100 \text{ mm towards the soffit}$$

Prestressing force = $P = 500$ kN

Span $AB = BC = L = 10$ m

By joining a line between the cables at end support and centre of span, the effective eccentricity at centre of span is obtained as $e = (100 + 100) = 200$ mm.

Using the tendon reaction method, the equivalent uniformly distributed load acting upwards is given by

$$w_e = \left[\frac{8\ Pe}{L^2}\right] = \left[\frac{8 \times 500 \times 0.2}{10^2}\right] = 8 \text{ kN/m}$$

Dead weight of the beam

$$g = (0.2 \times 0.5 \times 24) = 2.4 \text{ kN/m}$$

Live load on the beam $q = 5.6$ kN/m

Total load on the beam

$$(g + q) = 8.0 \text{ kN/m (acting downwards)}$$

Considering the two span continuous beam carrying an imposed load of 8 kN/m due to prestressing force, the resultant moment at mid support B is calculated as,

$$\text{R.M. at } B = 0.125\ w_e\ L^2 = (0.125 \times 8 \times 10^2) = 100 \text{ kN.m}$$

Primary moment (P.M) at $B = P.e = (500 \times 0.2) = 100$ kN.m

Hence, secondary moment at B is = [R.M. – P.M.] = [100 – 100] = 0

Hence, the cable is concordant.

The total downward load due to dead weight and live load are counterbalanced by the upward load developed due to the prestressing force. Hence, bending stresses are zero along the beam. However, direct stress due to prestressing force is computed as,

$$\text{Direct stress} = \left(\frac{P}{A}\right) = \left(\frac{500 \times 10^3}{200 \times 500}\right) = 5 \text{ N/mm}^2;$$

Pressure line coincides with centroidal axis.

Problem 10.18

A post-tensioned prestressed concrete continuous beam of two equal spans $AB = BC = 10$ m is prestressed by a continuous cable having a parabolic profile between the supports. The cable is concentric at all the three supports and has an eccentricity of 200 mm towards the soffit at the centre of spans. The beam is of rectangular section 500 mm wide by 1000 mm deep. The prestressing force in the cable is 500 kN.

a. Determine the position of thrust line in the beam due to prestressing at B.
b. If the cable is linearly transformed with a vertical shift of 200 mm towards the top at B, show that the position of thrust line is unchanged.

Solution:

Two span continuous beam ABC:

$$AB = BC = L = 10 \text{ m}$$

Prestressing force $P = 500$ kN

Eccentricity at supports $e = 0$

Eccentricity at mid spans $e = 200$ mm

Equivalent uniformly distributed load due to prestressing the parabolic cable is calculated as,

$$w_e = \left[\frac{8Pe}{L^2}\right] = \left[\frac{8 \times 500 \times 0.2}{10^2}\right]$$

$$= 8 \text{ kN/m (acting upwards)}$$

Resultant moment at mid support B

$$= M = \left[\frac{w_e L^2}{8}\right] = \left[\frac{8 \times 10^2}{8}\right] = 100 \text{ kN.m}$$

a. Shift of thrust line at $B = \left[\frac{M}{P}\right] = \left[\frac{100}{500}\right] = 0.2 \text{ m} = 200 \text{ mm}$

b. If the cable is linearly transformed with a vertical shift of 200 mm at B towards the top, the effective eccentricity of the parabolic cable at centre of spans will remain unchanged at 200 mm. Hence, the equivalent load also remains at 8 kN/m. Consequently, the position of thrust line at B will be the same as in case (a). Hence, the position of thrust line is unchanged due to linear transformation.

Problem 10.19

A continuous beam of two equal spans *ABC* ($AB = BC = 20$ m) with an overall depth of 1 m is prestressed by a continuous cable carrying an effective force of 500 kN. The cable profile is parabolic between the supports and it is concentric at the two end supports. The eccentricity at mid span sections is 200 mm towards the soffit and 400 mm towards the top fibre at mid support section. Evaluate the reactions developed at the supports due to prestress only and show that the cable is concordant.

Solution:

Prestressing force $P = 500$ kN

Eccentricity at end supports $= 0$

Eccentricity at mid support $e = 400$ mm

Eccentricity at mid spans $(0.5\,e) = 200$ mm

Span of the beam between supports $L = 20$ m

Using the tendon reaction method, the equivalent uniformly distributed load acting upwards is calculated using the relation,

$$w_e = \left[\frac{8Pe}{L^2}\right] = \left[\frac{8 \times 500 \times 0.4}{20^2}\right] = 4 \text{ kN/m}$$

Resultant moment at mid support *B*

$$\text{R.M} = \left[\frac{w_e L^2}{8}\right] = \left[\frac{4 \times 20^2}{8}\right] = 200 \text{ kN.m}$$

Secondary moment at *B*

$$\text{S.M} = [\text{R.M} - \text{P.M}] = [200 - (500 \times 0.4)] = 0$$

Hence, the cable is concordant.

Also, the slope of cable at support is given by

$$\theta = \left[\frac{4e}{L}\right] = \left[\frac{4 \times 0.4}{20}\right] = 0.08$$

Vertical component of prestressing force at end supports is given by

$$P \sin\theta = P\,\theta = [500 \times 0.08] = 40 \text{ kN (acting downwards)}$$

Vertical component at mid support $B = 2\,P\,\theta = 80$ kN

Reactions developed due to U.D.L of 4 kN/m acting upwards is 40 kN at *A* and *C* and 80 kN at *B*.

Hence, $R_A = R_B = R_C = 0$.

The cable is concordant since there are no redundant reactions.

Problem 10.20

A two span continuous beam *ABC* (*AB* = *BC* = 10 m) is prestressed by a parabolic cable which is concentric at all the support sections and having an eccentricity of 200 mm towards the soffit at mid spans. The effective force in the cable is 500 kN.

a. Calculate the secondary moment developed at mid support B.

b. Locate the pressure line in the beam when it supports a total (dead + live) load of 8 kN/m.

Solution

Span lengths;

$$AB = BC = L = 10 \text{ m}$$

Eccentricity of cable at supports $= 0$

Eccentricity of cable at mid spans $e = 200$ mm

Effective force in the cable $P = 500$ kN

Equivalent U.D.L on each of the spans

$$w_e = \left[\frac{8Pe}{L^2}\right] = \left[\frac{8 \times 500 \times 0.2}{10^2}\right] = 8 \text{ kN/m}$$

Moment due to equivalent load at *B*

$$M_B = \left[\frac{w_e L^2}{8}\right] = \left[\frac{8 \times 10^2}{8}\right] = 100 \text{ kN.m}$$

Secondary moment at *B*

$$\text{S.M} = [\text{R.M} - \text{P.M}]$$
$$= [100 - (500 \times 0)] = 100 \text{ kN.m}$$

Total load on the beam

= 8 kN/m (acting downwards)

Equivalent U.D.L = 8 kN/m (acting upwards)

Resultant load = (8 – 8) = 0

Hence, the pressure line coincides with the cable line along the length of the beam.

11

Prestressed Concrete Pipes and Tanks

Problem 11.1

Design a non-cylinder prestressed concrete pipe of internal diameter 500 mm to withstand a working pressure of 1 N/mm^2. High tensile wires of 2 mm diameter stressed to 1200 N/mm^2 at transfer are available for use. Permissible maximum stresses in concrete at transfer and working load are 13.5 and 0.8 N/mm^2 (compression) respectively. Loss ratio = 0.8, E_s = 210 kN/mm^2 and E_c = 35 kN/mm^2. Calculate (a) the minimum thickness of concrete for the pipe, (b) number of turns of wire per metre length of the pipe, (c) the test pressure required to produce a tensile stress of 0.7 N/mm^2 in the concrete when applied immediately after tensioning, and (d) the winding stress in the steel.

Solution

The cross section of the concrete pipe is shown in Fig. 11.1.

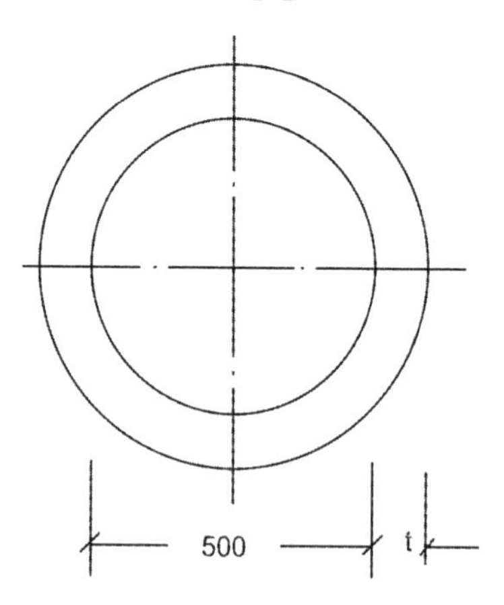

Fig. 11.1

a.
$$t > \left(\frac{N_d}{\eta f_{ct} - f_{\min w}}\right)$$
$$= \frac{1(500/2)}{(0.8 \times 13.5 - 0.8)} = 25 \text{ mm}$$

Number of turns of 2 mm wire stressed to 1200 N/mm^2 is given by

b.
$$n = \left(\frac{400\, t\, f_c}{d^2\, f_s}\right) = \left(\frac{4000 \times 25 \times 13.5}{2^2 \times 1200}\right) = 90 \text{ turns}$$

c. w_w = Test pressure required immediately after winding

($\eta = 1$)

$$f_c = \left[\frac{w_w D}{2t} + \frac{f_{\min w}}{\eta}\right]$$

$$\therefore \quad w_w = \frac{2t}{D}[f_c - f_{\min w}] = \frac{2 \times 25}{500}[13.5 - (-0.7)]$$
$$= 1.42 \text{ N/mm}^2$$

d. Winding stress in steel

f_{si} = winding stress in steel

$f_{si} = (1 + \alpha_e \cdot \rho) f_{se}$

$$x_e = 6 \text{ and } \quad \rho = \left(\frac{f_c}{f_s}\right) = \left(\frac{13.5}{1200}\right) = 0.01125$$

$$\therefore \quad f_{si} = (1 + 6 \times 0.01125)\ 1200 = 1281 \text{ N/mm}^2$$

Problem 11.2

A prestressed concrete cylinder pipe is formed by lining a steel cylinder of diameter 750 mm and thickness 2.5 mm with a layer of spun concrete 38 mm thick. If the pipe is required to withstand a hydraulic pressure of 0.85 N/mm^2, without developing any tensile stress in concrete, calculate (a) the required pitch of 4 mm wires, wound round the cylinder at a tensile stress of 980 N/mm^2, (b) the test pressure to produce a tensile stress of 1.4 N/mm^2, in the concrete immediately after winding, and (c) the approximate bursting pressure. Modular ratio = 6, tensile strength of wire = 1680 N/mm^2, yield stress of cylinder = 280 N/mm^2, loss ratio = 0.85.

Solution

$$t > \left(\frac{N_d}{\eta f_{ct} - f_{\min w}} - \alpha_e \cdot t_s\right)$$

$$> \left(\frac{0.85(750/2)}{0.85 \times 14 - 0} - 6 \times 2.5\right)$$

> 11.78 mm (provided thickness = 38 mm).

Hence, safe.

$$f_c = \left[\frac{N_d}{\eta(t + \alpha_e t_s)} + \frac{f_{\min w}}{\eta}\right]$$

$$= \left(\frac{0.85(750/2)}{0.85\,(38 + 6 \times 2.5)} + 0\right)$$

$= 7.07$ N/mm^2

$\therefore$ Number of turns of wire winding per metre length is

$$n = \left[\frac{4000(t + \alpha_e t_s) f_c}{d^2 \cdot f_s}\right]$$

$$= \left[\frac{4000(38 + 6 \times 2.5)7.07}{4^2 \times 980}\right]$$

$= 96$ turns/metre

a. $$\text{pitch} = \left(\frac{1000}{96}\right) = 10.4 \text{ mm}$$

Test pressure

b. $$w_w = \frac{2(t + \alpha_e t_s)}{D}[f_c - f_{\min w}]$$

$$= \frac{2(38 + 6 \times 2.5)}{750}[7.07 - (-1.4)]$$

$= 1.2$ N/mm^2

c. *Bursting pressure*

$$p_u = \left[\frac{0.00157\, d^2\, n\, f_{pu} + 2 t_s f_y}{D}\right]$$

$$= \frac{(0.00157 \times 4^2 \times 96 \times 1680) + 2 \times 2.5 \times 280}{750}$$

$= 7.2$ N/mm^2

Problem 11.3

A prestressed concrete pipe of 1.2 m diameter, having a core thickness of 75 mm is required to withstand a service pressure intensity of 1.2 N/mm^2. Estimate the pitch of 5 mm diameter hightensile wire winding if the initial stress is limited to 1000 N/mm^2. Permissible stresses in concrete being 12.5 N/mm^2 in compression and zero in tension. The loss ratio is 0.8, if the direct tensile strength of concrete is 2.5 N/mm^2, estimate the load factor against cracking.

Solution

Given:

$$D = 1.2\text{ m} \qquad d = 5\text{ mm}$$

$$t = 75\text{ mm} \qquad f_s = 1000\text{ N/mm}^2$$

$$p = 1.2\text{ N/mm}^2 \qquad n = \left(\frac{4000\,t\,f_c}{\pi d^2 \cdot f_s}\right)$$

$$f_c = \left[\frac{N_d}{\eta'} + \frac{f_{\min w}}{\eta}\right] = \left[\frac{1.2(1200/2)}{0.8 \times 75} + 0\right]$$

$$= 12\text{ N/mm}^2$$

$$\therefore \quad n = \left(\frac{4000 \times 75 \times 12}{\pi \times 5^2 \times 1000}\right) = 45.8\text{ turns/m}$$

$$\therefore \quad \text{pitch} = \left(\frac{1000}{45.8}\right) = 21.8\text{ mm}$$

Direct tensile strength of concrete = 2.5 N/mm^2

Hoop tension due to fluid pressure

$$= \left(\frac{pD}{2t}\right) = \left(\frac{1.2 \times 1200}{2 \times 75}\right) = 9.6\text{ N/mm}^2$$

Hoop compression due to prestress = 12 N/mm^2

$\therefore$ Resultant compressive stress in concrete

$$= (12 - 9.6) = 2.4\text{ N/mm}^2$$

Tensile strength of concrete = 2.5 N/mm^2

Additional fluid pressure required to develop a tensile stress of $(2.4 + 2.5) = 4.9$ N/mm^2 is given by

$$\left(\frac{2 \times 75 \times 49}{1200}\right) = 0.6125\text{ N/mm}^2$$

$\therefore$ Cracking fluid pressure = (1.2 + 0.6125) = 1.8125 N/mm^2

$\therefore$ Load factor against cracking = (0.8125/1.2) = 1.5

Problem 11.4

Design a cylindrical prestressed concrete water tank to suit the following data:

Capacity of the tank = 3.5×10^9 litres. Ratio of diameter to height = 4, maximum compressive stress in concrete at transfer not to exceed 14 N/mm^2 (compression). Minimum compressive stress under working load to be 1 N/mm^2. The prestress is to be provided by circumferential winding of 5 mm wires and by vertical cables of 12 wires of 7 mm diameter. The stress in wires at transfer = 1000 N/mm^2. Loss ratio = 0.75. Design the walls of the tank and details of circumferential wire winding and vertical cables for the following joint conditions at the base: (a) hinged, (b) fixed and (c) sliding base (Assume coefficient of friction as 0.5).

Solution

The salient dimensions of the tank is shown in Fig. 11.2.

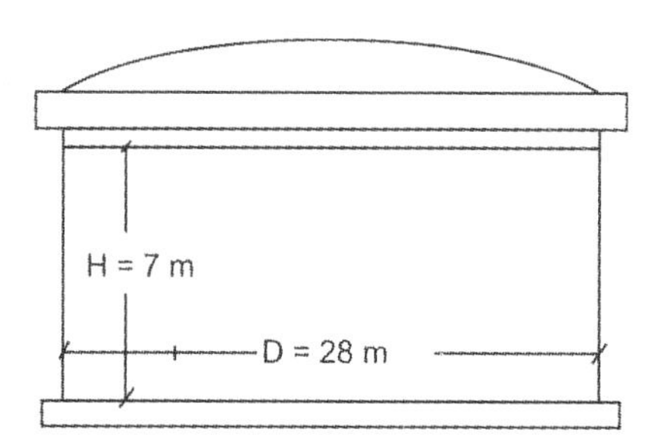

Fig. 11.2

a. *Hinged base*

Capacity of tank = 3.5×10^6 litres

Ratio of diameter to height $(D/H) = 4$

$$\left(\frac{\pi D^2}{4} \cdot H\right) = \left(\frac{3.5 \times 10^6 \times 10^3}{10^6}\right), \quad \text{But } H = (D/4)$$

Solving, $D = 28$ m $\qquad H = 7$ m

$t = 150$ mm

$$\left(\frac{H^2}{Dt}\right) = \left(\frac{7^2}{28 \times 0.15}\right) = 11.66$$

$$w_w = w \cdot H = (10 \times 7)\ \text{kN/m}^2 = 0.07\ \text{N/mm}^2$$

Table 11.1: Ring tension in cylindrical walls—hinged base free top (IS: 3370—Part IV); Ring tension N_d = (coefficient) × (wHR) kN/m; positive sign indicates tension

$\frac{H^2}{Dt}$	Coefficients at point									
	0.0H	0.1H	0.2H	0.3H	0.4H	0.5H	0.6H	0.7H	0.8H	0.9H
0.4	+0.474	+0.440	+0.395	+0.352	+0.305	+0.264	+0.215	+0.165	+0.111	+0.057
0.8	+0.423	+0.402	+0.381	+0.358	+0.330	+0.297	+0.249	+0.202	+0.145	+0.076
1.2	+0.350	+0.355	+0.361	+0.362	+0.358	+0.343	+0.309	+0.256	+0.186	+0.098
1.6	+0.271	+0.303	+0.341	+0.369	+0.385	+0.385	+0.362	+0.314	+0.233	+0.124
2.0	+0.205	+0.260	+0.321	+0.373	+0.411	+0.434	+0.419	+0.369	+0.280	+0.151
3.0	+0.074	+0.179	+0.281	+0.375	+0.449	+0.506	+0.519	+0.479	+0.375	+0.210
4.0	+0.017	+0.137	+0.253	+0.367	+0.469	+0.545	+0.579	+0.553	+0.447	+0.256
5.0	–0.008	+0.114	+0.235	+0.356	+0.469	+0.562	+0.617	+0.606	+0.503	+0.294
6.0	–0.011	+0.103	+0.223	+0.343	+0.463	+0.566	+0.639	+0.643	+0.547	+0.327
8.0	–0.015	+0.096	+0.208	+0.324	+0.443	+0.564	+0.661	+0.627	+0.621	+0.386
10.0	–0.008	+0.095	+0.200	+0.311	+0.428	+0.552	+0.666	+0.730	+0.678	+0.433
12.0	–0.002	+0.097	+0.197	+0.302	+0.417	+0.541	+0.664	+0.750	+0.720	+0.477
14.0	0.000	+0.098	+0.197	+0.299	+0.408	+0.531	+0.659	+0.761	+0.752	+0.531
16.0	+0.002	+0.100	+0.198	+0.299	+0.403	+0.521	+0.650	+0.764	+0.776	+0.543
	Coefficients at point									
	.75H	.80H	.85H	.90H	.95H					
20.0	+0.812	+0.817	+0.755	+0.603	+0.344					
24.0	+0.816	+0.339	+0.793	+0.647	+0.377					
32.0	+0.814	+0.861	+0.847	+0.721	+0.436					
40.0	+0.802	+0.866	+0.880	+0.778	+0.483					
48.0	+0.791	+0.854	+0.900	+0.820	+0.527					
56.0	+0.781	+0.859	+0.911	+0.852	+0.563					

Table 11.2: Moments in cylindrical walls—hinged base free at top (IS: 3370—Part IV); Moment M_w = (coefficient) (wH^3) kN/m; positive sign indicates tension at the outside

$\frac{H^2}{Dt}$	Coefficients at point									
	0.1*H*	0.2*H*	0.3*H*	0.4*H*	0.5*H*	0.6*H*	0.7*H*	0.8*H*	0.9*H*	10*H*
0.4	+.0020	+.0072	+.0151	+.0230	+.0301	+.0348	+.0357	+.0312	+.0197	0
0.8	+.0019	+.0064	+.0133	+.0207	+.0271	+.0319	+.0329	+.0292	+.0187	0
1.2	+.0016	+.0058	+.0111	+.0177	+.0237	+.0280	+.0296	+.0263	+.0171	0
1.6	+.0012	+.0044	+.0091	+.0145	+.0195	+.0236	+.0255	+.0232	+.0155	0
2.0	+.0006	+.0033	+.0073	+.0114	+.0158	+.0199	+.0219	+.0205	+.0145	0
3.0	+.0004	+.0018	+.0040	+.0063	+.0092	+.0127	+.0152	+.0153	+.0111	0
4.0	+.0001	+.0007	+.0016	+.0033	+.0057	+.0083	+.0109	+.0118	+.0092	0
5.0	.0000	+.0001	+.0006	+.0016	+.0034	+.0057	+.0080	+.0094	+.0078	0
6.0	.0000	.0000	+.0002	.0000	.0019	+.0039	+.0062	+.0078	+.0068	0
8.0	.0000	.0000	–.0002	.0000	+.0007	+.0020	+.0038	+.0057	+.0054	0
10.0	.0000	.0000	–.0002	–.0001	+.0002	+.0011	+.0025	+.0043	+.0045	0
12.0	.0000	.0000	–.0001	.0002	.0000	+.0005	+.0017	+.0032	+.0039	0
14.0	.0000	.0000	.0000	–.0001	.0001	.0000	+.0012	+.0026	+.0033	0
16.0	.0000	.0000	.0000	–.0001	–.0002	–.0004	+.0008	+.0022	+.0029	0

	Coefficients at point				
	.75*H*	.80*H*	.85*H*	.90*H*	.95*H*
20.0	+.0008	+.0014	+.0020	+.0024	+.0020
24.0	+.0005	+.0010	+.0015	+.0020	+.0017
32.0	.0000	+.0005	+.0009	+.0014	+.0013
40.0	.0000	+.0003	+.0006	+.0011	+.0011
48.0	.0000	+.0001	+.0004	+.0008	+.0010
56.0	.0000	.0000	+.0003	+.0007	+.0008

From Tables 11.1 and 11.2

The maximum ring tension and moments in tank walls are obtained as

$$N_d = (0.74 \times 10 \times 7 \times 14) = 725 \text{ kN/m} = 725 \text{ N/mm}$$

$$M_w = (0.0042 \times 10 \times 7^3) = 14.4 \text{ kN.m/m} = 14400 \text{ N.mm/mm}$$

Thickness of wall

$$t = \left[\frac{725}{(0.75 \times 14 - 1)}\right] = 76.3 \text{ mm}$$

Thickness adopted for practical considerations of housing cables of 30 mm ϕ = 150 mm

Net thickness available = (150 – 30) = 120 mm

Circumferential prestress

$$f_c = \left[\frac{725}{0.75 \times 120} + \frac{1}{0.75}\right] = 9.35 \text{ N/mm}^2$$

Spacing of 5 mm ϕ wires

$$s = \frac{2 \times 725}{0.07}\left(\frac{1000 \times 20}{9.35 \times 28 \times 10^3 \times 120}\right) = 13.18 \text{ mm}$$

No. of wires/metre = ((1000/13.18) = 76

Similarly, the number of wires required at the top of the tank = 16

Radial pressure due to prestress is

$$w_t = \left(\frac{2 \times 1000 \times 20}{13.18 \times 28 \times 10^3}\right) = 0.108 \text{ N/mm}^2$$

Max. vertical moment due to prestress

$$M_t = 14400\left(\frac{0.108}{0.07}\right) = 22217 \text{ N.mm/mm}$$

$$= 22.217 \times 10^6 \text{ N.mm/m}$$

Considering one metre length. The section modulus

$$Z = \left(\frac{1000 \times 150^2}{6}\right) = 375 \times 10^4 \text{ mm}^3$$

Vertical prestress required is

$$f_c = \left[\frac{1}{0.75} + \frac{22.217 \times 10^6}{375 \times 10^4}\right] = 5.92 \text{ N/mm}^2$$

Vertical prestressing force

$$= \left(\frac{5.92 \times 1000 \times 150}{1000}\right) = 888 \text{ kN}$$

Spacing of 12 wires of 7 mm ϕ cables

$$= \left(\frac{1000 \times 12 \times 38.4}{888}\right) = 518 \text{ mm}$$

b. *Fixed base*

The tank dimensions are those assumed in (a)

$D = 28$ m $\qquad H = 7$ m

$t = 150$ mm $= 0.75$

$$\left(\frac{H^2}{Dt}\right) = 11.66$$

$w_w = 0.07 \text{ N/mm}^2$

Referring to tables 11.3 and 11.4

$N_d = (0.61 \times 10 \times 7 \times 14) = 598$ kN/m

$= 598$ N/mm

$M_w = (0.011 \times 10 \times 7^3) = 37.8$ kN.m/m

$= 37800$ N.mm/mm

Minimum wall thickness obtained from the relation will be less. Hence, provide a minimum thickness of 150 mm. Using cable diameter = 30 mm net thickness of wall = (150 – 30) = 120 mm.

Required circumferential prestress is

$$f_c = \left[\frac{N_d}{\eta^t} + \frac{f_{\min w}}{\eta}\right] = \left[\frac{598}{0.75 \times 120} + \frac{1}{0.75}\right]$$

$= 9.35 \text{ N/mm}^2$

Spacing of 5 mm ϕ wire winding

$$S = \left(\frac{2N_d}{w_w} \cdot \frac{f_s A_s}{f_c Dt}\right)$$

$$= \left(\frac{2 \times 725}{0.07} \times \frac{1000 \times 20}{9.35 \times 28 \times 10^3 \times 120}\right)$$

$= 13.18$ mm

$\therefore$ No. of wires/metre = (1000/13.18) = 76

Table 11.3: Ring tension in cylindrical walls—Fixed base free top (IS: 3370—Part IV); Ring tension N_d = (coefficient) × (wHR) kN/m

$\frac{H^2}{Dt}$	Coefficients at point									
	0.0*H*	0.1*H*	0.2*H*	0.3*H*	0.4*H*	0.5*H*	0.6*H*	0.7*H*	0.8*H*	0.9*H*
0.4	+0.149	+0.134	+0.120	+0.101	+0.082	+0.066	+0.049	+0.029	+0.014	+0.004
0.8	+0.263	+0.239	+0.215	+0.190	+0.160	+0.130	+0.096	+0.063	+0.034	+0.010
1.2	+0.283	+0.271	+0.254	+0.234	+0.209	+0.180	+0.142	+0.099	+0.054	+0.016
1.6	+0.265	+0.268	+0.268	+0.266	+0.250	+0.226	+0.185	+0.134	+0.075	+0.023
2.0	+0.234	+0.251	+0.273	+0.285	+0.285	+0.274	+0.232	+0.172	+0.104	+0.031
3.0	+0.134	+0.203	+0.267	+0.322	+0.357	+0.362	+0.330	+0.262	+0.157	+0.052
4.0	+0.067	+0.164	+0.256	+0.339	+0.403	+0.429	+0.409	+0.334	+0.210	+0.073
5.0	+0.025	+0.137	+0.245	+0.346	+0.428	+0.477	+0.469	+0.398	+0.59	+0.092
6.0	+0.018	+0.119	+0.234	+0.344	+0.441	+0.504	+0.514	+0.447	+0.301	+0.112
8.0	–0.011	+0.104	+0.218	+0.335	+0.443	+0.534	+0.575	+0.530	+0.381	+0.151
10.0	–0.011	+0.098	+0.208	+0.323	+0.437	+0.542	+0.608	+0.589	+0.440	+0.179
12.0	–0.005	+0.097	+0.202	+0.312	+0.429	+0.543	+0.628	+0.633	+0.494	+0.211
14.0	–0.002	+0.098	+0.200	+0.306	+0.420	+0.539	+0.639	+0.666	+0.541	+0.241
16.0	–0.000	+0.099	+0.199	+0.304	+0.412	+0.531	+0.641	+0.687	+0.582	+0.265
	Coefficients at point									
	.75*H*	.80*H*	.85*H*	.90*H*	.95*H*					
20.0	+0.716	+0.654	+0.520	+0.325	+0.115					
24.0	+0.746	+0.702	+0.577	+0.372	+0.137					
32.0	+0.782	+0.768	+0.663	+0.459	+0.182					
40.0	+0.800	+0.805	+0.731	+0.530	+0.217					
48.0	+0.791	+0.828	+0.785	+0.593	+0.254					
56.0	+0.763	+0.838	+0.824	+0.636	+0.285					

Table 11.4: Moments in cylindrical walls—Fixed base free top (IS: 3370—Part IV);
Moment M_w = (coefficient) (wH^3) kN/m; positive sign indicates tension at the outside face

$\frac{H^2}{Dt}$	Coefficients at point									
	0.1*H*	0.2*H*	0.3*H*	0.4*H*	0.5*H*	0.6*H*	0.7*H*	0.8*H*	0.9*H*	10*H*
0.4	+.0005	+.0014	+.0021	+.0007	–.0042	–.0150	–.0302	–.0529	–.0816	–.1205
0.8	+.0011	+.0037	+.0063	+.0080	+.0070	+.0023	–.0068	–.0224	–.0465	–.0795
1.2	+.0012	+.0042	+.0077	+.0103	+.0112	+.0090	+.0022	–.0108	–.0311	–.0602
1.6	+.0011	+.0041	+.0075	+.0107	+.0121	+.0111	+.0058	–.0051	–.0232	–.0505
2.0	+.0010	+.0035	+.0068	+.0099	+.0120	+.0115	+.0075	–.0021	–.0185	–.0436
3.0	+.0006	+.0024	+.0047	+.0071	+.0090	+.0097	+.0077	+.0012	–.0119	–.0333
4.0	+.0003	+0.015	+.0028	+.0047	+.0066	+.0077	+.0069	+.0023	–.0080	–.0268
5.0	+.0002	+.0008	+.0015	+.0029	+.0046	+.0059	+.0059	+.0028	–.0058	–.0222
6.0	+.0001	+.0003	+.0008	+.0019	+.0032	+.0046	+.0051	+.0029	–.0041	–.0187
8.0	.0000	.0001	+.0002	+.0008	+.0016	+.0028	+.0038	+.0029	–.0022	–.0146
10.0	.0000	.0000	+.0001	+.0004	+.0007	+.0019	+.0029	+.0028	–.0012	–.0122
12.0	.0000	–.0001	+.0001	+.0002	+.0003	+.0013	+.0023	+.0026	–.0005	–.0104
14.0	.0000	.0000	.0000	.0000	.0001	+.0008	+.0019	+.0023	+.0001	–.0090
16.0	.0000	.0000	–.0001	–.0002	–.0001	+.0004	+.0013	+.0019	+.0001	–.0079
	Coefficients at point									
	.80*H*	0.85*H*	.90*H*	.95*H*	1.00*H*					
20.0	+.0015	+.0014	+.0005	–.0018	–.0063					
24.0	+.0012	+.0010	+.0007	–.0013	–.0053					
32.0	+.0007	+.0003	+.0007	–.0008	–.0040					
40.0	+.0002	+.0005	+.0005	–.0005	–.0032					
48.0	.0000	+.0001	+.0005	–.0003	–.0026					
56.0	.0000	.0000	+.0004	–.0001	–.0023					

Number of wires gradually reduced towards the top of the tank

Max. radial pressure due to prestress is

$$w_t = \left(\frac{2f_sA_s}{SD}\right) = \left(\frac{2\times1000\times20}{13.18\times28\times10^3}\right)$$

$$= 0.108 \text{ N/mm}^2$$

Max. vertical moment due to prestress is

$$M_t = M_w\left(\frac{w_t}{w_w}\right) = 37800\left(\frac{0.108}{0.07}\right)$$

$$= 58320 \text{ N.mm/mm}$$

$$= 58.32 \times 10^6 \text{ N.mm/m}$$

Considering one metre length of tank along circumference

$$Z = \left(\frac{1000\times150^2}{6}\right) = 375 \times 10^4 \text{ mm}^3$$

Vertical prestress required is

$$f_c = \left[\frac{f_{\min\cdot w}}{\eta} + \frac{M_t}{Z}\right] = \left[\frac{1}{0.75} + \frac{58.32\times10^6}{375\times10^4}\right]$$

$$= 16.83 \text{ N/mm}^2$$

Since, this stress exceeds the permissible value of $f_{ct} = 14$ N/mm^2, the thickness of the tank is increased to 200 mm at base.

$$Z = \left(\frac{1000\times200^2}{6}\right) = 666 \times 10^4 \text{ mm}^3$$

$$f_c = \left[\frac{1}{0.75} + \frac{58.32\times10^6}{666\times10^4}\right] = 10.08 \text{ N/mm}^2$$

Vertical prestressing force

$$= f_e \cdot A = (10.08 \times 1000 \times 200)/1000 = 2016 \text{ kN}$$

Using 7 mm ϕ – 12 nos cables

$$\text{Force/cable} = \left(\frac{\pi\times7^2}{4}\times\frac{1000}{1000}\right)12 = 461 \text{ kN}$$

$$\therefore \text{ Spacing of vertical cables} = \left(\frac{1000\times461}{2016}\right) = 228 \text{ mm}$$

Nominal reinforcements of 0.2 per cent of the cross-sectional area to be provided in the circumferential and longitudinal directions. This requirement will be fulfilled by providing 8 mm diameter mild steel bars at 300 mm spacing on both faces at a cover of 20 mm.

c. *Sliding base*

Assume, the tank walls are supported on elastomeric pads

Assume, coefficient of friction = 0.5

Dimensions of tank $D = 28$ m

$H = 7$ m

From table 11.5 and Fig. 11.3 the thickness of tank at base is taken as 230 mm gradually reducing to 150 mm at the top.

Hydrostatic pressure

$$w_w = wH = (10 \times 7) = 70 \text{ kN/m}^2 = 0.07 \text{ N/mm}^2$$

Max. ring tension

$$N_d = \left(10 \times 7 \times \frac{28}{2}\right) = 980 \text{ kN/m}$$

Table 11.5: Economic dimensional proportions for water tanks (*Preload Engineering Co., New York*) (Refer Fig. 11.3 for abbreviations)

Capacity, m^2	Dimensions, m								
	A	*B*	*C*	*D*	*E*	*F*	*G*	*H*	*I*
378	12.50	3.15	1.56	0.12	0.12	0.05	0.05	0.20	0.15
945	16.90	4.30	2.11	0.12	0.12	0.05	0.05	0.22	0.15
1890	21.35	5.35	2.67	0.12	0.12	0.05	0.05	0.30	0.17
2835	24.40	6.10	3.05	0.12	0.15	0.05	0.05	0.36	0.19
3780	26.95	6.70	3.36	0.12	0.18	0.05	0.05	0.38	0.22
5670	30.80	7.80	3.86	0.12	0.23	0.05	0.05	0.43	0.25
7570	33.85	8.55	4.23	0.12	0.24	0.05	0.06	0.48	0.27
9450	36.40	9.15	4.55	0.22	0.26	0.05	0.06	0.51	0.30
18900	46.00	11.45	5.75	0.22	0.44	0.05	0.10	0.69	0.38
37800	57.90	14.50	7.24	0.22	0.74	0.05	0.11	0.89	0.49
Economic proportion in U.S.A. $B : A = 1 : 4$									

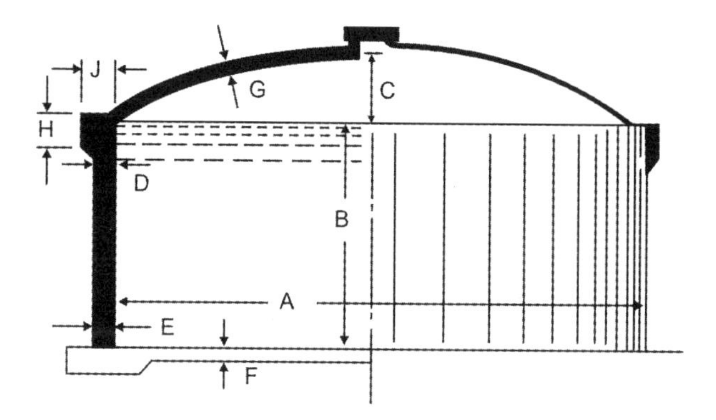

Fig. 11.3 Economic Dimensional Proportions

Self weight of wall $= \left(\dfrac{0.23+0.15}{2}\times 7\times 24\right) = 32$ kN/m

Frictional force at base $= N_0 = (0.5 \times 32) = 16$ kN/m

Minimum wall thickness at base $= \left[\dfrac{980}{(0.7\times 14)-1}\right] = 111$ mm

Net thickness available (allowing 30 mm for cables)

$= (230 - 30) = 200$ mm

Circumferential prestress

$$f_c = \left[\frac{980}{0.75\times 200}+\frac{1}{0.75}\right] = 7.86 \text{ N/mm}^2$$

Spacing of circumferential wire winding

$$S = \left(\frac{2\times 980}{0.07}\right)\times\left(\frac{1000\times 38.5}{14\times 28\times 10^3\times 200}\right)$$

$= 13.75$ mm

Number of wires/metre = (1000/13.75) = 73

The number of wires/metre at top is reduced depending upon the ring tension (10 to 12 wires/m)

Maximum radial pressure due to prestress at transfer

$$w_t = \left(\frac{2\times 1000\times 38.5}{7.7\times 28\times 10^3}\right) = 0.35 \text{ N/mm}^2$$

Maximum vertical moment due to working pressure

$$M_w = 0.247\, N_0\sqrt{Rt} \quad = 0.247\times 16\times\sqrt{14\times 0.23}$$

$$= 7.1 \text{ kN.m/m} \quad = 7100 \text{ N.mm/m}$$

Maximum vertical moment due to prestress is

$$M_t = 7100\left(\frac{0.35}{0.07}\right) = 35500 \text{ N.mm/mm}$$

$$= 35.5 \times 10^6 \text{ N.mm/m}$$

Considering one metre length of tank wall along the circumference

$$Z = \left(\frac{1000 \times 230^2}{6}\right) = 8.81 \times 10^6 \text{ mm}^3$$

The vertical prestress required is

$$f_c = \left[\frac{1}{0.75} + \frac{35.5 \times 10^6}{8.81 \times 10^6}\right] = 5.35 \text{ N/mm}^2$$

As per I.S. code, the minimum vertical prestress required to counteract the winding stress is

$$= (0.3 \times 14) = 4.2 \text{ N/mm}^2$$

$\therefore$ Vertical prestressing force

$$= \left(\frac{4.2 \times 1000 \times 230}{100}\right) = 966 \text{ kN}$$

Spacing of vertical cables

$$= \left(\frac{1000 \times 461}{966}\right) = 477 \text{ mm}$$

Ultimate tensile force in wires at base of tank

$$= \frac{(73 \times 38.5 \times 1500)}{1000} = 4215 \text{ kN}$$

$$\text{Load factor against collapse} = \left(\frac{4215}{980}\right) = 4.3$$

$$\text{Cracking load} = \frac{(1000 \times 230)[0.75 \times 14 + 1.7]}{1000}$$

(Assuming cracking stress in concrete = 1.7 N/mm^2)

$$= 2806 \text{ kN}$$

Factor of safety against cracking = (2806/980) = 2.8

Nominal reinforcements of 0.2 percent of the cross-section in the circumferential and vertical direction are arranged well distributed on each face.

Problem 11.5

A non cylinder prestressed concrete pipe 1 m internal diameter is to withstand a fluid pressure of 2 N/mm^2. Using 5 mm diameter H.T. wires initially tensioned to 1000 N/mm^2 and assuming f_{ct} = 16 N/mm^2, and loss ratio = 0.8, determine the pitch of circumferential wire winding. No tensile stresses are permitted at any stage. If the tensile strength of concrete is 2 N/mm^2, estimate the load factor against cracking.

Solution

Data: Diameter of pipe d = 1000 mm

Fluid pressure p = 2 N/mm^2

f_{ct} = 16 N/mm^2 f_{tw} = 0

Loss ratio η = 0.8

5 mm diameter high tensile wires initially tensioned to 1000 N/mm^2 are available for use.

Tensile strength of concrete f_t = 2 N/mm^2

Thickness of concrete pipe

$$t > \left[\frac{N_d}{\eta f_{ct} - f_{tw}}\right] > \left[\frac{2(1000/2)}{(0.8\times 16)-0}\right] > 78.12 \text{ mm}$$

Adopt 100 mm thick pipe for practical considerations

$$f_c = \left[\frac{2(1000/2)}{(0.8\times 100)}\right] = 12.5 \text{ N/mm}^2$$

Number of turns of 5 mm diameter high tensile wire initially stressed to 1000 N/mm^2 is computed as

$$n = \left[\frac{4000\, tf_c}{\Pi d^2 f_s}\right] = \left[\frac{(4000\times 100\times 12.5)}{\left(\Pi\pi\times 5^2\times 1000\right)}\right]$$

$$= 63.7 \text{ turns/m}$$

$$\text{Pitch of wire winding} = \left(\frac{1000}{63.7}\right) = 15.69 \text{ mm}$$

Adopt 5 mm diameter high tensile wire initially tensioned to a stress of 1000 N/mm^2 at a pitch of 15 mm.

Hoop tension due to fluid pressure

$$= \left[\frac{(2\times 1000)}{(2\times 100)}\right] = 10 \text{ N/mm}^2$$

Hoop compression due to prestress = 12.5 N/mm^2

Resultant compressive stress in concrete

$$= (12.5 - 10) = 2.5 \text{ N/mm}^2$$

Tensile strength of concrete = 2 N/mm^2

Additional fluid pressure required to develop a tensile stress of 4.5 N/mm^2 in concrete is given by

$$p = \left[\frac{(2 \times 100 \times 4.50)}{1000}\right] = 0.9 \text{ N/mm}^2$$

Cracking fluid pressure $= (2 + 0.9) = 2.9$ N/mm^2

Working pressure $= 2$ N/mm^2

Load factor against cracking $= \left(\dfrac{2.9}{2.0}\right) = 1.45$

Problem 11.6

A non cylinder prestressed concrete pipe of internal diameter 800 mm and thickness of concrete shell 74 mm conveys water at a working pressure of 1.5 N/mm^2. The length of each pipe is 4 m. The pipe are supported on pedestals at 4 m intervals. The longitudinal compressive prestress in the pipe due to prestressing is 3 N/mm^2. Check for the safety of the pipe for flexural stresses as per IS: 784 code considering the pipe as a hollow circular section spanning over 4 m.

Solution

Considering the pipe as a beam of hollow circular section spanning over 4 m.

Three times self weight of pipe $= (3 \times \pi \times 0.875 \times 0.075 \times 24)$

$= 14.83$ kN/m

Weight of water in pipe $= \dfrac{(\pi \times 0.8^2 \times 10)}{4} = 5.024$ kN/m

$\therefore$ Total U.D.L. on pipe $= 19.854$ kN/m

Maximum bending moment $= \dfrac{(19.854 \times 4^2)}{8} = 39.708$ kN.m

Second moment of area of cross section of pipe is computed as

$$I = \frac{(0.95^4 - 0.8^4)}{64} = 0.0198 \text{ m}^4$$

$$\text{Flexural tensile stress} = \left[\frac{(39.708 \times 10^6 \times 475)}{(0.0198 \times 10^{12})}\right]$$

$$= 0.95 \text{ N/mm}^2 \text{ (tension)}$$

Longitudinal compressive prestress = 3 N/mm^2

∴ Resultant compressive stress = (3 – 0.95) = 2.05 N/mm^2

The resultant flexural stress being compressive, the pipe is safe against cracking.

Longitudinal prestressing

Critical transient tensile stress at spigot end

= (0.6 × hoop stress)

= (0.6 × 12.5) = 7.5 N/mm^2

$$\text{Maximum permissible tensile stress} = 0.8\sqrt{f_{ci}}$$

$$= 0.8\sqrt{40} = 5 \text{ N/mm}^2$$

Hence, the tensile stress of (7.5 – 5) = 2.5 N/mm^2 should be counterbalanced by longitudinal prestressing.

Cross sectional area of pipe = (Π × 0.675 × 0.075) = 0.159 m^2

If P = Longitudinal prestressing force required, then

$$P = \left(\frac{0.159 \times 10^6 \times 2.5}{1000}\right) = 397.5 \text{ kN}$$

Using 5 mm diameter high tensile wires initially stressed to 1200 N/mm^2.

$$\text{Force in each wire} = \left[\frac{(\Pi \times 5^2 \times 1200)}{(4 \times 1000)}\right] = 23.55 \text{ kN}$$

$$\therefore \quad \text{Number of wires} = \left(\frac{397.5}{23.55}\right) = 16.87$$

Spacing of wires along the circumference is computed as

$$s = \left(\frac{\Pi \times 675}{16.87}\right) = 125 \text{ mm.}$$

Problem 11.7

A non cylinder prestressed concrete pipe of internal diameter 600 mm and thickness of concrete shell 75 mm is required to convey fluid at a pressure of 2 N/mm^2. The maximum and minimum compressive stress permissible in concrete are 16 and 2 N/mm^2 respectively. Loss ratio = 0.8

a. Design the circumferential wire winding using 5 mm diameter high tensile wires initially stressed to 1200 N/mm^2.

b. Design the longitudinal prestressing using 5 mm diameter high tensile wires tensioned to 1200 N/mm^2. The maximum permissible tensile stress under the critical transient loading (wire-wrapping at spigot end) should not exceed 0.8 f_{ci}, where f_{ci} is the cube strength of concrete at transfer having a value of 40 N/mm^2.

Solution

Given data:

t = 75 mm D = 600 mm

f_{ct} = 16 N/mm^2 d = 5 mm

$f_{\min w}$ = 2 N/mm^2 f_s = 1200 N/mm^2

η = 0.8 f_{ci} = 40 N/mm^2

a. *Circumferential wire winding*

$$f_c = \left[\frac{N_d}{\eta^t} + \frac{f_{\min \cdot w}}{\eta}\right] = \left[\frac{2(600/2)}{(0.8 \times 75)} + \frac{2}{0.8}\right]$$

$$= 12.5 \text{ N/mm}^2$$

Number of turns $$n = \left[\frac{400 t f_c}{\Pi d^2 f_s}\right] = \left[\frac{(4000 \times 75 \times 12.5)}{\Pi \times 5^2 \times 1200}\right]$$

$$= 39.8 \text{ turns/m}$$

Pitch of wire winding $$= \left(\frac{1000}{39.8}\right) = 25 \text{ mm}$$

Problem 11.8

A cylindrical tank wall of thickness of 100 mm is subjected to a design tensile force of 300 kN/m. If the compressive stress in concrete is limited to 15 N/mm^2 in compression and zero tension, design the pitch of circumferential wire winding using 5 mm diameter high tensile wires initially tensioned to 1200 N/mm^2. Assume a loss ratio of 0.8. If f_p = 1700 N/mm^2, determine the load factor against collapse.

Solution

Prestress required $$= \left(\frac{N_d}{\eta t}\right) = \left[\frac{(300 \times 10^3)}{(0.8 \times 100 \times 10^3)}\right]$$

$$= 3.75 \text{ N/mm}^2$$

$$\text{Prestressing force} = P = \left[\frac{3.75 \times 100 \times 10^3}{1000}\right] = 375 \text{ kN}$$

$$\text{Number of 5 mm diameter H.T. wires} = \left[\frac{375 \times 10^3}{19.6 \times 10^3}\right] = 19.1$$

$$\text{Pitch of wires} = \left[\frac{1000}{19.1}\right] = 52.3 \text{ mm}$$

$$\text{Ultimate tensile force} = \left(\frac{19.1 \times 19.6 \times 0.87 \times 1700}{1000}\right) = 553 \text{ kN}$$

$$\text{Load factor against collapse} = \left(\frac{553}{375}\right) = 1.47$$

Problem 11.9

A non cylinder concrete pipe 1.2 m diameter having a thickness of 75 mm and circumferential wire winding of high tensile wires of 5 mm diameter at a pitch of 30 mm, initially stressed to 1200 N/mm^2. Estimate the fluid pressure permissible if no tension is permitted in concrete and the loss ratio is 0.8.

Solution

$$\text{Number of turns/m} \quad n = \left(\frac{1000}{30}\right) = 33.33$$

$$\text{Compressive stress} \quad f_c = \left(\frac{n \Pi d^2 f_s}{4000\, t}\right)$$

$$= \left(\frac{33.33 \times \Pi \times 5^2 \times 1200}{4000 \times 75}\right)$$

$$= 10.46 \text{ N/mm}^2$$

$$\text{If } N_d = \text{Hoop tension} \quad = (f_c.\eta.t) = (10.46 \times 0.8 \times 75)$$

$$= 627.6 \text{ N}$$

$$\text{If } p = \text{fluid pressure} \quad p = \left(\frac{2N_d}{d}\right) = \left(\frac{2 \times 627.6}{1200}\right)$$

$$= 1.046 \text{ N/mm}^2$$

Problem 11.10

A prestressed concrete pipe is made up of a steel cylinder 1000 mm internal diameter and thickness 2 mm. Thickness of concrete pipe = 40 mm. Circumferential wire winding consists of 5 mm diameter high tensile wires initially tensioned to a stress of 1000 N/mm^2 at a pitch of 25 mm. Loss ratio = 0.8, f_{ct} = 16 N/mm^2. Modular ratio = 6. Estimate the maximum fluid pressure permissible in the pipe if no tension is permitted in the pipe.

Solution

Number of turns/m $n = \left(\frac{1000}{25}\right) = 40$ turns/m

$$f_c = \left[\frac{n \Pi d^2 f_s}{4000\,(t + mt_s)}\right] = \left[\frac{(40 \times \Pi \times 5^2 \times 1000)}{4000(40 + 6 \times 2)}\right]$$

$= 15$ N/mm^2

If N_d = Hoop tension $= (t + m\,t_s)\,(\eta\, f_{ct} - f_{\min\,w})$

$= (40 + 6 \times 2)\,(0.8 \times 15 - 0) = 624$ N

If p = fluid pressure or $p = \left(\frac{2N_d}{d}\right) = \frac{(2 \times 624)}{1000}$

$= 1.248$ N/mm^2

Problem 11.11

Design a non cylinder prestressed concrete pipe of 600 mm internal diameter to withstand a working hydrostatic pressure of 1.05 N/mm^2, using a 2.5 mm high tensile wire initially stressed to 1 kN/mm at transfer. Permissible maximum and minimum stresses in concrete at transfer and service loads are 14 and 0.7 N/mm. Assume 20 percent loss in prestress. Also calculate the test pressure required to produce a tensile stress of 0.7 N/mm^2 in concrete when applied immediately after tensioning and also the winding stress in steel, assuming the modulus of elasticity of steel and concrete as 210 and 35 kN/mm^2 respectively. Loss ratio = 0.8.

Solution

Internal diameter of pipe D = 600 mm

Hydrostatic pressure N_d = 1.05 N/mm^2

Diameter of H.T. wire d = 2.5 mm

Initial stress in wire = 1000 N/mm^2

Modulus of elasticity of steel

$$E_s = 210 \text{ kN/mm}^2$$

Modulus of elasticity of concrete $E_c = 35 \text{ kN/mm}^2$

Maximum permissible stress in concrete

$$f_{ct} = 14 \text{ N/mm}^2$$

Minimum permissible stress in concrete

$$f_{\text{min.w}} = 0.7 \text{ N/mm}^2$$

Initial stress in H.T wire $f_s = 1000 \text{ N/mm}^2$

Modular ratio $= [E_s/E_c] = 6$

The thickness of the pipe is computed using the equation

$$t > \left[\frac{N_d}{(\eta f_{ct} - f_{\text{min}.w})}\right] = \left[\frac{1.05(0.5 \times 600)}{\{(0.8 \times 14) - 0.7\}}\right] > 30 \text{ mm}$$

Using 30 mm thick concrete pipe, the actual compressive stress in concrete is $f_c = 14 \text{ N/mm}^2$. The number of turns of 2.5 mm H.T wire stressed to 1000 N/mm^2 is given by,

$$n = \left[\frac{4000\, t\, f_c}{\pi d^2 f_s}\right] = \left[\frac{4000 \times 30 \times 14}{\pi \times 2.5^2 \times 1000}\right] = 86 \text{ turns/m}$$

$$\text{Pitch of circumferential wire winding} = \left[\frac{1000}{86}\right] = 11.6 \text{ mm}$$

If W_w = test pressure required immediately after winding ($\eta = 1$), we have

$$f_c = \left[\frac{W_w D}{2\eta t} + \frac{f_{\text{min. }w}}{\eta}\right]$$

$$W_w = \frac{2t}{D}(f_c - f_{\text{min}.w}) = \left(\frac{2 \times 30}{600}\right)[14 - (-0.7)]$$

$$= 1.47 \text{ N/mm}^2$$

If $\quad f_{si}$ = winding stress in steel

$$f_{si} = [1 + a_e r] f_{se}$$

where, $\alpha_e = 6$ and $\rho = [f_c/f_s] = [14/1000] = 0.014$

$\therefore$ Winding stress in steel is $f_{si} = [1 + (6 \times 0.014)]\, 1000 = 1084 \text{ N/mm}^2$

Problem 11.12

A prestressed concrete cylinder pipe is to be designed using a steel cylinder of 1000 mm internal diameter and thickness 1.6 mm. The service internal hydrostatic pressure in the pipe is 0.8 N/mm^2. 4 mm diameter high tensile wires initially tensioned to a stress of 1 kN/mm^2 are available for circumferential winding. The yield stress of mild steel cylinder is 280 N/mm^2. The maximum permissible compressive stress in concrete at transfer is 14 N/mm^2 and no tensile stress is permitted under service load conditions. Determine the thickness of the concrete lining and the number of turns of circumferential wire winding and the factor of safety against bursting. Assume modular ratio as 6 and loss ratio as 0.8.

Solution

Hydrostatic pressure inside the pipe N_d = 0.8 N/mm^2

Internal diameter of steel cylinder pipe D = 1000 mm

Thickness of steel pipe t_S = 1.6 mm

Yield stress of mild steel pipe f_y = 280 N/mm^2

Permissible compressive stress in concrete at transfer

f_{ct} = 14 N/mm^2

Permissible tensile stress in concrete $f_{\min.w}$ = 0

Diameter of H.T wire winding = 4 mm

Initial stress in wires f_S = 1000 N/mm^2

Modular ratio $\alpha_e = 6$ and loss ratio $\eta = 0.8$

Ultimate tensile strength of wire f_{pu} = 1600 N/mm^2

The required thickness of the concrete pipe is evaluated using the relation,

$$t > \left[\frac{N_d}{(\eta f_{ct} - f_{\min.w})} - \alpha_e t_s\right] > \left[\frac{0.8\,(0.5 \times 1000)}{(0.8 \times 14) - 0} - (6 \times 1.6)\right] > 25.9\,\text{mm}$$

Use 26 mm thick concrete lining. Hence, f_c = 14 N/mm^2.

Number of turns H.T wire winding is given by the relation,

$$n = \left[\frac{4000\,(t + \alpha_e t_s) f_c}{\pi d^2 f_s}\right]$$

$$= \left[\frac{4000\,(26 + 6 \times 1.6)\,14}{(\pi \times 4^2 \times 1000)}\right] = 40\ \text{turns/metre}$$

Bursting pressure is estimated by the equation,

$$P_u = \left[\frac{0.00157\, d^2 n\, f_{pu} + 2 t_s f_y}{D}\right]$$

$$= \left[\frac{(0.00157 \times 4^2 \times 40 \times 1600) + (2 \times 1.6 \times 280)}{1000}\right]$$

$$= 2.516\ \text{N/mm}^2$$

Factor of safety against bursting

$$= \left[\frac{\text{Bursting pressure}}{\text{Working pressure}}\right] = \frac{2.516}{0.8} = 3.14.$$

Problem 11.13

A prestressed concrete pipe is to be designed to withstand a fluid pressure of 1.6 N/mm^2. The diameter of the pipe is 1200 mm and shell thickness is 100 mm. The maximum compressive stress in concrete at transfer is 16 N/mm^2. A residual compression of 2 N/mm^2 is expected to be maintained at service loads. Loss ratio is 0.8. High tensile wires of 5 mm diameter initially stressed to 1 kN/mm^2 are available for use. Determine

a. The number of turns of wire per metre length.

b. The pitch of wire winding.

Solution

Diameter of concrete pipe $D = 1200$ mm

Fluid pressure $N_d = 1.6$ N/mm^2

Thickness of concrete shell $t = 100$ mm

Diameter of high tensile wires $d = 5$ mm

Initial stress in wires $f_s = 1000$ N/mm^2

Compressive stress in concrete at transfer

$f_{ct} = 16$ N/mm^2

Residual compressive stress at service loads

$f_{\min.w} = 1$ N/mm^2

Loss ratio $\eta = 0.8$

The required thickness of the concrete pipe is evaluated using the relation,

$$t > \left[\frac{N_d}{(\eta f_{ct} - f_{\min.w})}\right] > \left[\frac{1.6\,(0.5 \times 1200)}{(0.8 \times 14) - 1}\right] > 94.1 \text{ mm}$$

Thickness provided = 100 mm. Hence, the compressive stress in concrete is given by,

$$f_c = \left[\frac{W_w D}{2\eta t} + \frac{f_{\min.w}}{\eta}\right] = \left[\frac{1.6 \times 1200}{2 \times 0.8 \times 100} + \frac{1}{0.8}\right]$$

$$= 13.25 \text{ N/mm}^2$$

Number of turns H.T wire winding is given by the relation,

$$n = \left[\frac{4000(t+\alpha_e t_s)\, f_c}{\pi d^2 f_s}\right] = \left[\frac{(4000\times 100\times 13.25)}{(\pi\times 5^2\times 1000)}\right]$$

$$= 68 \text{ turns/metre}$$

$$\text{Pitch of wire winding} = \left[\frac{1000}{68}\right] = 14.7 \text{ mm.}$$

Problem 11.14

A prestressed concrete pipe of 1200 mm diameter and shell thickness 100 mm has been designed to withstand a water pressure of 1.6 N/mm^2 with circumferential wire winding. The pipe is supported on pillars located at 6 m centres. Analysis of tensile stresses developed in the longitudinal direction indicates a tensile stress of 3 N/mm^2 due to circumferential wire winding near the spigot end of the pipe. Design suitable longitudinal prestressing using 7 mm diameter high tensile wires initially stressed to 1 kN/mm^2 to eliminate the tensile stresses in concrete. Also ensure that tensile stresses do not develop in the concrete when the pipe supports its self weight and water load over a span of 6 m.

Solution

Internal diameter of concrete pipe = 1200 mm

Thickness of concrete shell = 100 mm

Length of pipe unit = 6 m

Initial stress in high tensile wires = 1000 N/mm^2

Longitudinal tensile stress in concrete = 3.0 N/mm^2

Diameter of high tensile wires = 7 mm

Cross-sectional area of the pipe = $(\pi \times 1.4 \times 0.10)$

= 0.44 m^2

Longitudinal prestressing force required = $(0.44 \times 10^6 \times 3)$

= 1320×10^3 N

Using 7 mm diameter, H.T. wires initially stressed to 1000 N/mm^2

Force in each wire = (38.465×1000)

= 38465 N

$$\text{No. of longitudinal H.T wires} = \left[\frac{1320\times 10^3}{38465}\right] = 34.3$$

Provide 45 wires to produce excess precompression to resist the additional tensile stresses due to flexural effects of dead and fluid loads.

Hence, compressive stress induced in concrete

$$= \left[\frac{45 \times 38465}{0.44 \times 10^6}\right] = 393 \text{ N/mm}^2$$

Spacing of wires along the circumference of the pipe

$$= \left[\frac{\pi \times 1300}{34.3}\right] = 119 \text{ mm}$$

Flexural stresses should be checked when the pipe supports its self weight and water load supported over a span of 6 m.

Self weight of concrete pipe

$$= [\pi \times 1.3 \times 0.1 \times 24] = 9.8 \text{ kN/m}$$

Live load due to water in pipe

$$= \left[\left(\frac{\pi \times 1.2^2}{4}\right) \times 10\right] = 11.3 \text{ kN/m}$$

Total load w $= (9.8 + 11.3) = 21.1$ kN/m

Bending moment at centre of span

$$M = \left[\frac{wL^2}{8}\right] = \left[\frac{21.1 \times 6^2}{8}\right] = 94.95 \text{ kN.m}$$

Second moment of area of cross-section of pipe

$$I = \left[\frac{\pi(1.4^4 - 1.2^4)}{64}\right] = 0.086 \text{ m}^4$$

Section modulus of pipe cross-section

$$Z = \left[\frac{I}{y}\right] = \left[\frac{0.086}{0.7}\right] = 0.124 \text{ m}^3$$

Maximum bending stress at soffit of pipe

$$= \left[\frac{M}{Z}\right] = \left[\frac{94.95 \times 10^6}{0.124 \times 10^9}\right] = 0.765 \text{ N/mm}^2 \text{ (tension)}$$

Resultant stress at soffit of pipe

$$= [3.93 - 3.00 - 0.765]$$
$$= 0.165 \text{ N/mm}^2 \text{ (compression)}$$

Hence, the pipe section is free from tensile stresses due to service loads.

Problem 11.15

A prestressed concrete tank wall is subjected to a hoop tension of 1 kN/mm due to water pressure. The maximum and minimum permissible compressive stress in concrete under working pressure is 13 N/mm^2 and 1 N/mm^2 respectively. The loss of prestress due to various causes may be taken as 25 percent. High tensile wires of 5 mm diameter with an initial stress of 1000 N/mm^2 are available for circumferential wire winding. Design suitable thickness for the tank walls and the spacing of circumferential wire winding. Assume the diameter of the tank as 30 m and height of water in tank as 7.5 m.

Solution

Hoop tension in tank walls N_d = 1000 N/mm

Permissible maximum compressive stress in concrete

$$f_{ct} = 13 \text{ N/mm}^2$$

Permissible minimum compressive stress in concrete

$$f_{\min.w} = 1 \text{ N/mm}^2$$

Loss ratio $\eta = 0.75$

High tensile wires of 5 mm diameter are available for use.

Initial stress in wires = 1000 N/mm^2

Diameter of tank D = 30 m

Height of water in tank H = 7.5 m

Intensity of water pressure at base of tank

$$w_w = wH = (10 \times 7.5) = \text{kN/m}^2$$
$$= 0.075 \text{ N/mm}^2$$

Cross-sectional area of 5 mm H.T wires A_s = 19.6 mm^2

Minimum wall thickness required is given by the relation,

$$t = \left[\frac{N_d}{(\eta f_{ct} - f_{\min.w})}\right] = \left[\frac{1000}{(0.75 \times 13) - 1}\right] = 114.2 \text{ mm}$$

Adopting t = 120 mm, required circumferential prestress is computed as,

$$f_c = \left[\frac{N_d}{\eta t} + \frac{f_{\min.w}}{\eta}\right] = \left[\frac{1000}{(0.75 \times 120)} + \frac{1}{0.75}\right]$$
$$= 12.43 \text{ N/mm}^2$$

Spacing of 5 mm diameter wires initially stressed to 1000 N/mm^2 is given by,

$$s = \left[\left(\frac{2N_d}{w_w}\right)\left(\frac{f_s A_s}{f_c D t}\right)\right]$$

$$= \left[\left(\frac{2 \times 1000}{0.075}\right)\left(\frac{1000 \times 19.6}{12.43 \times 30 \times 10^3 \times 120}\right)\right] = 11.56 \text{ mm}$$

Number of wires per metre $= \left[\frac{1000}{11.56}\right] = 87.$

12

Prestressed Concrete Slabs and Grid Floors

Problem 12.1

The floor slab of an industrial structure, spanning over 8 m is to be designed as a one-way prestressed concrete slab with parallel post-tensioned cables. The slab is required to support a live load of 10 kN/m^2 with the compressive and tensile stress in concrete at any stage not exceeding 14 and zero N/mm^2 respectively. Design a suitable thickness for the slab and estimate the maximum horizontal spacing of the Freyssinet cables (12 of 5 mm diameter initially stressed to 1200 N/mm^2) and their position at mid span section. The loss ratio is 0.8.

Solution

Considering one metre width of slab. The dead load and live load moments are computed.

Live load = 10 kN/m^2, $\eta = 0.8$

$$\therefore \quad M_q = \left(\frac{10 \times 8^2}{8}\right) = 80 \text{ kN.m}$$

Let h = overall depth of slab

b = width of slab

$$M_g = \left(\frac{bh}{10^6} \times 24 \times \frac{8^2}{8}\right) = \left(\frac{192\,bh}{10^6}\right) \text{kN.m}$$

$$= 192\ bh \text{ N.mm}$$

$$f_{ct} = f_{cw} = 14 \text{ N/mm}^2$$

$$f_{tt} = f_{tw} = 0 \text{ N/mm}^2$$

Range of stress = $f_{br} = (0.8 \times 14 + 0) = 11.2$ N.mm^2

Hence, minimum section modulus required is given by

$$Z_b = \left(\frac{bh^2}{6}\right) = \frac{M_q + (1-\eta)M_g}{f_{br}}$$

$$\left(\frac{1000\,h^2}{6}\right) = \left[\frac{80\times 10^6 + (1-0.8)192\times 1000\,h}{11.2}\right]$$

Solving

$$h = 220 \text{ mm}$$

$$A = (220 \times 1000) = 22 \times 10^4 \text{ mm}^2$$

$$Z_b = Z_t = \left(\frac{1000\times 220^2}{6}\right) = 8.06 \times 10^6 \text{ mm}^3$$

$$M_g = (192 \times 1000 \times 220) = 4.22 \times 10^7 \text{ N.mm}$$

$$f_t = \left[0 - \frac{4.22\times 10^7}{8.06\times 10^6}\right] = -5.23 \text{ N/mm}^2$$

$$f_b = \left[0 + \frac{(80+42.2)\,10^6}{0.8\times 8.06\times 10^6}\right] = 18.95 \text{ N/mm}^2$$

The minimum prestressing force is

$$P = \frac{A}{2}(f_b + f_t) = \frac{22\times 10^4}{2}(18.95 - 5.23)$$

$$= 1509 \times 10^3 \text{ N} = 1509 \text{ kN}$$

$$e = \frac{Z(f_b - f_t)}{A(f_b + f_t)} = \left[\frac{8.06\times 10^6\,(18.95+5.23)}{22\times 10^4\,(18.95-5.23)}\right]$$

$$= 64.56 \text{ mm}$$

Spacing of cables

$$\text{Force in each cable} = \left(\frac{12\times 19.6\times 1200}{1000}\right) = 282.24 \text{ kN}$$

$$\therefore \quad \text{Spacing of cables} = \left(\frac{1000\times 282.24}{1509}\right) = 187 \text{ mm}$$

Problem 12.2

Design a post-tensioned prestressed concrete two-way slab, 6 m by 8 m in size to support a live load of 3 kN/m^2. If cables of four wires of 5 mm diameter stressed to 1000 N/mm^2 are available for use, determine the number of cables in the two principal directions. The stresses in concrete not to exceed 14 N/mm^2 in compression and tensile stresses are not permitted under service loads. The loss ratio is 0.8. Check for the limit states of serviceability and collapse.

Solution

$L_x = 6$ m, $L_y = 8$ m

Live load on slab $= 3$ kN/m^2

Force in each cable $= \left(\dfrac{4 \times 19.6 \times 1000}{1000}\right) = 78$ kN

$f_{ct} = f_{cw} = 15$ N/mm^2

$f_{tt} = f_{tw} = 0$

$\eta = 0.8$

Ratio $(L_y/L_x) = (8/6) = 1.33$

Thickness of slab $= \left(\dfrac{\text{span}}{50}\right) = \left(\dfrac{6000}{50}\right) = 120$ mm

Self weight of slab $= (0.12 \times 24 \times 1) = 2.88$ kN/m^2

Live load on slab $= 3.0$ kN/m^2

Finishes etc. $= 0.12$

$\therefore$ Total service load $= 6.00$ kN/m^2

Total ultimate design load

$$w_{ud} = (1.4 \times 3.00) + (1.6 \times 3.00) = 9.0 \text{ kN/m}^2$$

Referring to Table 12.1. Working moments in the middle strips are given by

$$M_x = (0.079 \times 6 \times 6^2) = 17.06 \text{ kN.m/m}$$

$$M_y = (0.056 \times 6 \times 6^2) = 12.10 \text{ kN.m/m}$$

Total moments in the middle strip (x-direction)

$$= (17.06 \times 0.75 \times 8) = 103 \text{ kN.m}$$

Using a minimum cover of 30 mm for the tendons at centre of slab, the distance between the top kern and the centroid of cable is

$$= (120 - 30 - 40) = 50 \text{ mm}$$

If P = Total prestressing force in the x-direction

$$10^3 \times P \times 50 = 103 \times 10^6$$

$\therefore$ $P = 2060$ kN

Force in each cable = 78 kN

$\therefore$ Number of cables in x-direction

(middle strip) $= \left(\dfrac{2060}{78}\right) = 27$

Table 12.1: Bending moment coefficients for rectangular panels supported on four sides with provision for torsion at corners (IS: 456)

No.	Type of panel and moments considered		B.M. coefficient for short span a_x values of L_x/L_x								B.M. coefficient a_y for long span (for all values of L_y/L_x)
			1.0	1.1	1.2	1.3	1.4	1.5	1.75 (or more)	2.0	
1	Interior panels	X	0.032	0.037	0.043	0.047	0.051	0.053	0.060	0.065	0.032
		Y	0.024	0.028	0.032	0.036	0.039	0.041	0.045	0.049	0.024
2	One short edge discontinuous	X	0.037	0.043	0.048	0.051	0.055	0.057	0.064	0.068	0.037
		Y	0.028	0.032	0.036	0.039	0.041	0.044	0.048	0.052	0.028
3	One long edge discontinuous	X	0.037	0.044	0.052	0.057	0.063	0.067	0.077	0.085	0.037
		Y	0.028	0.033	0.039	0.044	0.047	0.051	0.059	0.065	0.028
4	Two adjacent edges discontinuous	X	0.047	0.053	0.060	0.065	0.071	0.075	0.084	0.091	0.042
		Y	0.035	0.040	0.045	0.049	0.053	0.056	0.069	0.069	0.035
5	Two short edges discontinuous	X	0.045	0.049	0.052	0.056	0.059	0.060	0.065	0.069	—
		Y	0.035	0.037	0.040	0.043	0.044	0.045	0.049	0.052	0.035
6	Two long edges discontinuous	X	—	—	—	—	—	—	—	—	0.045
		Y	0.035	0.043	0.051	0.057	0.063	0.068	0.080	0.088	0.035
7	Three edges discontinuous	X	0.057	0.064	0.071	0.076	0.080	0.084	0.091	0.097	—
		Y	0.043	0.048	0.053	0.057	0.060	0.064	0.069	0.073	0.043
8	Three edges discontinuous	X	—	—	—	—	—	—	—	—	0.057
		Y	0.043	0.051	0.059	0.065	0.071	0.076	0.087	0.096	0.043
9	Four edges discontinuous	X	0.056	0.064	0.072	0.079	0.085	0.089	0.100	0.107	0.056

X–Negative moment at continuous edge, Y–Positive momet at mid span.

Spacing of cables $= \left(\dfrac{0.75 \times 8 \times 1000}{27}\right) = 222$ mm

Total moment in y-direction $= (12.1 \times 0.75 \times 6) = 55$ kN.m

Providing a cover of 40 mm to cables in y-direction. Distance between cable and top kern $= (120 - 40 - 40) = 40$ mm

prestressing force $= (55 \times 10^6)/(40 \times 10^3) = 1380$ kN

No. of cables (y-direction) $= \left(\dfrac{1380}{78}\right) = 18$

Spacing of cables $= \left(\dfrac{0.75 \times 6 \times 1000}{18}\right) = 250$ mm

The cable profile is parabolic with maximum eccentricity at centre and concentric at supports.

Problem 12.3

A simple flab slab, 10 m by 8 m overall size, is supported by four columns which are so placed as to form a symmetrical rectangular gird of 8 m by 6 m with cantilevers of 1 m on all sides. The imposed load on the slab is 1.5 kN/m^2. Prestressing cables consisting of four wires of 4 mm, stressed to 1000 N/mm^2 are available for use. Design the number of cables required and indicate their arrangement in the two principal directions.

Solution

Thickness of flat slab $= \left(\dfrac{6000}{40}\right) = 150$ mm

Self weight of slab $= (0.15 \times 24) = 3.6$ kN/m^2

Live load on slab $= 1.5$

Finishes etc. $= 0.9$

$\therefore$ Total load $= 6.0$ kN/m^2

Total load on four columns $= (6 \times 10 \times 8) = 480$ kN

Reaction on each column $= (480/4) = 120$ kN

The slab is analysed for +ve and –ve moments in the long span (x-direction) and short span (y-direction).

Moments in the direction of long span:

Positive moment (centre of slab)

$$M_{cp} = (240 \times 4) - (240 \times 2.5) = 360 \text{ kN.m}$$

Negative moment (supports) $= (1 \times 8 \times 6 \times 0.5) = 24$ kN.m

The prestressing force required is designed to resist the maximum moment of 360 kN.m

The cables are provided at a distance of 30 mm from the edge of the slab at critical sections. The cables profile is parabolic along the span so that the eccentricity is proportional to the moment at the section.

Total prestressing force required in the x-direction is given by

$$P = \left(\frac{360 \times 10^3}{70}\right) = 5143 \text{ kN}$$

$\therefore$ No. of cables in x-direction:

$$\text{Force in each cable} = \left(\frac{4 \times 12.56 \times 1000}{1000}\right) = 50 \text{ kN}$$

$$\therefore \quad \text{No. of cables} = \left(\frac{5143}{50}\right) = 103$$

In the x-direction, the width of column and middle strips each = (8/3) = 2.66 m. Similarly in the short span (y-direction) the maximum moment occurs at the centre of span and is given by,

$$M_{cp} = (240 \times 3 - 240 \times 2) = 240 \text{ kN.m}$$

$$\therefore \quad \text{No. of cables in } y\text{-direction} = \left(\frac{240 \times 10^3}{70 \times 50}\right) = 70$$

The cables are suitably arranged in the column and middle strips in the x- and y-directions.

Problem 12.4

A prestressed concrete continuous flab slab of overall size 16 m by 13 m is supported by nine columns arranged in three rows. The columns are spaced 7.5 and 6 m in the direction of long and short edges respectively with a cantilever of 0.5 m all round. The 20 cm thick slab, continuous over two bays in transverse directions supports a live load of 3 kN/m^2. Assuming the tensile stresses in the slab to be zero under full live load and the minimum cover to the centre of the cable to be 30 mm, determine the magnitude of the prestressing force in the direction of the long span. Using cables of 12 wires of 5 mm, stressed to 1000 N/mm^2, estimate the number of cables required for the long span direction and arrange them suitably in the column and middle strips.

Solution

The salient dimensions of prestressed concrete continuous flat slab is shown in Fig. 12.1.

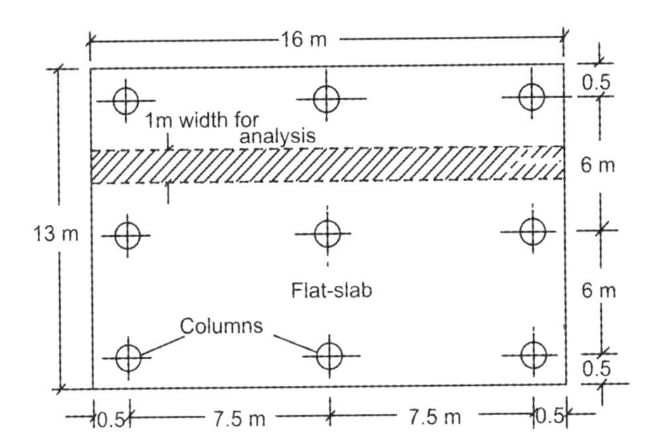

Fig. 12.1

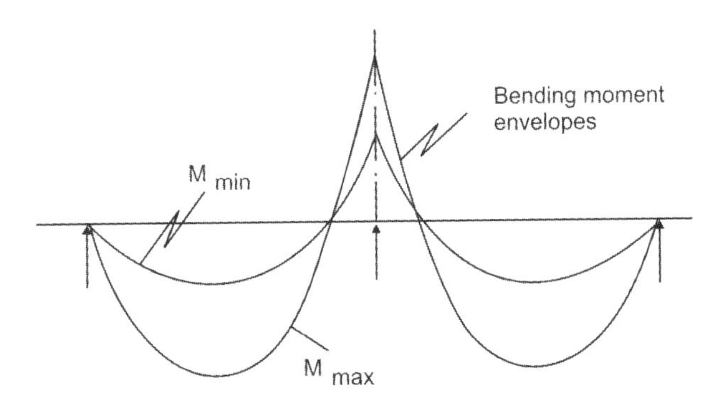

Fig. 12.2

$$\text{Thickness of slab} = \left(\frac{\text{span}}{40}\right) = \frac{(7.5 \times 1000)}{40} = 200 \text{ mm}$$

Self weight of slab $= (0.2 \times 24) = 4.8 \text{ kN/m}^2$

Live load on slab $= 3 \text{ kN/m}^2$

The design moments are evaluated by using coefficients provided in Reynolds R.C. Designer's hand book. The nature of maximum and minimum moments are shown in Fig. 12.2.

a. *Moments due to dead load* (M_g)

Span moment = (kgL^2) = $(0.062 \times 4.8 \times 7.5)^2 = 16.80$ kN.m

Mid support moment = $(-0.125 \times 4.8 \times 7.5^2) = -33.60$ kN.m

b. *Moments due to live load* (+ve and –ve)

Span moments

Positive (M_{Lp}) = $(0.093 \times 3 \times 7.5)^2 = 15.80$ kN.m

Negative (M_{Ln}) = $(-0.032 \times 3 \times 7.5^2) = -5.40$ kN.m

Mid support moments

Negative $M_{Ln} = (-0.125 \times 3 \times 7.5^2) = -21.20$ kN.m

The maximum and minimum moments at support and span are:

a. *At centre of span*

$$M_{max} = (M_{Lp} + M_g) = (15.80 + 16.80) = 32.60 \text{ kN.m}$$
$$M_{min} = (M_{Ln} + M_g) = (-5.40 + 16.80) = 11.40 \text{ kN.m}$$
$$\therefore \text{Range of moments} = (M_{max} - M_{min}) = 21.20 \text{ kN.m}$$

b. *At mid support*

$$M_{max} = M_{Lp} + M_g = (0 - 33.60) = -33.60 \text{ kN.m}$$
$$M_{min} = M_{Ln} + M_g = (-21.20 - 33.60) = -54.80 \text{ kN.m}$$
$$\therefore \text{Range of moments} = (M_{max} - M_{min}) = 21.20 \text{ kN.m}$$

Prestressing force

The absolute maximum moment occurs at the mid support section. Using a minimum cover of 30 mm to the cable. If no tensile stresses are permitted in the section, the distance between the cable and bottom kern is obtained as,

$$(70 + 33.3) = 103.3 \text{ mm}$$

If P = Prestressing force

$$P \times 103.3 = 54.80 \times 10^3$$
$$\therefore P = 530 \text{ kN}$$

Number of cables

Using 12 wires of 5 mm ϕ stressed to 1000 N/mm^2, the force in each cable = $(12 \times 20 \times 1000)/1000 = 240$ kN

Total prestressing force required for a width of 13 m

$$= (350 \times 13) = 6890 \text{ kN}$$

$\therefore$ Number of cables = $(6890/240) = 29$ cables

$$\text{Column strip} = \left(\frac{65}{100} \times 29\right) = 19$$

$$\text{Middle strip} = \left(\frac{35}{100} \times 29\right) = 10$$

Problem 12.5

A prestressed concrete waffle slab (grid floor) for a panel of 15 m by 12 m has the ribs arranged at 1.5 m centres in both directions. Design the grid floor assuming it as an orthotropic plate freely supported on four sides, and using the following data:

Live load on floor $= 3\ \text{kN/m}^2$

Thickness of slab $= 40$ mm

Overall depth of floor $= 400$ mm

Width of ribs $= 200$ mm

Loss ratio $= 0.85$

Permissible compressive stress in concrete at transfer and service load $= 15\ \text{N/mm}^2$. Tensile stresses are not permitted at transfer and working loads. Sketch the arrangements of cables for the central ribs in the transverse directions. Assume $f_{ck} = 40\ \text{N/mm}^2$.

Solution

The plan of the prestressed concrete waffle slab is shown in Fig. 12.3.

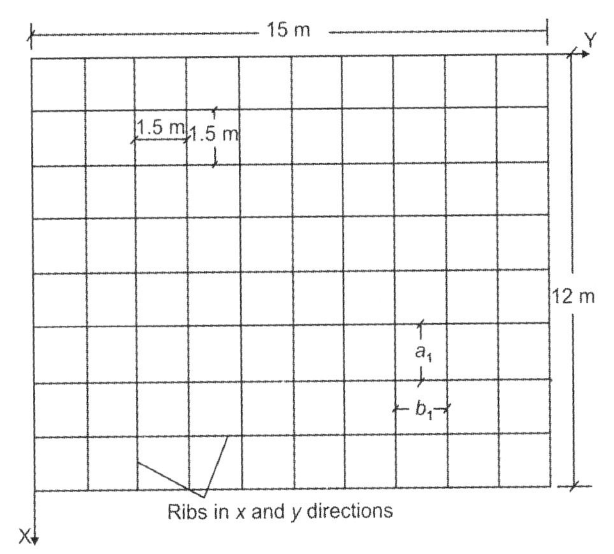

Fig. 12.3

The cross section of rib is shown in Fig. 12.4.

$$a_1 = b_1 = 1.5\ \text{m}$$

$$I_{xx} = I_{yy} = I$$

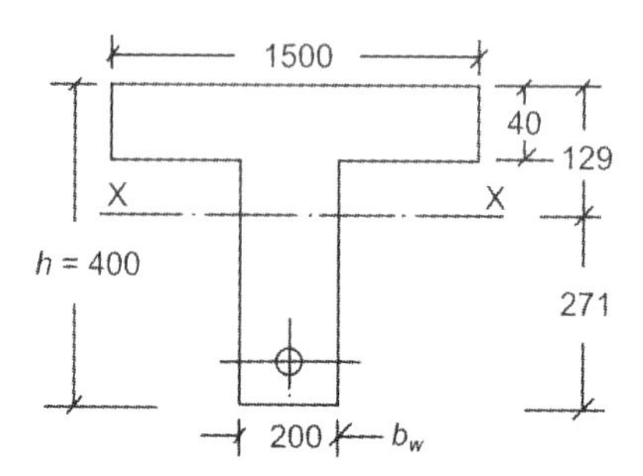

Fig. 12.4

Section properties

$I_{xx} = 2183 \times 10^6 \text{ mm}^4$

$Z_t = 16.92 \times 10^6 \text{ mm}^3$

$Z_b = 8.05 \times 10^6 \text{ mm}^3$

$A = 132000 \text{ mm}^2$

$E = 5700\sqrt{f_{CK}}$

$= 5700\sqrt{40}$

$= 360049 \text{ N/mm}^2$

$= 36.049 \times 10^6 \text{ kN/m}^2$

Flexural rigidity per unit length of plate is given by

$$D_x = \frac{EI}{b_1} = \left(\frac{2183 \times 10^6 E}{1.5 \times 10^{12}}\right) = 0.00145\,E = D_y$$

Torsional rigidity of the effective section is computed by using the coefficients recommended by Timoshenko.

For $\left(\frac{h}{b_w}\right) = \left(\frac{400}{200}\right) = 2.0,\ K_1 = 0.229$

$$\therefore \quad C_1 = C_2 = K_1 \cdot G \cdot (b_w)^3\, h$$

$$= 0.229\left[\frac{E}{2(1+0.15)}\right](200)^3\,(400)$$

$$= 3.18 \times 10^8\, E \text{ mm}^4 = 0.000318\, E \text{ m}^2$$

$$(C_x + C_y) = \left(\frac{C_1}{b_1} + \frac{C_2}{a_1}\right) \qquad \text{But } C_1 = C_2;\ a_1 = b_1$$

$$= 2\left[\frac{3.18 \times 10^8 E}{1.5 \times 10^{12}}\right] = 0.000424\, E \text{ m}^4$$

Computation of loads

Ribs in x-direction $= (9 \times 0.2 \times 0.36 \times 12 \times 24) = 186.6$ kN

Ribs in y-direction $= (7 \times 0.2 \times 0.36 \times 15 \times 24) = 181.4$ kN

$$\text{Slab (40 mm)} = (12 \times 15 \times 0.04 \times 24) = 173.0$$
$$\text{Floor finishes etc.} = 59.0$$
$$\therefore \quad \text{Total load} = 600.0 \text{ kN}$$

Total load

$$\text{Dead load/m}^2 = \left(\frac{600}{15 \times 12}\right) = 3.3 \text{ kN/m}^2$$
$$\text{Live load} = 3.0$$
$$\therefore \quad \text{Total design load } w_d = 6.3 \text{ kN/m}^2$$

The maximum vertical deflection at the centre of span is computed as

$$a = \int \frac{16 w_d}{\pi^6} \int \left\{ \frac{\sin\left(\frac{\pi x}{a_x}\right) \sin\left(\frac{\pi y}{b_y}\right)}{\left(\frac{D_x}{a_x}\right) + \left(\frac{C_x + C_y}{a_x^2 + b_y^2}\right) + \left(\frac{D_y}{b_y^2}\right)} \right\}$$

where, a = central deflection

$a_x = 12$ m $\quad b_y = 15$ m $\quad x = 6$ m

$E = 5700\sqrt{f_{ck}} = 5700\sqrt{40} \quad y = 7.5$ m

$= 36049 \text{ N/mm}^2 = 36.049 \times 10^6 \text{ kN/m}^2$

$$\sin\left(\frac{\pi_x}{a_x}\right) = \sin\left(\frac{\pi \times 6}{12}\right) = 1$$

$$\sin\left(\frac{\pi_y}{b_y}\right) = \sin\left(\frac{\pi \times 7.5}{15}\right) = 1$$

$$\left(\frac{16\, w_d}{6}\right) = \left(\frac{16 \times 6.3}{963}\right) = 0.1046$$

$$\left(\frac{D_x}{a_x^4}\right) = \left(\frac{0.00145\, E}{12^4}\right) = \left(\frac{0.00145 \times 36.049 \times 10^6}{12^4}\right)$$
$$= 2.52$$

$$\left(\frac{D_y}{b_y^4}\right) = \left(\frac{0.00145\, E}{15^4}\right) = \left(\frac{0.00145 \times 36.049 \times 10^6}{15^4}\right)$$
$$= 1.03$$

$$\left(\frac{C_x + C_y}{a_x^2\, b_y^2}\right) = \left(\frac{0.000424 \times 36.049 \times 10^6}{12^2 \times 15^2}\right) = 0.471$$

The central deflection is calculated using the expression given above.

$$a = 0.1046\left(\frac{1}{2.52 + 0.471 + 1.03}\right)$$

$$= 0.026 \text{ m} = 26 \text{ mm}$$

Moments in the central ribs

$$M_x = -D_x\left(\frac{\partial_a^2}{\partial_x^2}\right) = D_x\left(\frac{\pi}{a_x}\right)^2 \cdot a$$

$$= (0.00145 \times 36.049 \times 10^6)\left(\frac{\pi^2}{12^2}\right)0.026$$

$$= 93 \text{ kN.m}$$

$$M_y = D_y\left(\frac{\pi}{b_y}\right)^2 \cdot a$$

$$= (0.00145 \times 36.049 \times 10^6)\left(\frac{\pi^2}{15^2}\right)0.026$$

$$= 59 \text{ kN.m}$$

Dead load moment $= M_g = 0.52 \times$ total moment

Live load moment $= M_q = 0.48 \times$ total moment

Total moment resisted by effective section (x-direction)

$$= (1.5 \times 93) = 139.5 \text{ kN.m}$$

Total moment resisted by effective section (y-direction)

$$= (1.5 \times 59) = 88.5 \text{ kN.m}$$

Central rib in x-direction

Dead load moment $= M_g = (0.52 \times 139.5) = 72.5$ kN.m

Live load moment $= M_q = (0.48 \times 139.5) = 67$ kN.m

Assuming $\eta = 0.85$ $\quad f_{ct} = 15 \text{ N/mm}^2$, $\quad f_{tw} = 0$

$$f_{br} = (\eta f_{ct} - f_{tw}) = (0.85 \times 15) = 12.75 \text{ N/mm}^2$$

$$\therefore \quad Z_b \geq \left[\frac{M_q + (1-\eta)M_g}{f_{br}}\right]$$

$$\geq \left[\frac{67 \times 10^6 + (1 - 0.85)72.5 \times 10^6}{12.75}\right]$$

$$\geq 6.10 \times 10^6 < 8.05 \times 10^6 \text{ mm}^3 \text{ (provided)}$$

$$P = \frac{A(f_t Z_t + f_b Z_b)}{Z_t + Z_b}$$

$$f_b = \left(\frac{M_q + M_g}{\eta Z_b}\right) = \left(\frac{139.5 \times 10^6}{0.85 \times 8.05 \times 10^6}\right)$$

$$= 20.38 \text{ N/mm}^2$$

$$f_t = \left(-\frac{M_g}{Z_t}\right) = -\left(\frac{72.5 \times 10^6}{16.92 \times 10^6}\right)$$

$$= -4.28 \text{ N/mm}^2$$

$$\therefore \quad P = \left\{\frac{132000(-4.28 \times 16.92 \times 10^6 + 20.38 \times 8.05 \times 10^6)}{(16.92 + 8.05)\, 10^6}\right\}$$

$$= 486 \times 10^3 \text{ N} = 486 \text{ kN}$$

$$e = \frac{Z_t Z_b (f_b - f_t)}{A(f_t Z_t + f_b Z_b)}$$

$$= \left\{\frac{8.05 \times 16.92 \times 10^{12} (20.38 + 4.28)}{132000 \times 10^6 (-4.28 \times 16.92 + 20.38 \times 8.05)}\right\}$$

$$= 276 \text{ mm}$$

This eccentricity cannot be provided since $y_b = 271$ mm

Hence, the maximum possible eccentricity

$$= (271 - 101) = 170 \text{ mm}$$

The modified prestressing force

$$P = \left\{\frac{A f_b Z_b}{Z_b + Ae}\right\}$$

$$= \left\{\frac{132000 \times 20.38 \times 8.05 \times 10^6}{8.05 \times 10^6 + 132000 \times 170}\right\}$$

$$= 710 \times 10^3 \text{ N} = 710 \text{ kN}$$

Use Freyssinet cable of 18–7 mm ϕ stressed to 1100 N/mm^2 (x-direction)

$$\text{Force in cable} = \left(\frac{18 \times 38.4 \times 1100}{1000}\right) = 760 \text{ kN}$$

Central rib in y-direction

$$M_y = 88.5 \text{ kN.m}$$
$$M_g = (0.52 \times 88.5) = 46 \text{ kN.m}$$
$$M_q = (0.48 \times 88.5) = 42.5 \text{ kN.m}$$

$$f_b = \left(\frac{88.5 \times 10^6}{0.85 \times 8.05 \times 10^6}\right) = 12.9 \text{ N/mm}^2$$

$$f_t = \left(-\frac{46 \times 10^6}{16.92 \times 10^6}\right) = -2.7 \text{ N/mm}^2$$

Maximum possible eccentricity $e = 100$ mm

The modified prestressing force

$$P = \left(\frac{132000 \times 12.9 \times 8.05 \times 10^6}{8.05 \times 10^6 + 132000 \times 100}\right)$$
$$= 645 \times 10^6 \text{ N} = 645 \text{ kN}$$

In the y-direction for the central rib provide Freyssinet cable of 18–7 mm ϕ stressed to 1000 N/mm^2

$$\text{Force in each cable} = \left(\frac{18 \times 38.4 \times 1000}{1000}\right) = 691 \text{ kN}$$

Problem 12.6

A composite floor slab using precast prestressed plank units 63 mm deep in conjunction with *in situ* cast concrete is required to support a superimposed load of 8.75 kN/m^2 over an effective span of 5 m. Design the required thickness of the composite slab using the prestressed plank and cast *in situ* concrete and the required prestressing force in the precast units. Assume M-20 grade for *in situ* and M-42 for precast pre-tensioned units. Permissible stresses should conform to the provisions of the Indian standard code.

Solution

The cross-section of the composite floor slab is shown in Fig. 12.5.

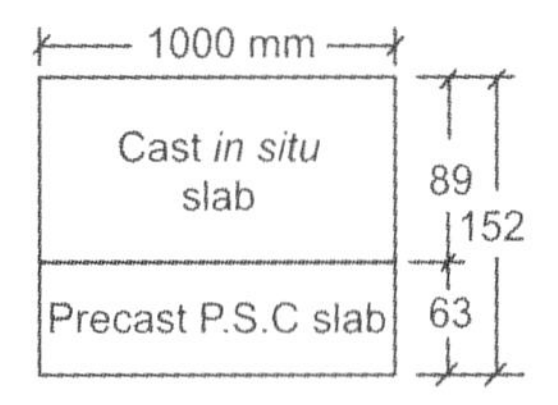

Fig. 12.5

Effective span = 5 m

Thickness of precast prestressed plank = 63 mm

Superimposed load = 8.75 kN/m^2

Referring to Table 12.1

Table 12.1: Prestressed concrete floor slabs

Section details		Solid prestressed plank units 63 mm deep; Total superimposed load, kN/m^2							
Total slab depth, mm	***In situ* topping depth, mm**	**3.0**	**4.0**	**5.0**	**6.0**	**7.50 span (m)**	**8.75**	**10.00**	**12.50**
100	38				4.25	3.83	3.65	3.45	3.15
114	50				4.75	4.35	4.05	3.80	3.45
125	63				5.15	4.70	4.40	4.15	3.80
138	76		5.85		5.10	5.00	4.75	4.45	4.10
152	88		6.20		5.80	5.30	5.00	4.75	4.35
165	100	7.00	6.55		6.10	5.60	5.30	5.00	4.55
178	114	7.30	6.40		6.40	5.85	5.60	5.25	4.85

Note: Spans to the left of the step-line are limited by deflection.

The thickness of cast *in situ* slab required to support a superimposed load of 8.75 kN/m^2 over a span of 5 m is read out as 89 mm

$\therefore$ Overall depth of slab = 152 mm

Properties of precast P.S.C. section

$$A = (1000 \times 63) = 63 \times 10^3 \text{ mm}^2$$

$$Z_t = Z_b = \frac{1000 \times 63^2}{6} = 661.5 \times 10^3 \text{ mm}^3$$

Section modulus of composite section

$$Z_b^1 = \left(\frac{1000 \times 152^2}{6}\right) = 385 \times 10^4 \text{ mm}^3$$

M = Moment acting on the precast part of a composite section during construction

Self weight of slab = (0.152 × 24) = 3.7 kN/m

$$M = (0.125 \times 3.7 \times 5^2) = 11.56 \text{ kN.m/m}$$

$$M^1 = (0.125 \times 8.75 \times 5^2) = 27.34 \text{ kN.m/m}$$

$$f_b = \left[\frac{f_{tw}}{\eta} + \frac{M}{\eta Z_b} + \frac{M^1}{\eta Z_b^1}\right]$$

Given, $f_{tw} = 0, \qquad \eta = 0.8$

$$f_b = \left[0 + \frac{11.56 \times 10^6}{0.8 \times 661.5 \times 10^3} + \frac{27.34 \times 10^6}{0.8 \times 385 \times 10^4}\right]$$

$$= 10.07 \text{ N/mm}^2$$

$$f_t = \left[f_t - \frac{M_{min}}{Z_t}\right] = 0 \qquad (M_{min} = 0)$$

$$\therefore \quad P = \frac{A(f_t Z_t + f_b Z_b)}{Z_t + Z_b}$$

$$= \left\{\frac{63 \times 10^3 (0 + 10.07 \times 661.5 \times 10^7)}{2 \times 661.5 \times 10^3}\right\}$$

$$= 317 \times 10^3 \text{ N} = 317 \text{ kN}$$

$$e = \frac{Z_t Z_b (f_b - f_t)}{A(f_t Z_t + f_b Z_b)}$$

$$= \left\{\frac{661.5^2 \times 10^6 (10.07 - 0)}{63 \times 10^3 (0 + 10.07 \times 661.5 \times 10^3)}\right\}$$

$$= 10.5 \text{ mm}$$

Problem 12.7

A prestressed concrete slab is to be designed as a one way slab spanning over 6 m. The permissible compressive stress in concrete is 15 N/mm^2 and no tension is permitted. Loss ratio = 0.8. The live load is 12 kN/m^2. Cable containing 12 wires of 5 mm diameter initially tensioned to 1200 N/mm^2 are available for use. Design the slab and determine the spacing of the cables.

Solution

Data: Live load $= q = 12$ kN/m^2 $\qquad f_{ct} = 15$ N/mm^2

$f_{tw} = 0 \qquad \eta = 0.8$

$L = 6$ m

Assume overall depth of the slab at 40 mm per metre span

$$h = (40 \times 6) = 240 \text{ mm}$$

Considering 1 m width of the slab,

$$A = (1000 \times 240) = 24 \times 10^4 \text{ mm}^2$$

$$Z_b = \left(\frac{bh^2}{6}\right) = \frac{(1000 \times 240^2)}{6} = 96 \times 10^5 \text{ mm}^3$$

$$g = (1 \times 0.24 \times 24) = 5.76 \text{ kN/m}$$

$$M_g = \frac{(5.76 \times 6^2)}{8} = 25.92 \text{ kN.m}$$

$$M_q = \frac{(12 \times 6^2)}{8} = 54 \text{ kN.m}$$

$$Z_b = \left[\frac{M_q + (1 - \eta)M_g}{\eta f_{ct} - f_{tw}}\right]$$

$$= \left[\frac{(54 \times 10^6) + (1 - 0.8)(25.92 \times 10^6)}{(0.8 \times 15 - 0)}\right]$$

$$= 49.32 \times 10^5 \text{ mm} < 96 \times 10^5 \text{ mm}^3, \text{ hence safe.}$$

$$f_t = f_{tt} - \left(\frac{M_g}{Z_b}\right) = 0 - \frac{(25.92 \times 10^6)}{(49.32 \times 10^5)}$$

$$= -5.25 \text{ N/mm}^2$$

$$f_b = \left[\frac{f_{tw}}{\eta} + \frac{(M_g + M_q)}{\eta Z_b}\right]$$

$$= \left[0 + \frac{(79.92 \times 10^6)}{(0.8 \times 49.32 \times 10^5)}\right] = 20.25 \text{ N/mm}^2$$

Minimum prestressing force is given by

$$P = \frac{A}{2}[f_t + f_b] = \frac{24 \times 10^4}{2}[-5.25 + 20.25]$$

$$= 1800 \times 10^3 \text{ N} = 1800 \text{ kN}$$

The eccentricity of the prestressing force is computed as

$$e = \left[\frac{Z(f_b - f_t)}{A(f_b + f_t)}\right]$$

$$= \left[\frac{49.32 \times 10^5\,(20.25 + 5.25)}{24 \times 10^4\,(20.25 - 5.25)}\right] = 34.93\text{ mm}$$

$$\text{Force in each cable} = \left[\frac{12 \times 19.6 \times 1200}{1000}\right] = 282.24\text{ kN}$$

$$\therefore \quad \text{Spacing of cables} = \left[\frac{1000 \times 282.24}{1800}\right] = 156.8\text{ mm}$$

Problem 12.8

A post tensioned prestressed concrete two way slab supported on walls all-around is 6 m by 6 m. The slab has to support a live load of 4 kN/m². Cables of four wires of 5 mm diameter high tensile wires initially stressed to 1000 N/mm² are available for use. Determine the number of cables if $f_{ct} = 15\text{ N/mm}^2$, $f_{tw} = 0$ and $\eta = 0.8$.

Solution

Data: $L_x = L_y = 6$ m Ratio of $\left(\frac{L_y}{L_x}\right) = 1$

$q = 4\text{ kN/m}^2$ $f_{ct} = 15\text{ N/mm}^2$

$f_{tw} = 0$ $\eta = 0.8$

$$\text{Force in each cable} = \left[\frac{(4 \times 19.6 \times 1000)}{1000}\right] = 78\text{ kN}$$

$$\text{Effective depth of slab} = \left(\frac{\text{span}}{50}\right) = \left(\frac{6000}{50}\right)$$

$$= 120\text{ mm}$$

Provide overall thickness $= 150$ mm

Self weight $g = (0.15 \times 24 \times 1) = 3.6\text{ kN/m}^2$

Load on slab

Live load $= 4.0\text{ kN/m}^2$

Self weight $= 3.6\text{ kN/m}^2$

Finishes $= 0.4\text{ kN/m}^2$

Total load $w = 8.0\text{ kN/m}^2$

Refer Table 12.1 and read out the moment coefficients for the

ratio $\left(\frac{L_y}{L_x}\right) = 1.0$

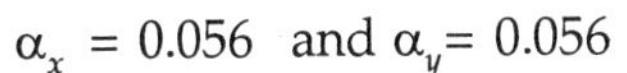

$$\alpha_x = 0.056 \text{ and } \alpha_y = 0.056$$

Working moments in x and y directions are computed as

$$M_x = M_y = (0.056 \times 8 \times 6^2) = 16.12 \text{ kN.m}$$

Total moment in the middle strips is computed as

$$M = (16.12 \times 0.75 \times 6) = 72.5 \text{ kN.m}$$

Number of cables

Using a minimum cover of 30 mm for cables, the distance between the top kern and centroid of cable is computed as

$$(150 - 30 - 50) = 70 \text{ mm}$$

If P = total prestressing force in x and y directions,

$$(P \times 10^3 \times 70) = 72.5 \times 10^6$$

Solving, $P = 1036$ kN

Force in each cable = 78 kN

Number of cables in each direction (middle strip) is

$$= \left(\frac{1036}{78}\right) = 13.28$$

$$\text{Spacing of cables} = \frac{(0.75 \times 6 \times 1000)}{13.28} = 338 \text{ mm}$$

Provide cables at a spacing of 338 mm in the middle strip in each direction.

Problem 12.9

A prestressed concrete grid floor of size 12 m by 7 m is made up of slab and T-beam crossing each other in both directions. The slab is 87.5 mm thick and the beams have a rib of width 200 mm and depth 212.5 mm. The effective width of flange of T-beam may be taken at 1250 mm. The analysis of moments due to dead and live loads indicates the maximum moments in the central T-beams of long span (x-direction) and short span (y-direction) as follows:

Type of moments	Short span (x-direction)	Long span (y-direction)
Dead load moments (M_g)	31.5 kN.m	14.2 kN.m
Live load moments (M_q)	27.0 kN.m	12.1 kN.m

The permissible compressive stress in concrete is 15 N/mm^2 and no tension is permitted at any stage. Assume the loss ratio as 0.85. Using Freyssinet cables containing 18 wires of 5 mm diameter initially stressed to 1100 N/mm^2, check the adequacy of the section provided and determine the prestressing force and eccentricity required for the cables in the short and long span directions.

Solution

T-beam and slab dimensions:

Width of flange = 1250 mm
Thickness of flange = 87.5 mm
Depth of rib = 212.5 mm
Width of rib = 200 mm
Overall depth of section = 300 mm

Permissible compressive stress in concrete $= f_{ct} = 15 \text{ N/mm}^2$

Tensile stresses are not permitted:

$$f_{tw} = 0$$

Initial stress in H.T wires $= 1100 \text{ N/mm}^2$

Loss ratio $\eta = 0.85$

The cross-sectional area of the T-section

$$A = 157200 \text{ mm}^2$$

The second moment of area of the T-section

$$I = (92 \times 10^7) \text{ mm}^4$$

The centroidal axis is located at a distance of 85 mm from the top.

Hence $y_t = 85$ mm and $y_b = 215$ mm

Section modulus is computed as

$$Z_t = \left[\frac{I}{y_t}\right] = \left[\frac{(92 \times 10^7)}{85}\right] = (10.8 \times 10^6) \text{ mm}^3$$

$$Z_b = \left[\frac{I}{y_b}\right] = \left[\frac{(92 \times 10^7)}{215}\right] = (4.2 \times 10^6) \text{ mm}^3$$

The minimum section modulus required is calculated as

$$Z_b \geq \left[\frac{M_q + (1-\eta) M_g}{f_{br}}\right]$$

$$f_{br} = [\eta f_{ct} - f_{tw}] = [0.85 \times 15) - 0] = 12.75 \text{ N/mm}^2$$

$$Z_b \geq \left[\frac{27 + (1-0.85)31.5}{12.75}\right] 10^6 \geq (2.5 \times 10^6) \text{ mm}^3$$

$$< (4.25 \times 10^6) \text{ mm}^3$$

Hence, the section can safely resist the applied moments.

The minimum prestressing force required is obtained from the expression,

$$P = \frac{A(f_t Z_t + f_b Z_b)}{(Z_t + Z_b)}$$

where, $$f_b = \left[\frac{M_q + M_g}{\eta Z_b}\right] = \left[\frac{58.5 \times 10^6}{0.85 \times 4.25 \times 10^6}\right] = 16.3 \text{ N/mm}^2$$

$$f_t = -\frac{M_g}{Z_t} = \left[\frac{31.5 \times 10^6}{1.08 \times 10^6}\right] = -2.9 \text{ N/mm}^2$$

$$P = \left[\frac{157200 \times (16.3 \times 4.2 - 2.9 \times 10.8)10^6}{15.05 \times 10^6}\right]$$

$$= (395 \times 10^3) \text{ N} = 395 \text{ kN}$$

Using Freyssinet cables of 18–5 mm diameter stressed to 1100 N/mm^2, force in each cable

$$= \left[\frac{(18 \times 20 \times 1100)}{1000}\right] = 396 \text{ kN}$$

One cable is provided in the central rib (x-direction) at an eccentricity given by the relation,

$$e = \frac{Z_t Z_b (f_b - f_t)}{A(f_t Z_t + f_b Z_b)}$$

$$= \frac{(10.8 \times 4.2)(10^{12})(16.3 + 2.9)}{[157200 \times (16.3 \times 4.2 - 2.9 \times 10.8)10^6]} = 150 \text{ mm}$$

In the central rib in y-direction, we have the moments as,

$$M_g = 14.2 \text{ kN.m and } M_q = 12.1 \text{ kN.m}$$

$$f_b = \left[\frac{26.3 \times 10^6}{0.85 \times 4.2 \times 10^6}\right] = 7.3 \text{ N/mm}^2$$

$$f_t = -\left[\frac{14.2 \times 10^6}{10.8 \times 10^6}\right] = -1.32 \text{ N/mm}^2$$

$$P = \left[\frac{157200 \times (7.3 \times 4.2 - 1.32 \times 10.8)10^6}{15.05 \times 10^6}\right]$$

$$= (175 \times 10^3) \text{ N} = 175 \text{ kN}$$

The corresponding eccentricity is obtained as

$$e = \left[\frac{10.8 \times 4.2 \times 10^{12}(7.3 + 1.32)}{2.65 \times 10^{12}}\right] = 150 \text{ mm}$$

This eccentricity is not practicable as it obstructs the cables in the x-direction. Hence, the maximum possible eccentricity is obtained as,

$$e = [150 - 20 - 40 - 20] = 70 \text{ mm}$$

The modified prestressing force located at this reduced eccentricity is computed as,

$$P = \left[\frac{Af_b Z_b}{z_b + Ae}\right]$$

$$= \left[\frac{157200 \times 7.3 \times 4.2 \times 10^6}{(4.2 \times 10^6) + (157200 \times 70)}\right]$$

$$= (320 \times 10^3) \text{ N} = 320 \text{ kN}$$

Using Freyssinet cable (18 – 5) initially stressed to 900 N/mm^2 will yield a force of 324 kN.

Problem 12.10

A highway bridge deck slab spanning 10 m is to be designed as a one way prestressed concrete slab with parallel post-tensioned cables carrying an effective force of 620 kN. The deck slab is required to support a uniformly distributed live load of 25 kN/m^2. The permissible stresses in concrete should not exceed 15 N/mm^2 in compression and no tension is permitted at any stage. Design the spacing of the cables and their position at mid span section. Assume loss of prestress as 20 percent.

Solution

Span of the deck slab = 10 m

Distributed working live load = 25 kN/m^2

Force in the cable = 620 kN

Permissible compressive stress in concrete

$$f_{ct} = 15 \text{ N/mm}^2$$

Permissible tensile stress in concrete

$$f_{\min.w} = 0$$

Loss ratio $\eta = 0.85$

The live and dead load moments are computed considering one metre width of the slab.

$$M_g = \left[\frac{25 \times 10^2}{8}\right] = 312.5 \text{ kN.m}$$

Let h = overall depth of the slab

b = width of the slab

$$\therefore \quad M_g = \left[\left(\frac{bh}{10^6}\right) \times 24 \times \left(\frac{10^2}{8}\right)\right]$$

$$= \left[\frac{300bh}{10^6}\right] \text{kN.m} = 300\, bh \text{ N.mm}$$

Range of stress at bottom fibre

$$f_{br} = [\eta f_{ct} - f_{\min.\omega}] = [(0.85 \times 15) - 0] = 12 \text{ N/mm}^2$$

Hence, the minimum section modulus is given by the expression,

$$Z_b > \left[\frac{bh^2}{6}\right] > \left[\frac{M_q + (1-\eta) M_g}{f_{br}}\right]$$

$$\therefore \left[\frac{1000 \times h^2}{6}\right] > \left[\frac{(312.5 \times 10^6) + (1-0.8)\, 300 \times 1000 \times h}{12}\right]$$

or $h^2 - 30h + 156250 = 0$

Solving, $h = 410$ mm

$$\therefore \quad A = (1000 \times 410) = (41 \times 10^4) \text{ mm}^2$$

$$Z_b = Z_t = \left[\frac{1000 \times 410^2}{6}\right] = (28 \times 10^6) \text{ mm}^3$$

and $M_g = (300\, bh) = (300 \times 1000 \times 410) = (123 \times 10^6)$ N.mm

$$f_t = \left[0 - \left(\frac{123 \times 10^6}{0.8 \times 28 \times 10^6}\right)\right] = -4.4 \text{ N/mm}^2$$

$$f_b = \left[0 + \frac{(312.5 + 123)10^6}{(0.8 \times 28 \times 10^6)}\right] = 19.4 \text{ N/mm}^2$$

The minimum prestressing force required is computed as,

$$P = \frac{A}{2}(f_b + f_t) = \frac{(41 \times 10^4)}{2}(19.4 - 4.4)$$

$$= (3075 \times 10^3) \text{ N} = 3075 \text{ kN}$$

The eccentricity is calculated as,

$$e = \frac{Z(f_b - f_t)}{A(f_b + f_t)}$$

$$= \left[\frac{28 \times 10^4 (19.4 + 4.4)}{41.4 \times 10^4 (19.4 - 4.4)}\right] = 109 \text{ mm}$$

Spacing of the cables

$$= \left[\frac{(1000 \times 620)}{3075}\right] = 201 \text{ mm}$$

Adopt a spacing of 200 mm centres for the cables in the span direction.

13

Prestressed Concrete Shell and Folded Plate Structures

Problem 13.1

A concrete cylindrical shell roof covering an area of 10 m by 30 m is to be designed with prestressed edge beams. Using the beam approximation, prepare a preliminary design for the edge beams using the following data:

Radius of the shell	= 7.5 m
Thickness of shell	= 75 m
Semi central angle	= 40°
Width of edge beam	= 150 mm
Depth of edge beam	= 1500 mm
Imposed load on shell	= 1 kN/m^2

The centroid of the cables (12 wires of 7 mm diameter stressed to 1000 N/mm^2) is located at a distance of 300 mm from the soffit of the beam. Calculate the prestressing force necessary in each beam for zero stress at the soffit under:

a. dead load + half live load; and
b. dead load + live load
c. Also find the number of cables required in each beam for case (a)

Solution:

The dimensions of the shell with edge beams as shown in Fig. 13.1 are first assumed and the analysis is made by using beam theory.

The dimensions assumed are as follows:

Semi central angle	α = 40° = 0.6981 radians
Span of shell,	L = 30 m
Chord width	= 10 m
Radius of shell	= 7.5 m

Thickness of shell $= 0.075$ m
Width of edge beam $= B = 0.15$ m

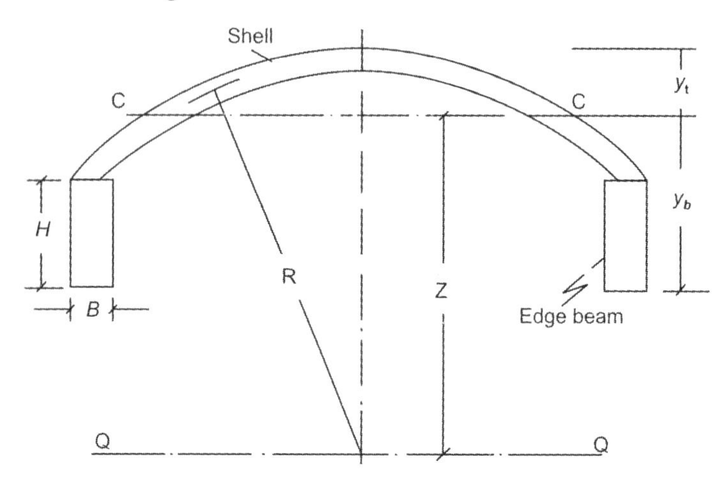

Fig. 13.1

Depth of edge beam $H = 1.5$ m
Rise of shell $= R(1 - \cos\alpha)$
$= 7.5(1 - 0.766)$
$= 1.76$ m

$$Z = \left\{\frac{tR^2 \sin\alpha + [R\cos\alpha - (H/2)]\, H \cdot B}{Rt\alpha + HB}\right\}$$

$$= \left\{\frac{(0.075 \times 7.5^2 \times 0.6428) + (7.5 \times 0.766 - 0.75)\, 1.5 \times 0.15}{(7.5 \times 0.075 \times 0.6981) + (1.5 \times 0.15)}\right\}$$

$= 6.2$ m

$\therefore \quad y_t = 1.338$ m
$y_b = 1.922$ m

Second moment of area of the shell about QQ axis is given by

$$I_{SQ} = tR^3\left(\alpha + \frac{1}{2}\sin 2\alpha\right)$$

$$= 0.075 \times 7.5^3\left(0.6981 + \frac{1}{2} \times 0.9848\right)$$

$$= 37.6 \text{ m}^4$$

Second moment of area of beams about QQ-axis is

$$I_{BQ} = 2\left(\frac{BH^3}{12}\right) + 2BH\left(R\cos\alpha - \frac{H}{2}\right)^2$$

$$= 2\left(\frac{0.15 \times 1.5^3}{12}\right) + 2 \times 0.15 \times 1.5\left(7.5 \times 0.7660 - \frac{1.5}{2}\right)^2$$

$$= 11.31 \text{ m}^4$$

Cross sectional area of shell and edge beams

$$A = 2(Rt\alpha + HB)$$

$$= 2(7.5 \times 0.075 \times 0.6981 + 1.5 \times 0.15)$$

$$= 1.234 \text{ m}^2$$

The second moment of area of the shell and edge beams about centroidal axis C-C is given by

$$I_C = (I_{SQ} + I_{BQ} - AZ^2)$$

$$= (37.6 + 11.31 - 1.234 \times 6.2^2)$$

$$= 1.48 \text{ m}^4$$

Calculation of loads on shell

Self weight = $(0.075 \times 24) = 1.8$ kN/m^2

Live load = 1.0

Total load = 2.8 kN/m^2 of surface area

Load on shell per meter of span

$$= (2.8 \times 7.5 \times 2 \times 0.6981) = 29.3 \text{ kN/m}$$

Self weight of edge beams = $2(1.5 \times 0.15 \times 24)$ = 10.8 kN/m

Live load on top of edge beam = $(2 \times 0.15 \times 1)$ = 0.3 kN/m

$\therefore$ Total load $w_d = (29.3 + 10.8 + 0.3)$ = 40.4 kN/m

$$\text{Maximum moment } M = \left(\frac{w_d L^2}{8}\right) = \left(\frac{40.4 \times 30^2}{8}\right)$$

$$= 4545 \text{ kN.m}$$

Bending stresses in the shell and edge beams are obtained as follows:

a. At top of shell $= \left\{\frac{(4545 \times 10^6)(1338)}{1.48 \times 10^{12}}\right\} = 4.10$ N/mm^2

(compression)

b. At bottom of shell or top of edge beam

$$= \left\{\frac{(4545 \times 10^6 \times 460)}{1.48 \times 10^{12}}\right\} = 1.41 \text{ N/mm}^2$$

(tension)

c. At soffit of edge beam $= \left\{\frac{(4545 \times 10^6 \times 1922)}{1.48 \times 10^{12}}\right\}$

$= 5.90 \text{ N/mm}^2$ (tension)

Prestressing force

Locating the centroid of the prestressing force at a distance of 400 mm from the soffit of the edge beams, the available eccentricity is given by

$$e = (1922 - 400) = 1522 \text{ mm}$$

Case a: *Dead load + half live load*

Load on shell/m of span = (2.3 × 7.5 × 2 × 0.6981) = 24 kN/m

Live load on top of edge beam = (2 × 0.15 × 0.5) = 0.15 kN/m

Total load w_d = (24 + 10.8 + 0.15) = 34.95 kN/m

$$\text{Maximum moment} = \left(\frac{34.95 \times 30^2}{8}\right) = 3932 \text{ kN.m}$$

Bending stress at soffit of beam

$$= \left[\frac{(3932 \times 10^6)\,1922}{1.48 \times 10^{12}}\right]$$

$= 5.10 \text{ N/mm}^2$ (tension)

If the bending stress at soffit of the beam is zero under dead load + half live load,

$$\left[\frac{P}{A} + \frac{PeY_b}{I}\right] = 5.10$$

$$P\left[\frac{1}{1.234 \times 10^6} + \frac{1522 \times 1922}{1.48 \times 10^{12}}\right] = 5.10$$

Solving, $P = 1834 \times 10^3$ and $N = 1834$ kN

∴ Prestressing force required for each beam

$$= \left(\frac{1834}{2}\right) = 917 \text{ kN}$$

Case b: *Dead load + live load*

$$\left[\frac{P}{A}+\frac{PeY_b}{I}\right] = 5.90$$

$$P\left[\frac{1}{1.234\times10^{6}}+\frac{1522\times1922}{1.48\times10^{12}}\right] = 5.90$$

$\therefore$ $$P = 2122 \text{ kN}$$

$\therefore$ Prestressing force required for each beam

$$= (2122)\ 0.5 = 1061 \text{ kN}$$

Using Freyssinet cables of 12–7 mm ϕ stressed to 1000 N/mm^2,

Force in each cable = 460 kN

$\therefore$ Number of cables $= \left(\frac{917}{460}\right) = 2$

Problem 13.2

A pre-tensioned hyperboloidal shell is required to cover the roof of a bus depot 30 m by 60 m in size. The loading should conform to IS: 875. M-50 grade concrete and high tensile wires conforming to IS: 1785 are available for use. Design the shell units and check for load factors against collapse in flexure and shear. Also check for stresses under serviceability limit state.

Solution

The cross-section of pre-tensioned hyperboloidal shell is shown in Fig. 13.2.

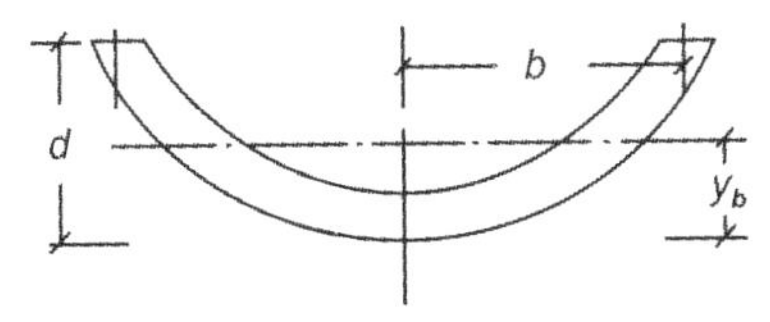

Fig. 13.2

The following dimensions are assumed:

Span		= 30 m
Effective span c/c of walls		= 30 m
Overall length of unit		= 31 m
Semi width of element	b	= 2.1 m
Thickness of shell	t	= 100 mm

Assuming the cross section of shell to correspond to a parabolic curve weight of shell unit

$$= (2.2942 \times 2.1 \times 31 \times 0.1 \times 24) = 358 \text{ kN}$$

Width of elemental unit $= 4.2$ m

Area of concrete $A_c = (2.2942 \times 2100 \times 100) = 481782 \text{ mm}^2$

Centroidal distance from soffit is given by

$$y_b = (0.1831 \times 2100) = 384.5 \text{ mm}$$

Depth of shell $d = \left(\frac{b}{2}\right) = \left(\frac{2100}{2}\right) = 1050 \text{ mm}$

$\therefore$ $y_t = (1050 - 384.5) = 665.5$ mm

Second moment of area of section about neutral axis:

$$I_c = 0.0536\, b^3\, t$$
$$= 0.0536 \times (2100)^3 \times 100$$
$$= 496 \times 10^8 \text{ mm}^4$$

Section modulus

$$Z_b = 0.2928 \times b^2 \times t \quad \text{or} \quad (I_c/y_b)$$
$$= 0.2928 \times (2100)^2 \times 100$$
$$= 129 \times 10^6 \text{ mm}^3$$
$$Z_t = (0.1693 \times b^2 \times t)$$
$$= (0.1693 \times (2100)^2 \times 100)$$
$$= 74.6 \times 10^6 \text{ mm}^3$$

Loads and moments

Self weight of unit $= (0.481 \times 24) = 11.54$ kN/m

Live load on roof $= 0.75 \text{ kN/m}^2$

Total live load $= (0.75 \times 4.2) = 3.15$ kN/m

$\therefore$ Total design load $= (11.54 + 3.15) = 14.49 = 14.50$ kN/m

Maximum B.M. at mid span

$$M_d = (14.50 \times 15.5 \times 15) - (0.5 \times 14.50 \times 15.5^2)$$
$$= 1630 \text{ kN.m}$$

Eccentricity 'e' at mid span is given by

$$e = y_b - (\text{cover} + \text{diameter of H.T. wire})$$
$$= 384.5 - (25 + 7) = 352.5 \text{ mm}$$

Effective prestressing forces is

$$P = \left[\frac{M_d}{\left(\frac{Z_b}{A_c}\right) + e}\right] = \left[\frac{1630 \times 10^6}{\left(\frac{129 \times 10^6}{481782}\right) + 352.5}\right]$$

$$= 2630 \times 10^3 \text{ N} = 2630 \text{ kN}$$

Initial prestressing force

$$P_i = (1.25 \times 2630) = 3287 \text{ kN}$$

Number of 7 mm ϕ high tensile wires stressed to 1100 N/mm^2 is given by

$$\text{No. of wires} = \left(\frac{3287 \times 10^3}{38.4 \times 1100}\right) = 78 \text{ wires}$$

The wires are arranged in two layers at the centre of span, spread over two bands towards the supports.

Problem 13.3

Design a V-shaped pertensioned folded plate roof to cover an industrial warehouse measuring 20 m by 50 m. Loading is as per IS: 875. M-45 grade concrete and 5 mm H.T. wires are available for use, check for stresses under working loads and for load factors required as per IS: 1343 in flexure. Sketch the details of wires in the plates and the valley junction cable in the cross-section of the pre-tensioned units.

Solution

The cross-section of V-shaped pre-tensioned folded plate is shown in Fig. 13.3.

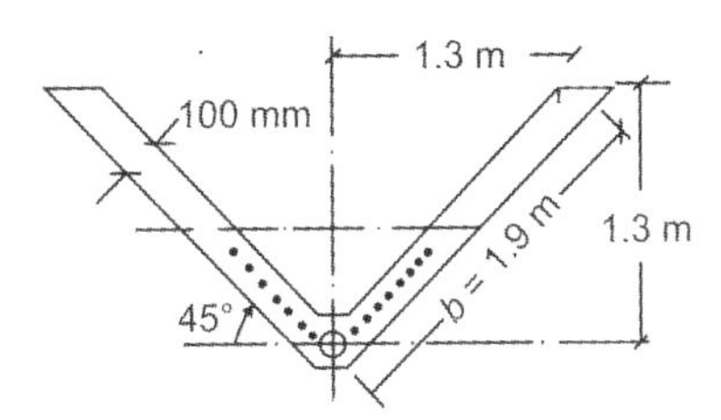

Fig. 13.3

Span of folded plate		= 20 m
Thickness of plate	t	= 100 mm
Depth of V-plate unit		= (1/15) × 20 = 1.3 m
Width of plate	b	= $1.3\sqrt{2}$ = 1.9 m
Angle of inclination	θ	= 45°

Sectional properties

Cross-sectional area $A = (1.9 \times 0.1) = 0.19 \text{ m}^2$

Second moment of area

$$I_c = \left[\frac{4t \sin\theta(b/2)^3}{3}\right] = \left[\frac{4 \times 0.1 \times 0.707(0.95)^3}{3}\right]$$

$$= 0.081 \text{ m}^4$$

Distance of extreme fibres from the centroidal axis

$$y_t = y_b = (1.3/2) = 0.65 \text{ m}$$

Section modulus $Z = Z_b = Z_t$

$$= \left(\frac{I}{y_t}\right) = \left(\frac{0.081}{0.65}\right) = 0.1246 \text{ m}^3$$

Loads

Self weight $= (0.1 \times 2 \times 1.9 \times 24) = 9.12$ kN/m

Live load $= (1 \times 2 \times 1.9) = 3.80$

(1 kN/m²)

Finishes etc. $= 1.08$

Total load w_d $= 14.00$ kN/m

$$\text{Max. moment } M = \left(\frac{14 \times 20^2}{8}\right) = 700 \text{ kN.m}$$

Longitudinal bending stress at soffit

$$\left(\frac{M}{Z_b}\right) = \left(\frac{700 \times 10^6}{0.1246 \times 10^9}\right)$$

$$= 5.6 \text{ N/mm}^2$$

If the centre of gravity of the high tensile wires is kept at the bottom kern point, eccentricity,

$$e = \left(\frac{I}{Ay_t}\right) = \left(\frac{0.081 \times 10^{12}}{0.19 \times 10^6 \times 0.65 \times 10^3}\right)$$

$$= 655 \text{ mm}$$

Since, this is not practicable

$$\text{Adopt} \qquad e = \left(\frac{d}{6}\right) = \left(\frac{1.9 \times 10^3}{6}\right) = 316 \text{ mm}$$

If P = Prestress force required to impart a compressive stress f_b at the soffit, then

$$P = \left(\frac{f_b A Z_b}{Z_b + Ae}\right)$$

$$= \left[\frac{5.6 \times 0.19 \times 10^6 \times 0.1246 \times 10^9}{(0.1246 \times 10^9) + (0.19 \times 10^6 \times 316)}\right]$$

$$= 717 \times 10^3 \text{ N} = 717 \text{ kN}$$

Using 5 mm wires stressed to 1200 N/mm^2

$$\text{Number of wires} = \left(\frac{717 \times 10^3}{19.6 \times 1200}\right) = 32 \text{ wires}$$

Provide 10 wires in each plate and one of 12–5 mm cable at the valley junction.

Problem 13.4

A spherical dome is to be designed to cover a circular tank of 36 m diameter. The rise of dome is one-eighth of the diameter, the thickness of dome is 1/500 diameter and the live load on dome is 1.5 kN/m^2. Design a suitable prestressed ring girder for the dome given the permissible compressive stress in the concrete as 14 N/mm^2. The loss ratio is 0.8. Use 7 mm diameter high tensile wires initially stressed to 1000 N/mm^2 to provide the necessary prestress.

Solution

The salient dimensions of the spherical dome is shown in Fig. 13.4.

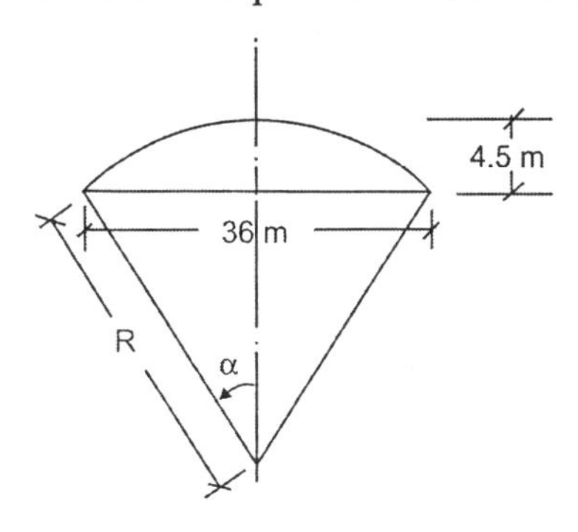

Fig. 13.4

$$\text{Diameter of base} = 36 \text{ m}$$

$$\text{Rise of dome} = \left(\frac{1}{8} \times 36\right) = 4.5 \text{ m}$$

Thickness of dome $= \left(\frac{1}{500} \times 36 \times 10^3\right) = 72$ mm

Adopt $t = 75$ mm

Radius of the shell dome

$(R - 4.5)^2 + 18^2 = R^2$

$\therefore \quad R = 38.25$ m

Semi central angle

$\alpha = 28° \, 4'$

$\cos\alpha = 0.8823$

$\cot\alpha = 1.88$

Self weight $= (0.075 \times 24) = 1.8$ kN/m^2

Live load $= 1.5$

Total load $w_d = 3.3$ kN/m^2

$$N_a = \left(\frac{w_d R}{1 + \cos\alpha}\right) = \left(\frac{3.3 \times 38.25}{1 + 0.8823}\right)$$

$= 67$ kN/m

Max. meridional compressive stress

$= (67 \times 10^3)/(1000 \times 75) = 0.893$ N/mm^2

Nominal reinforcement of 0.25 per cent of concrete section is provided in the meridional and circumferential directions

Total load $W = 2\pi R^2 w_d (1 - \cos\alpha)$

$= 2\pi \times 38.25^2 \times 3.3 (1 - 0.8823)$

$= 3570$ kN

Hoop tension in ring beam

$$N = \left(\frac{W}{2\pi}\right)\cot\alpha$$

$$= \left(\frac{3570}{2\pi}\right) \times 1.88 = 1068 \text{ kN}$$

$\therefore$ Initial prestressing force

$$P = \left(\frac{1068}{0.8}\right) = 1335 \text{ kN}$$

Cross sectional area of ring beam is given by

$$A_c = \left(\frac{1335 \times 10^3}{0.8 \times 14}\right) = 119196 \text{ mm}^2$$

Provide a ring beam of 250 mm wide by 500 mm deep.

($A_c = 125000$ mm^2)

Number of 7 mm high tensile wires

$$= \left(\frac{1424 \times 10^3}{38.4 \times 1000}\right) = 37 \text{ wires}$$

Problem 13.5

A pre-tensioned V-shaped folded plate roof is to be designed for a factory having a span of 25 m. The cross section of the unit is shown in Fig. 13.5. The thickness of plate unit is 100 mm. Live load is 1 kN/m^2. 7 mm diameter high tensile wires with a ultimate tensile strength of 1500 N/mm^2 are available for use. Design the number of wires required in the folded plate.

Solution

The cross section of the folded plate is shown in Fig. 13.5.

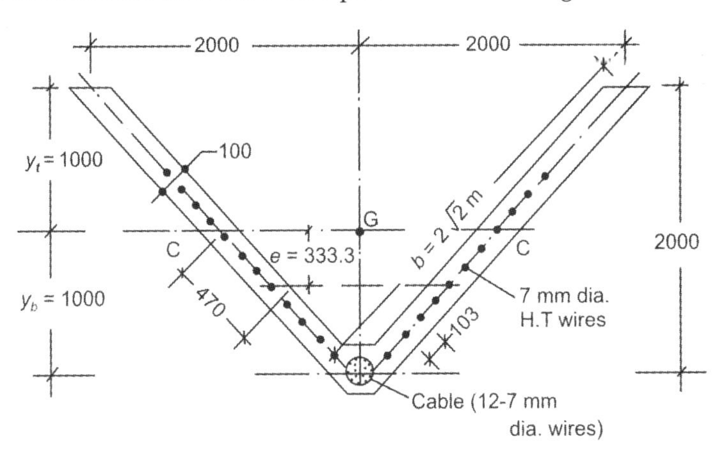

Fig. 13.5

Span = 25 m, Thickness of plates = 100 mm (t)

Width of plate $b = 2\sqrt{2}$

Cross sectional area $A = \left(2 \times 2\sqrt{2} \times 0.1\right) = 0.568 \text{ m}^2$

Second moment of area $I_C = \left[\dfrac{4t\sin^2\theta(b/2)^3}{3}\right]$

$$= \left[\frac{4 \times 0.1 \times 0.5\left(\sqrt{2}\right)^3}{3}\right] = 0.190 \text{ m}^4$$

Distance of extreme fibre $y_t = y_b = 1$ m

Section modulus $Z_t = Z_b = \left(\dfrac{I_c}{y_b}\right) = \left(\dfrac{0.190}{1}\right) = 0.190\ \text{m}^3$

Loads on the folded plate are computed as

Self weight = (2 × 2 × 0.1 × 24) = 13.80 kN/m

Live load = (2 × 2 × 2 × 1) = 5.65

Finishes etc. = 0.55

Total load = 20.00 kN/m

$$\text{Maximum bending moment} = M = \left(\frac{20 \times 25^2}{8}\right)$$

$$= 1562.5\ \text{kN.m}$$

Longitudinal bending stress at soffit is obtained as

$$f_b = \left(\frac{M_b}{Z_b}\right) = \left(\frac{1562.5 \times 10^6}{0.190 \times 10^9}\right) = 8.22\ \text{N/mm}^2$$

If the centre of gravity of the high tensile wires is kept at the bottom kern point, then

$$\text{Eccentricity}\quad e = \left(\frac{I}{A y_t}\right) = \left(\frac{0.190 \times 10^2}{0.568 \times 10^6 \times 10^3}\right) = 333.3\ \text{mm}$$

If P = Prestressing force required to impart a compressive stress of f_b at the soffit is computed as

$$P = \left[\frac{A f_b Z_b}{Z_b + Ae}\right]$$

$$= \left[\frac{0.568 \times 10^6 \times 8.22 \times 0.190 \times 10^9}{(0.190 \times 10^9) + (0.568 \times 10^6 \times 333.3)}\right]$$

$$= 2.34 \times 10^6\ \text{N} = 2340\ \text{kN}$$

Using 7 mm diameter high tensile wires initially stressed to 1200 N/mm^2, the number of wires required is computed as

$$\text{Number of wires} = \left[\frac{2340 \times 10^3}{38.5 \times 1200}\right] = 52\ \text{wires}$$

Provide one cable of 12–7 mm at the valley junction. The remaining 40 wires are distributed between the two plates.

$$(40 \times e) + (12 \times 1420) = (52 \times 470)$$

$$\therefore \quad e = 185 \text{ mm}$$

$$\text{Spacing of wires} = \left[\frac{2(1420 - 185)}{20}\right] = 123 \text{ mm}$$

Provide 20 high tensile wires in each plate at a spacing of 123 mm in the plane of the plate.

Problem 13.6

A spherical dome of an R.C. circular tank of 20 m diameter has a rise of 4 m and the thickness of the dome is 75 mm. The live load on the dome is 1.5 kN/m^2. Design a suitable prestressed concrete ring beam assuming the permissible compressive stress in concrete as 15 N/m^2. The loss ratio is 0.8. Adopt 5 mm diameter high tensile wires initially stressed to 1200 N/mm^2 to provide the necessary prestress.

Solution

The salient dimensions of the spherical dome is shown in Fig. 13.6.

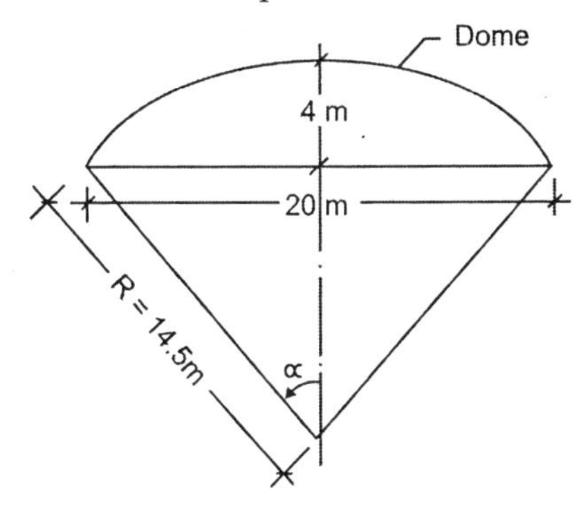

Fig. 13.6

Diameter at base = 20 m

Rise of the dome = 4 m

Thickness of the dome = 75 mm

Let R = radius of the shell dome

Then $(R - 4)^2 + 10^2 = R^2$

Solving, $R = 14.5$ m

If α = semicentral angle

$$\sin \alpha = \left(\frac{10}{14.5}\right) = 0.689$$

$\therefore \qquad \alpha = 43° - 30'$

$\cos\alpha = 0.725$

$\cot\alpha = 1.053$

Self weight of dome $= (0.075 \times 24) = 1.8\ \text{kN/m}^2$

Live load $= 1.5$

Finishes $= 2.0$

Distributed load $w_d = 3.5\ \text{kN/m}^2$

Total load on dome $w = 2\Pi R^2 w_d (1 - \cos\alpha)$

$= 2\Pi \times 14.5^2 \times 3.5\,(1 - 0.75) = 1271\ \text{kN}$

Hoop tension in ring beam $= \left(\dfrac{W}{2\Pi}\right)\cot\alpha$

$= \left(\dfrac{1271}{2}\right)1.053 = 213\ \text{kN}$

Initial prestressing force $P = \left(\dfrac{213}{0.8}\right) = 267\ \text{kN}$

Cross sectional area of ring beam $A_c = \left[\dfrac{213 \times 10^3}{0.8 \times 15}\right]$

$= 17750\ \text{mm}^2$

Adopt a ring beam of size 150 mm by 150 mm

Number of 5 mm diameter high tensile wires

$= \left[\dfrac{267 \times 10^3}{20 \times 1200}\right] = 11.125$

Provide one cable containing 12 wires of 5 mm diameter at the centre of the ring beam.

Problem 13.7

The concrete dome of a circular water tank 30 m diameter is supported by a ring beam at the base of the tank. Analysis indicates the maximum hoop tension in the ring beam as 420 kN. The permissible compressive stress in concrete is 14 N/mm^2. The loss of prestress may be taken as 25 percent. Adopting Freyssinet cables containing 12 wires of 5 mm diameter initially stressed to 1200 N/mm^2, design a suitable ring beam and the number of cables required to resist the service load ring tension.

Solution

Hoop tension in ring beam = 420 kN

Initial stress in H.T wires = 1200 N/mm^2

Loss of prestress η = 0.75

Freyssinet cables consisting 12 wires of 5 mm diameter are available for use.

Permissible compressible stress in concrete

$$f_{ct} = 14 \text{ N/mm}^2$$

Initial prestressing force required to resist hoop tension

$$P = \left[\frac{420}{0.75}\right] = 560 \text{ kN}$$

Cross-sectional area of ring beam

$$= \left[\frac{560 \times 10^3}{14}\right] = 40000 \text{ mm}^2$$

Adopt a ring beam of size 200 mm by 200 mm.

$$\text{Force in each cable} = \left[\frac{20 \times 12 \times 1200}{1000}\right] = 288 \text{ kN}$$

$$\text{Number of cables required} = \left[\frac{560}{288}\right] = 2$$

Provide 2 Freyssinet cables each containing 12 wires of 5 mm diameter.

Problem 13.8

The edge beam of a cylindrical shell roof is to be designed as a prestressed concrete girder spanning over 40 m to resist the following loads. The shell roof has a radius of 7.5 m and thickness of the shell is 100 mm. The width and depth of edge beams are 150 and 1500 mm respectively. The cross-sectional area of shell and edge beams are computed as 1.5 m^2. The centroidal axis lies at a distance of 1.2 m from the top of shell and 2 m from the soffit of the edge beams. The second moment of area about centroidal axis is 1.5 m^4. If the total dead and live loads per metre run of the shell roof is 40 kN/m.

a. Estimate the tensile stresses developed at the soffit of the edge beams.

b. Design the prestressing force necessary to nullify the tensile stresses in the edge beams.

c. Calculate the number of cables containing 12 – 8 mm diameter H.T wires initially stressed to 1100 N/mm^2 required in the edge beams.

Solution

Span of edge beams L = 40 m

Width of edge beam = 150 mm

Depth of edge beam = 1500 mm

Distance of soffit of beam from centroidal axis y_b = 2000 mm

Total load of shell roof + live load w = 40 kN/m

Cross-sectional area of shell with edge beams A = 1.5 m^2

Second moment of area of section I = 1.5 m^4

Initial stress in cable = 1100 N/mm^2

Maximum bending moment at the centre of span

$$M = \left[\frac{wL^2}{8}\right] = \left[\frac{40 \times 40^2}{8}\right] = 8000 \text{ kN.m}$$

Bending stress at soffit of edge beam

$$= \left[\frac{My_b}{I}\right] = \left[\frac{(8000 \times 10^6 \times 2000)}{(1.5 \times 10^{12})}\right]$$

$$= 10.66 \text{ N/mm}^2 \text{ (tension)}$$

If P = prestressing force required to eliminate the tensile stress, then we have the relation

$$\left[\frac{P}{A} + \frac{Pey_b}{I}\right] = P\left[\frac{1}{A} + \frac{ey_b}{I}\right] = 10.66$$

$$P\left[\frac{1}{(1.5 \times 10^6)} + \frac{(1500 \times 2000)}{(1.5 \times 10^{12})}\right] = 10.66$$

Solving, the prestressing force,

$$P = [3.81 \times 10^6]\ N = 3810 \text{ kN}$$

$$\text{Force in each cable} = \left[\frac{12 \times 50 \times 1100}{1000}\right] = 660 \text{ kN}$$

$$\text{Number of cables} = \left[\frac{3810}{660}\right] = 5.77 \cong 6$$

Provide 3 cables each in the two edge beams at an eccentricity of 1500 mm.

Problem 13.9

A prestressed concrete V-shaped folded plate proof of an industrial structure span over 30 m and is built up of plates of thickness 100 mm and width 2.82 m. The plates are joined by a valley cable and they are inclined at an angle of 45° to the horizontal. The total cross-sectional area of each V-unit is 0.568 m^2 and the second moment of area about the centroidal axis is 0.190 m^4. The extreme fibre distance of the unit from centroidal axis is 1 m. The total load inclusive of self weight, live load and finishes amounts to 20 kN/m. Estimate

a. The maximum bending tensile stress at the soffit of the V-shaped folded plate.
b. The effective prestressing force, located at the bottom kern point, required to counteract the tensile stress at the soffit.
c. The required number of 8 mm diameter H.T. wires initially stressed to 1200 N/mm^2.

Solution

Span of the fold plate roof $L = 30$ m

Thickness of plate = 100 mm

Width of plate = 2.82 m

Angle of inclination of the plate unit = 45°

Cross-sectional area of V-unit = 0.568 m^2

Depth of V-unit = 2 m

Distance of extreme fibres from centroidal axis

$$y_t = y_b = y = 1 \text{ m}$$

Second moment of area of cross-section = $I = 0.190 \text{ m}^4$

Total load of the structural system $w = 20$ kN/m

Section modulus of the V-unit

$$= \left[\frac{I}{y}\right] = \left[\frac{0.190}{1}\right] = 0.190 \text{ m}^3$$

Maximum bending moment at the centre of span

$$M = \left[\frac{wL^2}{8}\right] = \left[\frac{20 \times 30^2}{8}\right] = 2250 \text{ kN.m}$$

Bending tensile stress at soffit

$$= \left[\frac{M}{Z_b}\right] = \left[\frac{(2250 \times 10^6)}{(0.190 \times 10^9)}\right]$$

$$= 12.8 \text{ N/mm}^2 \text{ (tension)}$$

If P = effective prestressing force required to nullify the tensile stress at soffit, it is computed using the relation,

$$P = \left[\frac{Af_b Z_b}{Z_b + Ae}\right]$$

where f_b = compressive stress required at soffit to counteract the tensile stress = 12.8 N/mm^2

and $$e = \left[\frac{I}{Ay_t}\right] = \left[\frac{0.190 \times 10^{12}}{0.568 \times 10^6 \times 10^3}\right] = 333.3 \text{ mm}$$

$\therefore$ $$P = \left[\frac{0.568 \times 10^6 \times 12.8 \times 0.190 \times 10^6}{(0.190 \times 10^9) + (0.568 \times 10^6 \times 333.3)}\right]$$

$$= (3600 \times 10^3) \text{ N} = 3600 \text{ kN}$$

Using 8 mm diameter H.T. wires stressed to 1200 N/mm^2,

Force in each wire $= \left[\frac{50 \times 1200}{1000}\right] = 60$ kN

Number of wires required $= \left[\frac{3600}{60}\right] = 60$

Using one cable containing 12 wires of 8 mm diameter at the valley junction, the remaining 48 wires are distributed between the two plates.

Problem 13.10

A prestressed concrete hyperboloidal shell is to be designed to cover a roof of a factory spanning over 25 m. It is proposed to use pre-tensioned shell units of semi-width 1.6 m and span length 25 m. The thickness of the shell is 100 mm. Self weight of shell is 5.28 kN/m. Cross-sectional area of concrete is 367072 mm^2. Centroidal distance from soffit is 293 mm. Distance of top fibre from the centroidal axis is 507 mm. Second moment of area of the shell section about neutral axis is 21950 × 10^6 mm^4. The section modulus of top and bottom fibres is 433 × 10^5 mm^3 and 749 × 10^5 mm^3 respectively. Live load on the roof is 0.75 kN/m^2. Design the required prestressing force and the number of 7 mm diameter H.T. wires required.

Solution

Effective span of shell roof L = 25 m

Thickness of shell roof t = 100 mm

Semi-width of shell $= 1.6$ m

Self weight of shell $g = 5.28$ kN/m

Live load on shell $q = 0.75$ kN/m

Cross-sectional area of concrete $A_c = 367072 \text{ mm}^2$

Distance of top and bottom fibres from centroidal axis:

$$y_t = 507 \text{ mm and } y_b = 293 \text{ mm}$$

Second moment of area of shell about centroidal axis

$$I = 21950 \times 10^6 \text{ mm}^4$$

Section modulus:

$$Z_t = (433 \times 10^5) \text{ mm}^3 \text{ and } Z_b = (749 \times 10^5) \text{ mm}^3$$

H.T. wires of 7 mm diameter stressed to 1000 N/mm^2 are available for use.

Total live load $= (0.75 \times 3.2) = 2.4$ kN/m

$\therefore$ Total design load $w = [5.28 + 2.4] = 7.67$ kN/m

Maximum bending moment at mid span is computed as,

$$M_d = \left[\frac{wL^2}{8}\right] = \left[\frac{7.68 \times 25^2}{8}\right] = 600 \text{ kN.m}$$

Eccentricity of the prestressing force at mid span is computed as,

$$e = [y_b - (\text{cover} + \text{diameter of H.T. wire})$$
$$= [293 - (25 + 5)] = 263 \text{ mm}$$

Effective prestressing force is given by the relation,

$$P = \left[\frac{M_d}{\left(\frac{Z_b}{A_c}\right) + e}\right] = \left[\frac{600 \times 10^6}{\left(\frac{749 \times 10^5}{367072}\right) + 263}\right]$$

$$= (1280 \times 10^3) \text{ N} = 1280 \text{ kN}$$

Number of 7 mm diameter H.T. wires (stressed to 1000 N/mm^2) required is

$$= \left(\frac{1280000}{35 \times 1200}\right) = 33 \text{ wires}$$

The wires are arranged in two layers at the centre of span spread over two bands towards the supports.

14

Prestressed Concrete Poles and Piles

Problem 14.1

A pre-tensioned prestressed concrete pole is to be designed to suit the following data:

Height of pole above ground = 10 m

Wind force on wires acting at

a height of 8 m from base = 2 kN

Wind force on pole = 1.6 kN

Permissible compressive stress in concrete f_{cw} = 16 N/mm^2

No tension is permitted under working loads.

Loss ratio = 0.8

High tensile wires of 8 mm diameter initially stressed to 1200 N/mm^2 are available for use.

Design a suitable section for the pole at base and the number of wires required in the pole.

Solution

Referring to Fig. 14.1

Maximum working moment at base of pole is computed as

$$M_d = [(2 \times 8) + (1.6 \times 5)] = 24 \text{ kN.m}$$

Section modulus required at base is

$$Z_t = Z_b \geq \left[\frac{2M_d}{(f_{cw} - f_{tw})}\right]$$

$$\geq \left[\frac{2 \times 24 \times 10^6}{(16 - 0)}\right] \geq (3 \times 10^6) \text{ mm}^3$$

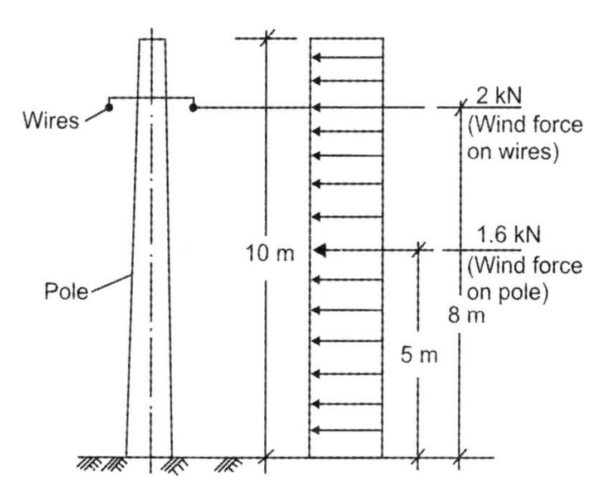

Fig. 14.1

Adopt a rectangular section with a width b = 200 mm

Overall depth $$h = \sqrt{\frac{6Z}{b}} = \sqrt{\frac{6\times3\times10^6}{200}} = 300 \text{ mm}$$

Hence, adopt a rectangular section 200 mm wide by 300 mm deep at base gradually tapering to 200 mm by 200 mm at the top of the pole.

Prestress in member $$\left(\frac{M_d}{\eta Z}\right) = \left(\frac{24\times10^6}{0.8\times3\times10^6}\right) = 10 \text{ N/mm}^2$$

Initial prestressing force $$= \left(\frac{10\times200\times300}{1000}\right) = 600 \text{ kN}$$

Force in each wire $= (50 \times 1200) = 60000$ N

Number of wires required $$= \left(\frac{600\times10^3}{60000}\right) = 10 \text{ wires}$$

Adopt 10 high tensile wires at the base section with 5 wires equally distributed in the compression and tension zones with a clear cover of 30 mm.

Problem 14.2

A prestressed concrete pole is to be designed to suit the following data:

Height of pole above ground = 12 m

Wind force on wires acting at 2 m from top = 1.5 kN

Wind force on pole acting at mid height = 2.5 kN

Permissible stresses in concrete in compression and tension are 16 N/mm^2 and 4 N/mm^2 respectively.

Loss ratio = 0.8

High tensile wire of 8 mm diameter initially stressed to 1200 N/mm^2 are available for use.

Design a suitable rectangular section for the pole and the number of high tensile wires at the base section.

Solution

Referring to Fig. 14.2.

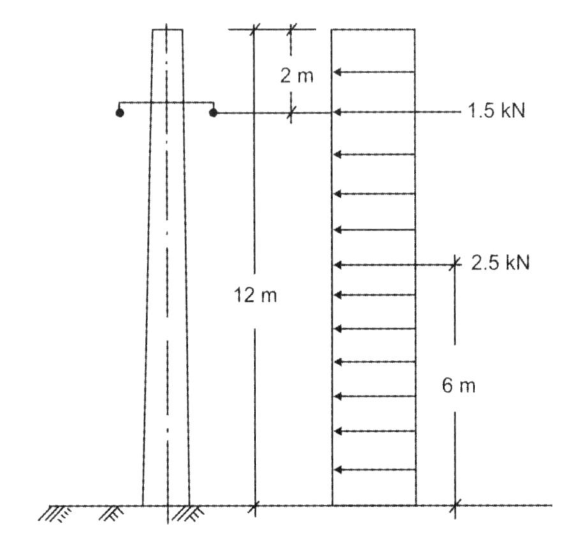

Fig. 14.2

Maximum moment at base of pole

$$M_d = [(1.5 \times 10) + (2.5 \times 6)] = 30 \text{ kN.m}$$

Section modulus required at base is given by

$$Z_t = Z_b \geq \left[\frac{2M_d}{(f_{cw} - f_{tw})}\right] \geq \left[\frac{2 \times 30 \times 10^6}{(16 - (-4))}\right]$$

$$\geq 3 \times 10^6 \text{ mm}^3$$

Adopt a rectangular section with a width of 200 mm

Overall depth $h = \sqrt{\frac{6Z}{b}} = \sqrt{\frac{6 \times 3 \times 10^6}{200}} = 300 \text{ mm}$

Hence, adopt a section rectangular in shape 200 mm wide by 300 mm deep at base gradually tapering to 200 mm by 200 mm at the top of the pole.

$$\text{Prestress in member} \left(\frac{M_d}{\eta Z}\right) = \left(\frac{30 \times 10^6}{0.8 \times 30 \times 10^6}\right)$$

$$= 12.5 \text{ N/mm}^2$$

$$\text{Initial prestressing force} = \left(\frac{12.5 \times 200 \times 300}{1000}\right) = 750 \text{ kN}$$

$$\text{Force in each wire} = (50 \times 1200) = 60000 \text{ N}$$

$$\text{Number of wires required} = \left(\frac{750 \times 10^6}{60000}\right) = 12.5$$

Adopt 14 high tensile wires of 8 mm diameter at base section with 7 wires equally distributed in the compression and tension zones.

Problem 14.3

The columns of an industrial building are supported on prestressed concrete piles of 8 m length. Each pile is subjected to an axial load of 4000 kN. The permissible effective prestress is not to exceed 5 N/mm^2. Design a suitable pile of square cross section. Also design the number of strands of 7–15.2 mm, required for the piles, if the ultimate tensile strength of the strand is 260 kN. Assume cylinder compressive strength of concrete as 40 N/m^2.

Solution

Data: $N = 4000$ kN $\quad f_{cy} = 40$ N/mm^2

$f_{cp} = 5$ N/mm^2

If N = Axial load on the pile

$$N = A_c\left[0.33\, f_{cy} - 0.27\, f_{cp}\right]$$

$$(4000 \times 10^3) = A_c[0.33 \times 40 - 0.27 \times 5]$$

Area of concrete $A_c = 338 \times 10^3$ mm^2

Referring to Table 14.1, a square pile of size 600 mm by 600 mm is adopted

$$A_c = 360 \times 10^3 \text{ mm}^2$$

$$\text{Prestressing force} = \left[\frac{5 \times 360 \times 10^3}{1000}\right] = 1800 \text{ kN}$$

Effective force in each strand (7--15.2 mm) under
service load = $(0.6 \times 260) = 156$ kN

Number of strands = $\left(\frac{1800}{156}\right) = 11.53$

Adopt 12 strands in a square pile of 600 mm size.

Problem 14.4

The columns of a multistoried building are to be supported on prestressed concrete pile foundations. It is proposed to use 350 mm solid octagonal piles having an effective length of 4.5 m to support a total axial service load of 1000 kN together with a moment of 25 kN.m. Design the high tensile steel strands of 7–12.5 mm wires in the piles to support the load assuming a load factor of 2 against collapse. The pile is to be designed to be lifted at any point along its length for installation. Assume cylinder compressive strength of concrete as 56 N/mm^2 and loss ratio is 0.85.

Solution

Data: $L = 4.5$ m $\quad f_{cy} = 56$ N/mm^2

$M = 25$ kN.m $\quad h = 350$ mm

$\eta = 0.85 \quad N_u = (2 \times 1200) = 2400$ kN

$N = 1200$ kN $\quad M_u = (2 \times 25) = 50$ kN.m

Referring to the interaction diagram (Fig. 14.3) and Table 14.1 and using 350 mm solid octagonal pile the section properties of the pile are as follows:

$A = 1.02 \times 10^5$ mm^2 $\quad w_d = 2.45$ kN/m

$Z = 4.70 \times 10^6$ mm^3 $\quad i = 90$ mm

Slenderness ratio $\left(\frac{L}{i}\right) = \left[\frac{(4.5 \times 1000)}{90}\right] = 50$

Eccentricity $e = \left(\frac{M_u}{N_u}\right) = \left[\frac{(50 \times 1000)}{2400}\right] = 20.83$

$\therefore$ Ratio of $\left(\frac{e}{h}\right) = \left(\frac{20.83}{350}\right) = 0.0595$

The ultimate moment and load capacity of the 350 mm solid octagonal pile for $M_u = 80$ kN.m and $(L/i) = 50$ interpolated from Fig. 14.3 is 2500 kN

Hence, $P_u = 2500$ kN

Effective prestress $= 8.4$ N/mm^2

$$\therefore \quad \text{Initial prestressing force} = \left(\frac{8.4 \times 1.02 \times 10^5}{0.85 \times 10^3}\right) = 1030 \text{ kN}$$

Use eight strands of 7–15.2 mm wires, initially stressed to 1120 N/mm^2.

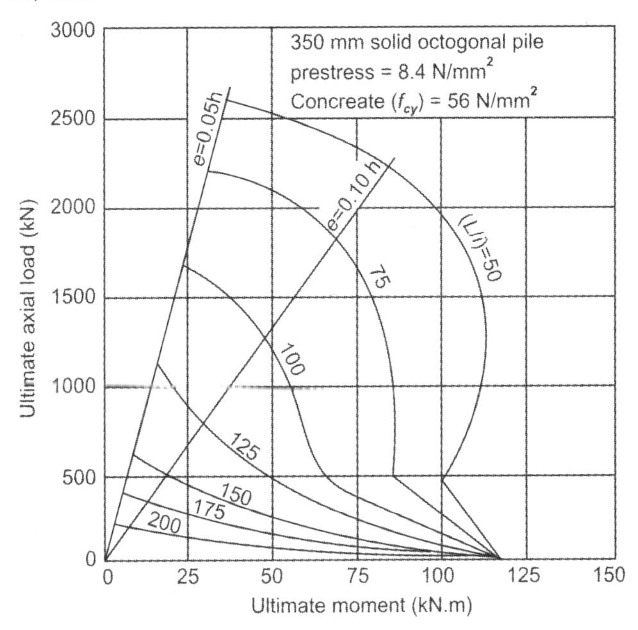

Fig. 14.3 Load moment characteristics of prestressed piles

Handling stresses

Self weight of pile = 2.45 kN/m

Impact allowance (50%) = 1.23 kN/m

w_d = 3.68 kN/m

Maximum moment in pile when lifted at centre is computed as

$$M = \left(\frac{3.68 \times 2.25^2}{2}\right) = 9.315 \text{ kN.m}$$

$$\text{Tensile bending stress} = \left(\frac{9.315 \times 10^6}{4.70 \times 10^6}\right) = 1.98 \text{ N/mm}^2$$

$$\text{Resultant stress} = (8.4 - 1.98) = 6.24 \text{ N/mm}^2$$

(compression)

Hence, the pile is safe against handling stresses.

Table 14.1: Section properties and allowable service loads of prestressed concrete piles

Size; Wire spiral; Strand; Square solid — Size; Square hollow — Core diameter; Size; Octagonal solid or hollow — Size; Round hollow

Size, mm	Core dia., mm	Area, mm^2 × 10^3	Weight, kN/m	Moment of inertia, mm^4 × 10^6	Section modulus, mm^3 × 10^6	Radius of gyration, mm	Perimeter, mm	Allowable concrete service load, kN — Cylinder strength, f_{cy} (N/mm^2) 35	42	49	56
(1)	(2)	(3)	(4)	(5)	(6)	(7)	(8)	(9)	(10)	(11)	(12)
					Square piles						
250	Solid	62.5	1.50	325	2.60	72	1000	652	790	950	1100
300	Soild	90.0	2.16	675	4.50	87	1200	940	1160	1360	1580
350	Soild	122.5	9.95	1250	7.15	101	1400	1280	1570	1870	2150
400	Soild	160.0	3.85	2130	10.60	118	1600	1680	2050	2440	2830
450	Soild	202.5	4.87	3400	15.20	132	1800	2130	2600	3090	3560
500	Soild	250.0	6.00	5200	20.70	146	2000	2630	3200	3800	4400
500	275	174.5	4.20	4950	19.80	161	2000	2000	2440	2900	3340
600	Soild	360.0	8.65	10800	36.00	174	2400	3760	4600	5480	6300
600	300	270.0	6.50	10400	34.70	190	2400	3030	3700	4400	5080
600	350	237.5	5.70	10100	33.60	196	2400	2760	3390	4000	4630

(Contd.)

(Contd.)

(1)	(2)	(3)	(4)	(5)	(6)	(7)	(8)	(9)	(10)	(11)	(12)
					Octagonal piles						
600	375	220.0	5.30	9800	32.70	199	2400	2610	3200	3800	4380
250	Solid	52.0	1.25	216	1.73	65	830	535	660	770	900
300	Soild	74.8	1.69	433	2.87	77	1000	770	948	1110	1290
350	Soild	102.0	2.45	822	4.70	90	1160	1050	1290	1520	1760
400	Soild	133.0	3.20	1400	7.00	103	1330	1370	1690	1980	2300
450	Soild	168.0	4.05	2230	9.90	116	1500	1730	2130	2500	2900
500	Soild	207.0	4.98	3420	13.70	129	1660	2140	2630	3100	3580
500	275	135.0	3.25	3130	12.50	146	1660	1390	1710	2000	2320
550	Soild	253.0	6.08	5020	18.30	141	2000	2600	3200	3750	4350
600	325	156.0	3.75	4480	16.30	164	2000	1610	1980	2320	2680
600	Soild	300.0	7.20	7100	23.60	154	2000	3100	3800	4450	5180
600	375	165.0	3.95	6120	20.40	181	2000	1700	2090	2450	2840
					Round piles						
900	650	303.0	7.28	23400	52.0	280	2830	3130	3830	4500	5220
1200	950	420	10.10	61800	102.5	382	3780	4320	5300	6250	7250
1350	1100	480.0	11.50	91000	135.0	435	4250	4950	6080	7150	8300

Problem 14.5

A pre-tensioned prestressed concrete pole of rectangular section 150 mm wide by 400 mm deep at the base is proposed for a pole of height 10 m. The analysis of wind loads on pile face and wires indicate a maximum design moment of 25 kN.m at the base section. The permissible compressive stress in concrete is 14 N/mm^2 and no tension is permitted under working loads. The loss of prestress may be taken as 30 percent. 5 mm high tensile wires initially stressed to 1500 N/mm^2 are available for use. Check the adequacy of the section and determine the number of wires required in the section.

Solution

Cross-section of pole at the base:

Width $b = 150$ mm

Depth $h = 400$ mm

Maximum working moment at base section

$$M_d = 25 \text{ kN.m}$$

Permissible compressive stress in concrete

$$f_{cw} = 14 \text{ N/mm}^2$$

Tensile stresses are not permitted:

$$f_{tw} = 0$$

Loss ratio $\eta = 0.7$

H.T wires of 5 mm diameter, initially stressed to 1500 N/mm^2 are available for use. The required section modulus is given by the expression,

$$Z_t = Z_b > \frac{2M_d}{(f_{cw} - f_{tw})} = \frac{(2 \times 25 \times 10^6)}{(14 - 0)}$$

$$= 357 \times 10^4 \text{ mm}^3$$

Section modulus of the rectangular section provided

$$= \left[\frac{bh^2}{6}\right] = \left[\frac{150 \times 400^2}{6}\right] = (400 \times 10^4) \text{ mm}^3$$

The section modulus provided is greater than the required value, hence, the section is adequate.

Prestress in member

$$= \left[\frac{f_{tw}}{\eta} + \frac{M_d}{\eta Z}\right] = \left[0 + \frac{(25 \times 10^6)}{(0.7 \times 400 \times 10^4)}\right]$$

$$= 8.92 \text{ N/mm}^2$$

Initial prestressing force

$$= \left[\frac{(8.92 \times 150 \times 400)}{1000}\right] = 535 \text{ kN}$$

Permissible force in 5 mm wire

$$= [(19.6 \times 0.8 \times 1500)/1000] = 23.52 \text{ kN}$$

Number of wires required

$$= \left[\frac{535}{23.52}\right] = 22.7$$

Provide a total of 24 wires with 12 wires on each face. The wires in each face are arranged in two rows of 6 each.

Problem 14.6

A prestressed concrete pole has a cross-section of 150 mm by 200 mm at the top and a large section at the ground level. The pole has been designed for flexure due to wind loads with the required number of high tensile wires. The pole is subjected to an ultimate torque of 4.5 kN.m at the top due to skew snapping of wires. Design suitable reinforcements to resist the torsion at the top section. Assume the grade of concrete as M-50 and Fe-415 grade HYSD bars.

Solution:

Cross-sectional dimensions:

$$h_{max} = 200 \text{ min}$$
$$h_{min} = 150 \text{ mm}$$

Ultimate torsional moment $T = 4.5$ kN.m

The section at the top of the pole will be checked for torsional shear stresses.

$$\text{Torsional shear stress } \tau_t = \left\{\frac{2T}{h_{min}^2\left[h_{max} - \left(\frac{h_{min}}{3}\right)\right]}\right\}$$

$$= \left\{\frac{2 \times 4.5 \times 10^6}{150^2\left[200 - \left(\frac{150}{3}\right)\right]}\right\} = 2.66 \text{ N/mm}^2$$

According to the revised IS: 1343 code, the torsional shear stress does not exceed the value of 4.8 N/mm^2 for M-50 grade concrete.

Hence, longitudinal and transverse reinforcements are to be designed according to the specifications prescribed in the Indian standard code IS: 1343.

Using 12 mm diameter two-legged links, the spacing is given by

$$S = \left[\frac{A_{sv}0.8x_1y_1 0.87 f_{yv}}{T}\right],$$

where, x_1 and y_1 are the shorter and longer dimensions of the links.

$$= \left[\frac{2\times113\times0.8\times110\times160\times0.87\times415}{2\times4.5\times10^6}\right] = 127 \text{ mm}$$

The cross-sectional area of longitudinal reinforcement is given by

$$A_s = \left(\frac{A_{sv}}{S}\right)\left(\frac{f_{yv}}{f_y}\right)(x_1+y_1)$$

$$= \left(\frac{2\times113}{127}\right)\left(\frac{415}{415}\right)(110+160) = 488 \text{ mm}^2$$

Four longitudinal bars of 12 mm diameter are provided as corner bars along with the legged links at 125 mm centres.

Problem 14.7

The foundation for the column of an industrial structure is made up a prestressed octagonal pile with an effective length of 5 m. The axial service load and bending moment transmitted to the pile is 1100 kN and 37.5 kN.m respectively. Design a 350 mm solid octagonal pile assuming a load factor of 2 against collapse. The pile is to be designed to be lifted at any point along its length for installation. Assume a loss ratio of 0.85.

Solution:

Axial load N = 1000 kN

Bending moment M = 37.5 kN.m

Octagonal pile of overall width = 350 mm

Effective length of pile = 5 m

Ultimate design loads on pile are computed as

$$N_u = (2\times1100) = 2200 \text{ kN}$$

$$M_u = (2\times37.5) = 75 \text{ kN.m}$$

Referring to the Table 14.1, the section properties of the pile are listed as,

$$A = (1.02\times10^5) \text{ mm}^2$$

$$g = 2.45 \text{ kN/m}$$

$$Z = 4.70\times10^6 \text{ mm}^3$$

Radius of gyration i = 90 mm

Slenderness ratio $\left[\frac{L}{i}\right] = \left[\frac{5 \times 1000}{90}\right] = 55.5$

Eccentricity $e = \left[\frac{M_u}{N_u}\right] = \left[\frac{75 \times 1000}{2200}\right] = 34$ mm

$\therefore$ Ratio $\left(\frac{e}{h}\right) = \left(\frac{34}{350}\right) = 0.097$

The ultimate moment and load capacity of the 350 mm solid octagonal pile for $M_u = 75$ kN.m and $(L/i) = 55.5$. From Fig. 14.3, P_u is obtained as 2250 kN.

Effective prestress = 8.4 N/mm^2

$\therefore$ Initial prestressing force = $\left[\frac{8.4 \times 1.02 \times 10^5}{0.85 \times 1000}\right] = 1030$ kN

Use eight strands of 7–12.5 mm wire initially stressed to 1120 N/mm^2.

The pile is checked for handling stresses to be lifted at any point along its length for installation.

Self weight of pile = 2.43 kN/m
Impact allowance (50%) = 1.23 kN/m
Total load = 3.68 kN/m

Maximum bending moment in the pile when lifted at centre of pile is given by,

$$M = \left[\frac{3.68 \times 2.5^2}{2}\right] = 11.5 \text{ kN.m}$$

Tensile bending stress = $\left[\frac{(11.5 \times 10^6)}{(4.70 \times 10^6)}\right] = 2.45$ N/mm^2

Resultant stress = [8.4 – 2.45] = 5.95 N/mm^2 (compression)

Hence, the pile is safe against handling stresses.

Problem 14.8

Design the number of high tensile strands required for an Indian RDSO/T sleeper for a broad gauge track. The sleeper is 2.75 m long and has a trapezoidal cross-section with the following dimensions and properties.

Width at top = 150 mm
Width at bottom = 250 mm
Height of sleeper = 210 mm

Cross-sectional area $A = 42000 \text{ mm}^2$

Section modulus of top fibre $Z_t = 1403 \times 10^3 \text{ mm}^3$

Section modulus of bottom fibre $Z_b = 1543 \times 10^3 \text{ mm}^3$

The sleeper is subjected to maximum and minimum bending moments of magnitude 13 and 2 kN.m at different sections in the beam. The permissible compressive stress in concrete is 14 N/mm^2 at transfer and 24 N/mm^2 at service loads. No tension is allowed at any stage. High tensile strands of 6.3 mm nominal diameter initially stressed to 1200 N/mm^2 are available for use. Check the adequacy of the section and design the number of strands required in the sleeper.

Solution

$$f_{ct} = 14 \text{ N/mm}^2 \quad \text{and} \quad f_{tw} = 0$$

$$f_{br} = (\eta f_{ct} - f_{tw}) = \{(0.8 \times 14) - 0\} = 11.2 \text{ N/mm}^2$$

The minimum section modulus required is given by the relation,

$$Z_b = \left[\frac{M_q + (1-\eta)M_g}{f_{br}}\right] = \left[\frac{\{13 + (1-0.80)2\}10^6}{11.2}\right]$$

$$= 1375 \times 10^3 \text{ mm}^3 < 1403 \times 10^3 \text{ mm}^3$$

Bending stress due to moment

$$f = \left[\frac{M}{Z}\right] = \left[\frac{13 \times 10^6}{1403 \times 10^3}\right] = 9.2 \text{ N/mm}^2$$

(compression or tension)

Prestressing force required to eliminate this tension is calculated as

$$P = [A f] = [42000 \times 9.2] = 386400 \text{ N} = 386.4 \text{ kN}$$

Using high tensile strands of 6.3 mm nominal diameter initially stressed to 1200 N/mm^2 having a cross-sectional area of 23.2 mm^2, the force in each strand is computed as,

Force in each strand $= [1200 \times 23.2] = 27840 \text{ N}$

$= 27.84 \text{ kN}$

Number of strands required $= \left[\dfrac{386.4}{27.84}\right] = 13.8$

Provide a total of 16 strands with 8 strands distributed in two rows towards the top and bottom faces of the sleeper.

Problem 14.9

A two lane highway 7.5 m wide by 100 m long resting on a sub-base is to be prestressed in the longitudinal and transverse directions. The thickness of the concrete slab is 150 mm. The coefficient of friction between the slab and sub-base is estimated to be 1.5. Freyssinet cables containing 12–5 mm H.T. wires are available for use at site. A minimum longitudinal prestress of 2 N/mm^2 should be ensured. Design the spacing of cables in both the directions.

Solution

Coefficient of friction $\mu = 1.5$

Dimension of concrete slab = 100 m by 7.5 m

(thickness = 150 mm)

The longitudinal prestress required at the ends of slabs after allowing for loss due to friction is given by the relation,

$$f_c = [2 + 0.5\, \mathrm{m} D_c L]$$

$$= 2 + \left[\frac{0.5 \times 1.5 \times 24 \times 1000 \times 100}{10^6}\right]$$

$$= 3.80 \text{ N/mm}^2$$

Total prestressing force

$$= P = [A f_c] = \left[\frac{(7.5 \times 1000 \times 150 \times 3.8)}{1000}\right]$$

$$= 4300 \text{ kN}$$

Force in each cable containing 12 wires of 5 mm diameter initially stressed to 1200 N/mm^2 is computed as,

$$\text{Cable force} = \left[\frac{19.6 \times 1200 \times 12}{1000}\right] = 288 \text{ kN}$$

Spacing of longitudinal cables

$$= \left[\frac{7.5 \times 1000 \times 288}{4300}\right] = 500 \text{ mm}$$

Transverse prestress $= (0.7 \times 3.80) = 2.66$ N/mm^2

$$\text{Force/metre} = \left[\frac{2.66 \times 1000 \times 150}{1000}\right] = 399 \text{ kN/m}$$

$$\text{Spacing of transverse cables} = \left[\frac{288 \times 1000}{399}\right] = 720 \text{ mm}$$

Problem 14.10

An airport runway pavement is 150 m long and 20 m wide. Design an oblique cable system using Freyssinet cables comprising 12 wires of 5 mm diameter. The coefficient of friction between the pavement and subgrade may be assumed as 1.0. The thickness of the pavement is 200 mm and the minimum longitudinal and transverse prestress should not be less than 2 and 1.5 N/mm^2 respectively. Determine the obliquity and spacing of the cables.

Solution

$$L = 150 \text{ m} \qquad t = 200 \text{ mm}$$
$$B = 20 \text{ m} \qquad \mu = 1.0$$
$$f_{cL} = 2 \text{ N/mm}^2 \qquad f_{cT} = 1.5 \text{ N/mm}^2$$

Obliquity of the cables is determined using the relation,

$$\tan\alpha = \sqrt{\frac{2f_{cT} + \mu D_c B}{2f_{cL} + \mu D_c L}},$$

where, α is the inclination of the cable.

$$\mu D_C B = [1.0 \times 24 \times 10^3 \times 20]/10^6 = 0.48 \text{ N/mm}^2$$
$$\mu D_C L = [1.0 \times 24 \times 10^3 \times 150]/10^6 = 3.60 \text{ N/mm}^2$$

$$\tan\alpha = \sqrt{\frac{(2\times 1.5)+0.48}{(2\times 2)+3.60}} = 0.68$$

$$\therefore \qquad \alpha = 34° \, 13'$$

Spacing of cables
$$s = \left[\frac{4P\sin\alpha}{t(2f_{ct} + \mu D_c B)}\right]$$
$$= \left[\frac{4\times 288\times 10^3 \times 0.5622}{200\,(2\times 15 + 0.48)}\right] = 930 \text{ mm}$$

Force in each cable
$$P = \left[\frac{(20\times 1200\times 12)}{1000}\right] = 288 \text{ kN}$$

15 Prestressed Concrete Bridges

Problem 15.1

Design a post-tensioned prestressed concrete slab bridge deck for a national highway crossing to suit the following data:

Clear span = 8 m
Width of bearings = 400 mm
Clear width of the roadway = 7.5 mm
Foot paths 1 m on either side
Wearing coat thickness = 80 mm
Live load: I.R.C. class AA tracked vehicle
Type of structure: Class 1 type (no tensile stresses)

Materials; M-40 grade concrete and 7 mm diameter high tensile wires with an ultimate tensile strength of 1500 N/mm^2 housed in cables with 12 wires and anchored by Freyssinet anchorages of 150 mm diameter.

For supplementary reinforcement adopt Fe-415 grade HYSD bars. Compressive strength of concrete at transfer

$$f_{ci} = 35 \text{ N/mm}^2$$

Loss ratio $\eta = 0.8$

The design should conform to the specifications of codes IRC-6, IRC-18, IRC-21.

Solution

1. *Permissible stresses*

Permissible compressive stresses in concrete at transfer and working loads as specified in IRC-18 are as follows:

$$f_{ct} \leq 0.45 f_{ci} \leq (0.45 \times 35) \leq 15.75 \text{ N/mm}^2$$

Adopt $f_{ct} = 15 \text{ N/mm}^2$

$$f_{cw} \leq 0.33 f_{ck} \leq (0.33 \times 40) \leq 13.2 \text{ N/mm}^2$$

$$f_{tt} = f_{tw} = 0 \text{ (since class-1 type structure)}$$

2. *Depth of slab and effective span*

Assuming the thickness of slab at 50 mm per metre of span for highway slab bridge decks,

Overall thickness of slab = (50 × 8) = 400 mm

Width of bearing = 400 mm

∴ Effective span = (8 + 0.4) = 8.4 m

The cross section of the deck slab is shown in Fig. 15.1

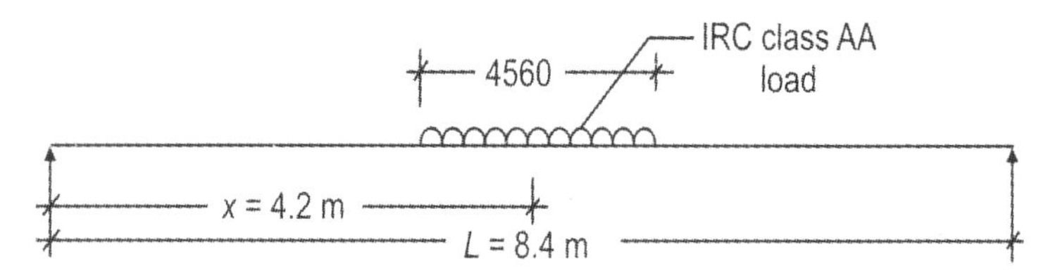

Fig. 15.1: Cross section of deck slab

3. *Dead load bending moments*

Dead weight of slab = (0.4 × 24) = 9.60 kN/m^2

Dead weight of W.C. = (0.08 × 22) = 1.76 kN/m^2

Dead weight of foot:

Path etc. (Lumpsum) = 0.64 kN/m^2

Total dead load = 12.00 kN/m^2

Hence, dead load bending moment is computed as

$$M_g = \frac{(12 \times 8.4^2)}{8} = 105.84 \text{ kN.m}$$

4. *Live load bending moments*

Generally the bending moment due to live load will be maximum for IRC class AA tracked vehicle.

Impact factor for the class *AA* tracked vehicle is 25 per cent for 5 m span linearly decreasing to 10 per cent for 9 m span.

∴ Impact factor = 12.5 per cent for 8.4 m span

The tracked vehicle is placed symmetrically on the span.

Effective length of load = [3.6 + 2(0.4 + 0.08)] = 4.56 m

Effective width of slab perpendicular to the span is expressed as

$$b_e = k \cdot x\left(1 - \frac{x}{L}\right) + b_w$$

Referring to Fig. 15.2, $x = 4.2$ m, $L = 8.4$ m and B = 9.5 m

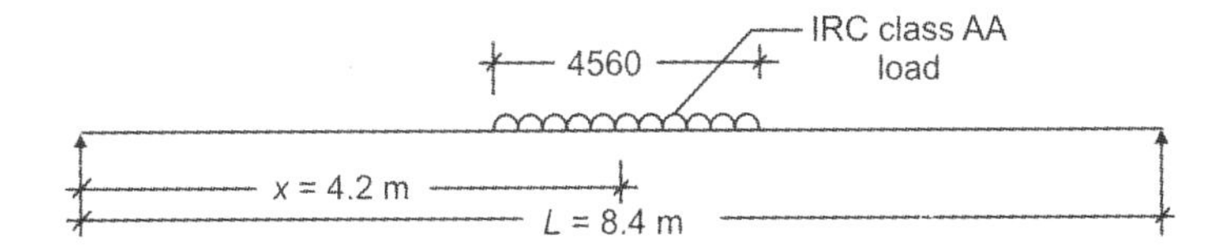

Fig. 15.2: Position of load for maximum bending moment

$\therefore$ The ratio $\left(\frac{B}{L}\right) = \left(\frac{9.5}{8.4}\right) = 1.13$ and

$$b_w = (0.85 + 2 \times 0.08) = 1.01 \text{ m}$$

Table 15.1: Values of constant k (IRC: 21–1987)

$\left(\frac{B}{L}\right)$	Values of k		$\left(\frac{B}{L}\right)$	Values of k	
	For simply supported slabs	For continuous slabs		For simply supported slabs	For continuous slabs
0.1	0.40	0.40	1.1	2.60	2.28
0.2	0.80	0.80	1.2	2.64	2.36
0.3	1.16	1.16	1.3	2.72	2.40
0.4	1.48	1.44	1.4	2.80	2.48
0.5	1.72	1.68	1.5	2.84	2.48
0.6	1.96	1.84	1.6	2.88	2.52
0.7	2.12	1.96	1.7	2.92	2.56
0.8	2.24	2.08	1.8	2.96	2.60
0.9	2.36	2.16	1.9	3.00	2.60
1.0	2.48	2.24	2.0 and above	3.00	2.60

From Table 15.1 for $\left(\frac{B}{L}\right) = 1.13$ and simply supported slabs, read out the value of k = 2.61.

$$\therefore \quad b_e = (2.61 \times 4.2)\left[1 - \left(\frac{4.2}{8.4}\right)\right]1 + 1.01$$

$$= 6.491 \text{ m}$$

The IRC class *AA* tracked vehicle is placed close to the kerb with the required minimum clearance as shown in Fig. 15.3

Net effective width of dispersion = 7.921 m

Total load of two tracks with impact = 700 × 1.25 = 787.5 kN.

$$\text{Average intensity of load} = \left[\frac{787.5}{4.56 \times 7.92}\right] = 21.8 \text{ kN/m}^2$$

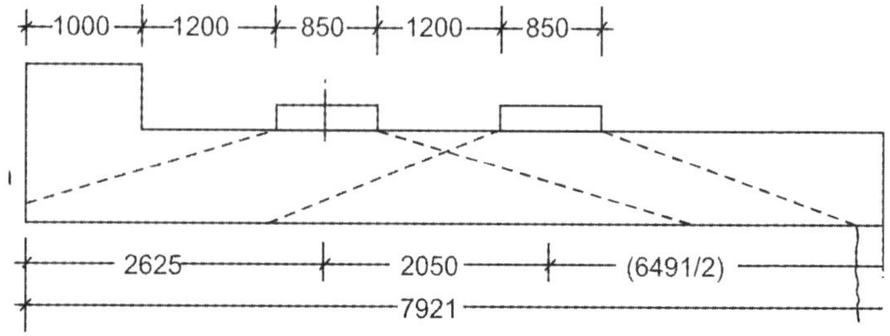

Fig. 15.3: Effective width of dispersion for IRC class *AA* tracked vehicle

Maximum bending moment due to live load is computed as

$$M_q = [(21.8 \times 4.56)0.5 \times 4.2] - [(21.8 \times 4.56)0.5 \times 0.25 \times 4.56]$$

$$= 153 \text{ kN.m}$$

5. *Shear due to class AA tracked vehicle*

For maximum shear force at the support section, the IRC class *AA* tracked vehicle is arranged as shown in Fig. 15.4.

Effective width of dispersion is given by

$$be = \text{k} \cdot x\left(1 - \frac{x}{L}\right) + b_w$$

where, $x = 2.28$ m, $L = 8.4$ m, $B = 9.5$ m, $b_w = 1.01$ m

and $$\left(\frac{B}{L}\right) = \left(\frac{9.5}{8.4}\right) = 1.13$$

From Table 15.1, for $\left(\frac{B}{L}\right) = 1.13$, the value of $k = 2.61$

$$\therefore \quad b_e = \left[2.61 \times 2.28\left(1 - \frac{2.28}{8.4}\right) + 1.01\right] = 5.35 \text{ m}$$

Referring to Fig. 15.3

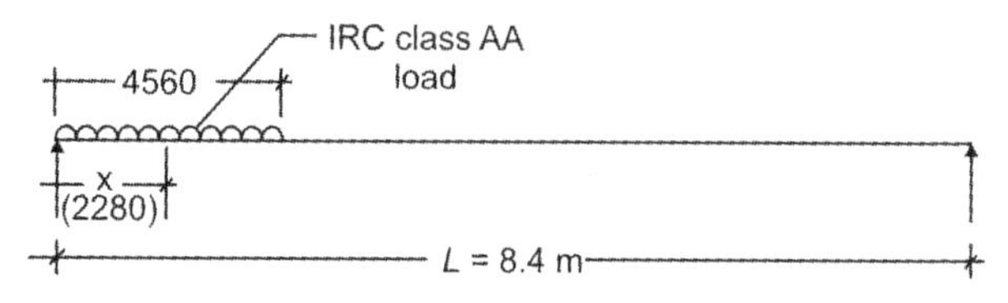

Fig. 15.4: Position of load for maximum shear

Width of dispersion for two tracks is computed as,

$$\text{Dispersion width} = \left[2625 + 2050 + \left(\frac{5350}{2}\right)\right] = 7350 \text{ mm}$$

$$\text{Intensity of load} = \left[\frac{787.5}{(4.56 \times 7.35)}\right] = 23.5 \text{ kN/m}^2$$

$$\text{Shear force} \quad V_A = \left[\frac{(23.5 \times 4.56 \times 6.12)}{8.4}\right] = 78.1 \text{ kN}$$

Dead load shear $= (0.5 \times 12 \times 8.4) = 50.4$ kN

$\therefore$ Total design shear force $= (78.1 + 50.4) = 128.5$ kN

6. *Check for minimum section modulus*

Dead load moment $M_g = 105.84$ kN.m

Live load moment $M_q = 153.00$ kN.m

$$\text{Section modulus} \quad Z_t = Z_b = Z = \left[\frac{(1000 \times 400^2)}{6}\right]$$

$$= 26.66 \times 10^6 \text{ mm}^3$$

The permissible stress in concrete at transfer (f_{ct}) is obtained from IRC-18, as

$$f_{ct} = 15 \text{ N/mm}^2, \qquad f_{tw} = 0, \qquad \eta = 0.8$$

Range of stress at bottom fibre $= f_{br} = (\eta f_{ct} - f_{tw})$

$= (0.8 \times 15 - 0) = 12$ N/mm²

The minimum section modulus required is given by

$$Z_b \geq \left[\frac{M_q + (1 - \eta) M_g}{f_{br}}\right]$$

$$\geq \left[\frac{153 + (1 - 0.8)105.84}{12}\right] \times 10^6$$

$$\geq 14.5 \times 10^6 \text{ mm}^3 < 26.66 \times 10^6 \text{ mm}^3 \text{ (provided)}$$

Hence, the section selected is adequate to resist the service loads without exceeding the permissible stresses.

7. *Minimum prestressing force*

The minimum prestressing force required is computed using the relation,

$$P = \left[\frac{A(f_t Z_t + f_b Z_b)}{Z_t + Z_b}\right]$$

where, $f_t = \left(f_{tt} - \frac{M_g}{Z_t}\right) = \left[0 - \frac{105.84 \times 10^6}{26.66 \times 10^6}\right]$

$= -3.97 \text{ N/mm}^2$

and $f_b = \left[\frac{f_{tw}}{\eta} + \frac{M_g + M_q}{\eta Z_b}\right]$

$= \left[0 + \frac{(105.84 + 153)10^6}{0.8 \times 26.66 \times 10^6}\right]$

$= 12.13 \text{ N/mm}^2$

$\therefore \quad P = \left[\frac{(1000 \times 400)(26.66 \times 10^6)(12.13 - 3.97)}{2 \times 26.66 \times 10^6}\right]$

$= 1632 \times 10^3 \text{ N} = 1632 \text{ kN}$

Using Freyssinet cables containing 12 wires of 7 mm diameter initially stressed to 1200 N/mm^2,

Force in each cable $= \left[\frac{12 \times 38.5 \times 1100}{1000}\right] = 508 \text{ kN}$

$\therefore$ Spacing of cables $= \left[\frac{1000 \times 508}{1632}\right] = 311 \text{ mm}$

Adopt cables at 300 mm centres.

8. *Eccentricity of cables*

The eccentricity of the cables at the centre of span is obtained from the relation,

$$e = \left[\frac{Z_t Z_b (f_b - f_t)}{A(f_t Z_t + f_b Z_b)}\right]$$

$$= \left[\frac{(26.66^2) \times 10^{12} (12.13 + 3.97)}{1000 \times 400 \times 26.66 \times 10^6 (12.13 - 3.97)}\right]$$

$= 131.5 \text{ mm}$

The cables are arranged in a parabolic profile with a maximum eccentricity of 131.5 mm at centre of span and reducing to zero at supports.

The cross section of slab deck with position of cables at centre of span section is shown in Fig. 15.5

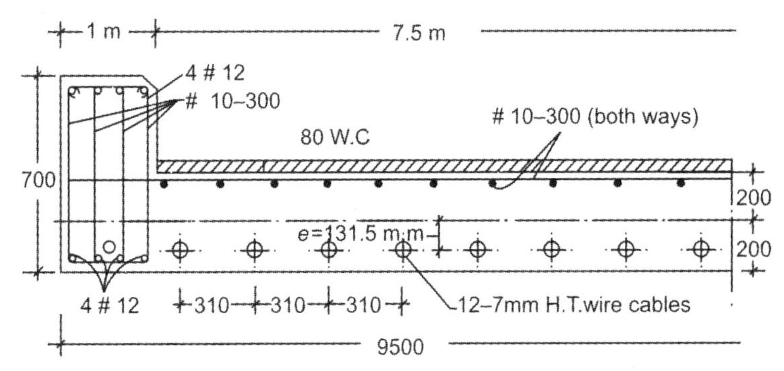

Fig. 15.5: Cross section of deck slab at centre of span

Problem 15.2

Design a post-tensioned prestressed concrete beam to suit the following data:

Effective span = 30 m

Live load = 11 kN/m

M-50 grade concrete

$$f_{ck} = 50\ \text{N/mm}^2, \qquad f_{ci} = 41\ \text{N/mm}^2$$

$$f_t = 0.24\sqrt{f_{ck}} = 0.24\sqrt{50} = 1.7\ \text{N/mm}^2$$

$$E_C = 5700\sqrt{f_{ck}} = 5700\sqrt{50} = 40.3\ \text{N/mm}^2$$

Loss ratio $\eta = 0.85$

High tensile wires of 8 mm diameter– 12 nos in each cable with Freyssinet anchorages are available for use.

Initial stress in the wires = 1100 N/mm^2

Ultimate tensile strength of wires f_p = 1500 N/mm^2

Design the beam as class-1 type structure to conform to IS: 1343 code specifications.

Solution:

1. *Cross sectional dimensions*

Unsymmetrical tee section is selected

Depth of girder = 40 to 50 mm/m of span

= (40 × 30) = 1200 mm

= (50 × 30) = 1500 mm

Adopt overall depth h = 1300 mm

Top width of flange = 0.4 to 0.5 h

= (0.4 × 1300) to (0.5 × 1300)

= 520 to 650 mm

Adopt top width of flange $b = 600$ mm

Thickness of top flange $0.20\,h = 0.20 \times 1300 = 250$ mm

Thickness of web = minimum = 150 mm to house 50 mm diameter cables with a clear cover of 50 mm on either side.

The dimensions of bottom flange is based on number of cables and the number of each cables in a row. Providing for 3 cables in a row, 350 mm wide by 300 mm deep flange is adopted.

The cross sectional dimensions of the girder is shown in Fig. 15.6.

2. *Sectional properties*

$A = 367500\ \text{mm}^2$, $y_t = 570$ mm,

$y_b = 730$ mm $h = 1300$ mm,

$I = 72490 \times 10^6\ \text{mm}^4$ $Z_t = 127 \times 10^6\ \text{mm}^3$,

$Z_b = 99 \times 10^6\ \text{mm}^3$

3. *Design moments and shear forces*

Self weight $g = (0.3675 \times 24) = 8.8$ kN/m

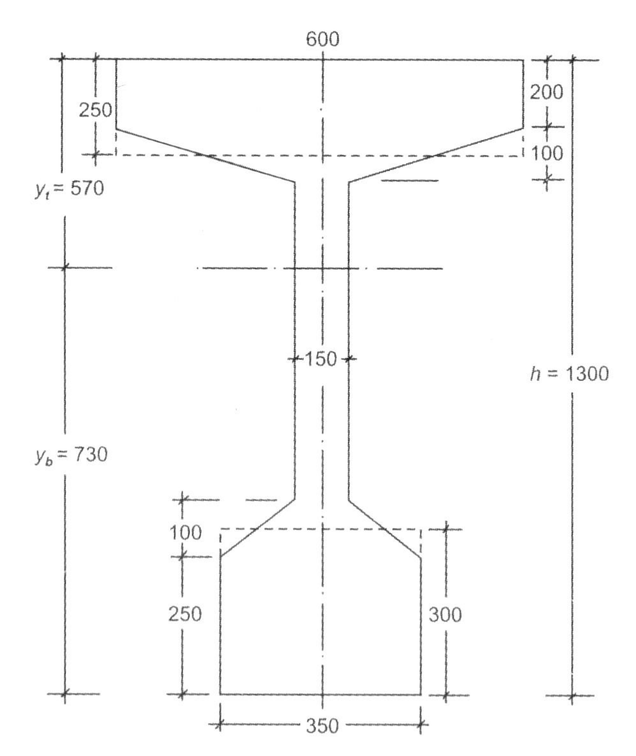

Fig. 15.6: Cross sectional dimensions

Self weight moment $M_g = 0.125 \times 8.8 \times 30^2 = 990$ kN.m

Live load $q = 11$ kN/m

Live load moment $M_g = 0.125 \times 11 \times 30^2 = 1237.5$ kN.m

Total shear force $V = V_g + V_q$

$= 0.5(8.8 + 11)\, 30 = 297$ kN

4. *Permissible stresses* (as per IS: 1343–2012)

$$f_{ck} = 50 \text{ N/mm}^2, \qquad f_{ci} = 41 \text{ N/mm}^2$$

Since, the beam is post-tensioned, referring to Fig. 8 of IS: 1343 code, the permissible compressive stress in concrete is read out as,

$$f_{ct} = 0.427, \qquad f_{ci} = (0.427 \times 41) = 17.5 \text{ N/mm}^2$$

For class-1 type member, $f_{tt} = f_{tw} = 0$, since no tension is permitted at transfer and working loads.

The permissible compressive stress in concrete under working loads is interpolated from Fig. 7 of IS: 1343.

$$f_{cw} = 0.37, \qquad f_{ck} = 0.37 \times 50 = 18.5 \text{ N/mm}^2$$

5. *Check for minimum section modulus*

$$Z_b \geq \left[\frac{M_q + (1-\eta) M_g}{(\eta f_{ct} - f_{tw})}\right]$$

$$\geq \left[\frac{(1237.5 \times 10^6) + (1 - 0.85) 990 \times 10^6}{(0.85 \times 17.5 - 0)}\right]$$

$\geq 93.17 \times 10^6 \text{ mm}^3 < 99 \times 10^6 \text{ mm}^3$ (provided)

6. *Prestressing force and eccentricity*

$$f_t = \left[f_{tt} - \frac{M_g}{Z_t}\right] = \left[0 - \frac{990 \times 10^6}{127 \times 10^6}\right]$$

$= -7.8 \text{ N/mm}^2$

$$f_b = \left[\frac{f_{tw}}{\eta} + \frac{M_g + M_q}{\eta Z_b}\right] = \left[0 + \frac{(2227.5 \times 10^6)}{(0.85 \times 99 \times 10^6)}\right]$$

$= 26.5 \text{ N/mm}^2$

$$\text{Eccentricity} = e = \left[\frac{Z_b Z_t (f_b - f_t)}{A(f_t Z_t + f_b Z_b)}\right]$$

$$= \left[\frac{(99 \times 127) 10^{12} (26.5 + 7.8)}{(367500)\left(-7.8 \times 127 \times 10^6\right) + \left(26.5 \times 99 \times 10^6\right)}\right]$$

$= 718.33$ mm which is impracticable

Hence, maximum possible eccentricity = [730 – 150] = 580 mm

Modified prestressing force is computed as

$$P = \left[\frac{A\,f_b\,Z_b}{(Z_b + A.e)}\right]$$

$$= \left[\frac{(367500 \times 26.5 \times 99 \times 10^6)}{(99 \times 10^6) + (367500 \times 580)}\right]$$

$$= 3088 \times 10^3 \text{ N} = 3088 \text{ kN}$$

Using Freyssinet cables 12 nos of 8 mm diameter initially stressed to 1100 N/mm^2,

$$\text{Number of cables} = \left[\frac{3080 \times 10^3}{12 \times 50 \times 1100}\right] \cong 5$$

7. *Permissible tendon zone*

   a. **At centre of span**

$$e \le \left[\left(\frac{Z_b\,f_{ct}}{P}\right) - \left(\frac{Z_b}{A}\right) + \left(\frac{M_g}{P}\right)\right]$$

$$\le \left[\frac{(99 \times 10^6 \times 17.5)}{(3088 \times 10^3)} - \frac{(99 \times 10^6)}{(367500)} + \frac{(990 \times 10^6)}{(3088 \times 10^3)}\right]$$

$$\le 612 \text{ mm}$$

$$e \ge \left[\frac{Z_b f_{tw}}{P} - \frac{Z_b}{A} + \frac{M_g + M_q}{P}\right]$$

$$\ge \left[0 - \frac{99 \times 10^6}{367500} + \frac{2227.5 \times 10^6}{0.85 \times 3088 \times 10^3}\right]$$

$$\ge 580 \text{ mm}$$

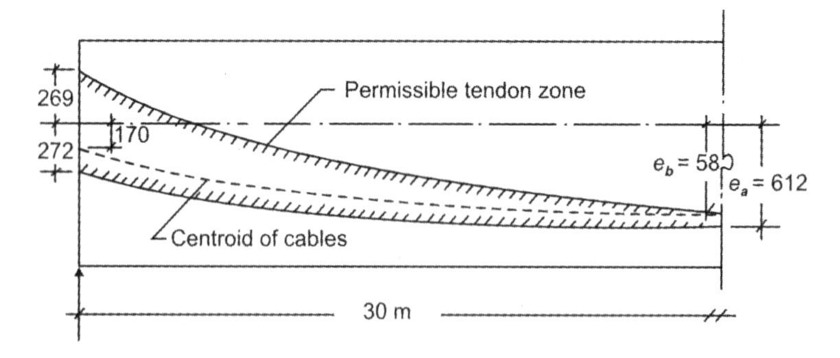

Fig. 15.7: Permissible tendon zone

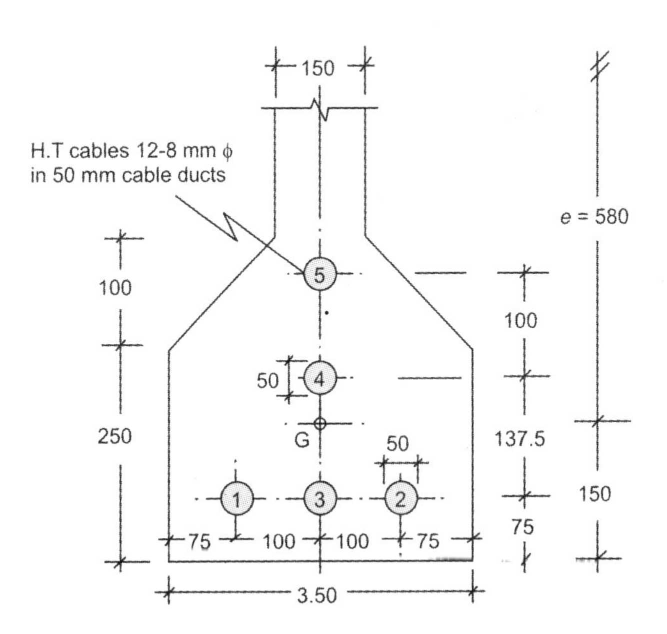

Fig. 15.8: Arrangement of cables at centre of span

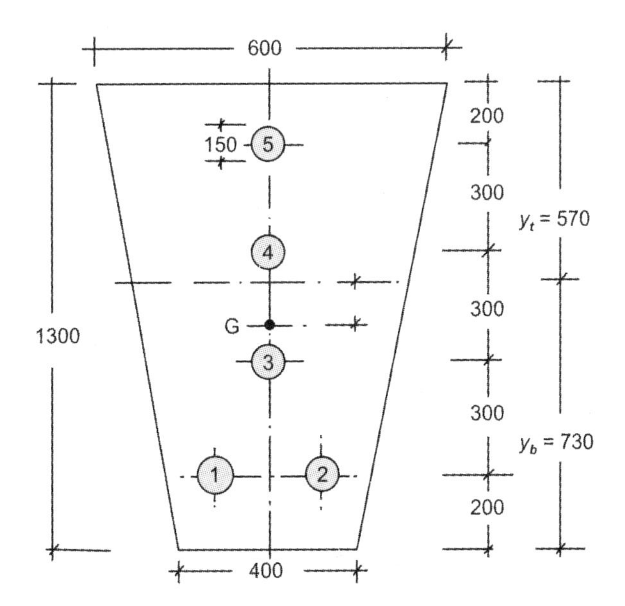

Fig. 15.9: Arrangement of cables of supports

b. **Support section**

$$e \le \left[\frac{Z_b f_{ct}}{P} - \frac{Z_b}{A} + 0\right] \le 272 \text{ mm}$$

$$e \ge \left[0 - \frac{Z_b}{A} + 0\right] \ge 269 \text{ mm}$$

The permissible tendon zone is shown in Fig. 15.7.

The arrangement of cables at centre of span and support sections are shown in Figs 15.8 and 15.9.

8. *Check for ultimate flexural strength*

At centre of span section, the section properties are as follows:

$A_p = 3000 \text{ mm}^2$, $d = 1150$ mm
$f_{ck} = 50 \text{ N/mm}^2$, $b_w = 150$ mm
$f_p = 1500 \text{ N/mm}^2$, $b = 600$ mm
$t = 250$ mm

Design $M_u = 1.5\,(M_g + M_q)$
$= 1.5[990 \times 10^6 + (1237.5 \times 10^6)]$
$= 3341.25$ kN.m

$A_p = A_{pw} + A_{pf}$
$A_{pf} = 0.45\, f_{ck}\,(b - b_w)(t/f_p)$

$$= 0.45 \times 50(600 - 150)\left(\frac{250}{1500}\right) = 1680 \text{ mm}^2$$

$\therefore$ $A_{pw} = [3000 - 1680] = 1320 \text{ mm}^2$

$$\left(\frac{A_{pw} f_p}{f_{ck} b_w d}\right) = \left(\frac{1320 \times 1500}{50 \times 150 \times 1150}\right) = 0.23$$

Referring to Table-11 of IS: 1343, read out the value of

$$\left(\frac{f_{pu}}{0.87 f_p}\right) = 0.92 \quad \text{and} \quad \left(\frac{x_u}{d}\right) = 0.45$$

$\therefore$ $f_{pu} = (0.92 \times 0.87 \times 1500) = 1200.6 \text{ N/mm}^2$ and
$x_u = (0.45 \times 1150) = 517.5$ mm
$M_u = A_{pw} f_{pu}\,(d - 0.42\, x_u) + 0.45\, f_{ck}\,(b - b_w)\, t(d - 0.5t)$
$= (1320 \times 1200.6)(1150 - 0.42 \times 517.5)$
$+ (0.45 \times 50)(600 - 150)\, 250(1150 - 125)$
$= 4072 \times 10^6$ N.mm
$= 4072$ kN.m > 3341.25 kN.m

Hence, the section satisfies the limit state of ultimate flexural strength.

9. *Check for ultimate shear strength at support section*

Required ultimate shear force

$$V_u = 1.5\left(V_g + V_q\right) = 1.5(297) = 445.5 \text{ kN}$$

$$f_t = 1.7 \text{ N/mm}^2$$

$$f_{cp} = \left(\frac{\eta P}{A}\right) = \left(\frac{0.85 \times 3088 \times 10^3}{367500}\right) = 7.14 \text{ N/mm}^2$$

$$\text{Slope of cable } \theta = \left(\frac{4e}{L}\right) = \left[\frac{(4 \times 410)}{(30 \times 1000)}\right] = 0.0547$$

$$V_{co} = 0.67 b_w D\sqrt{f_t^2 + 0.8 f_{cp} f_t} + \eta P \sin\theta$$

$$= 0.67 \times 150 \times 1300 \frac{\sqrt{1.7^2 + 0.8 \times 7.14 \times 1.7}}{1000} + (0.85 \times 3088 \times 0.0547)$$

$$= 607 \text{ kN} > 445.5 \text{ kN, Hence safe.}$$

Minimum shear reinforcement

Using 10 mm diameter two legged stirrups (Fe-415 HYSD bars)

$$\text{Spacing} \quad S_v = \left(\frac{A_{sv} 0.87 f_y}{0.4 b_w}\right)$$

$$= \left(\frac{2 \times 79 \times 0.87 \times 415}{0.4 \times 150}\right) = 950 \text{ mm}$$

Spacing $\not> 0.75\, d_t$ or $(4 \times b_w)$ whichever is smaller

$\not> (0.75 \times 1250)$ or $(4 \times 150) = 600$ mm

Adopt 10 mm diameter two legged stirrups at 600 mm centres throughout the span.

10. *Check for deflection at serviceability limit state*

Eccentricity of cable at centre of span $e_1 = 580$ mm

Eccentricity of cable at support section $e_2 = 170$ mm

Deflection at centre of span due to prestressing force is

$$a_p = -\left[\frac{P \cdot L^2 (5e_1 + e_2)}{48 EI}\right] \text{ (upward direction)}$$

$$= -\left[\frac{\left(3088 \times 10^3 \times 30^2 \times 10^6\right)(5 \times 580 + 170)}{48 \times 40.3 \times 10^3 \times 72490 \times 10^6}\right]$$

$$= -61 \text{ mm (upwards)}$$

Deflection due to dead and live loads are computed as

$$a_{(g+q)} = \left[\frac{5(g+q)L^4}{384E \cdot I}\right]$$

$$= \left[\frac{5 \times (8.8+11)(30 \times 10^3)^4}{384 \times 40.3 \times 10^3 \times 72490 \times 10^6}\right]$$

$$= 71.50 \text{ mm (downwards)}$$

If creep coefficient $\phi = 1.6$

Long term deflection is evaluated as

$$= (1+\phi)[\eta a_p + (a_{g+q})]$$

$$= (2.6)[0.85 \times (-61) + 71.50]$$

$$= 51 \text{ mm (downwards)}$$

Maximum permissible deflection $\not> \left(\frac{\text{span}}{250}\right)$

$$= \left(\frac{30 \times 10^3}{250}\right) = 120 \text{ mm}$$

Hence, deflection is within permissible limits.

11. *End block design*

Equivalent prisms on which the anchorage forces are considered to be effective are shown in Fig. 15.10.

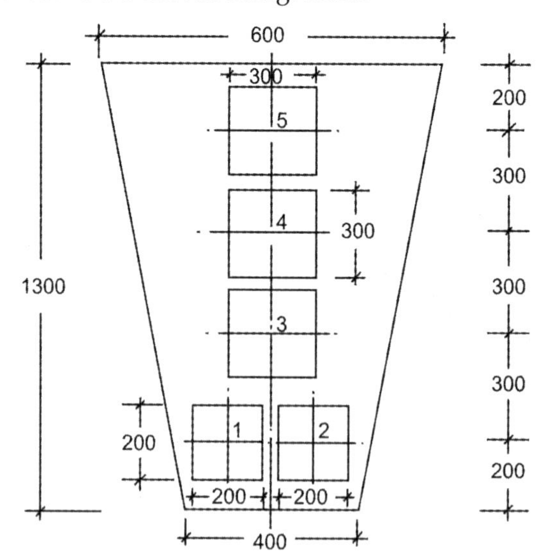

Fig. 15.10: Equivalent prisms for anchorage forces

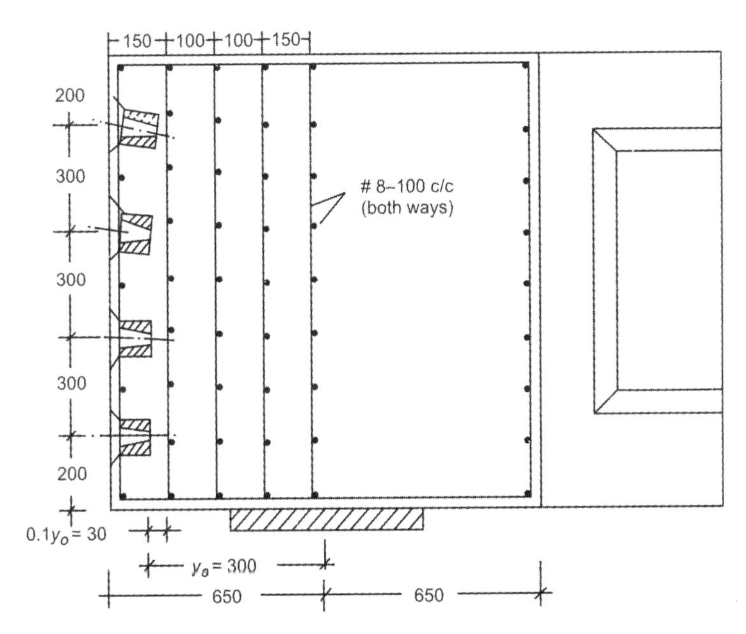

Fig. 15.11: End block reinforcements

For cables 1 and 2

$$P_k = \left(\frac{3088}{5}\right) = 618 \text{ kN}$$

$$y_{po} = 150 \text{ mm} \qquad y_o = 200 \text{ mm}$$

Hence, the ratio $\left(\dfrac{y_{po}}{y_o}\right) = \left(\dfrac{150}{200}\right) = 0.75$

$\therefore$ Bursting tension $F_{bst} = P_k\left[0.32 - 0.3\left(\dfrac{y_{po}}{y_o}\right)\right]$

$$= 618[0.32 - (0.3 \times 0.75)] = 58.7 \text{ kN}$$

For cables 3, 4 and 5,

$$y_{po} = 150 \text{ m m} \quad \text{and} \quad y_o = 300 \text{ mm}$$

Ratio $\left(\dfrac{y_{po}}{y_o}\right) = \left(\dfrac{150}{300}\right) = 0.5$

Bursting tension $= F_{bst} = 618[0.32 - (0.3 \times 0.5)] = 105$ kN

Using Fe-250 grade mild steel bars,

$$A_{st} = \left[\frac{(105 \times 10^3)}{(0.87 \times 250)}\right] = 483 \text{ mm}^2$$

Number of 8 mm diameter bars= $\left(\frac{483}{50}\right) = 10$

Provide 8 mm diameter bars at 100 mm centres both in the horizontal and vertical direction as shown in Fig. 15.11.

12. *Supplementary reinforcement*

According to Indian standard code IS: 1343 specifications the minimum longitudinal reinforcement should be not less than 0.15 percent of the cross sectional area.

$$A_{st} = \left(\frac{0.15 \times 367500}{100}\right) = 551.25 \text{ mm}^2$$

Provide 16 mm diameter HYSD bars as shown in Fig. 15.12.

Minimum web reinforcement = 0.15 per cent of web area in plan

$$= \left[\frac{(0.15 \times 1000 \times 150)}{100}\right] = 225 \text{ mm}^2$$

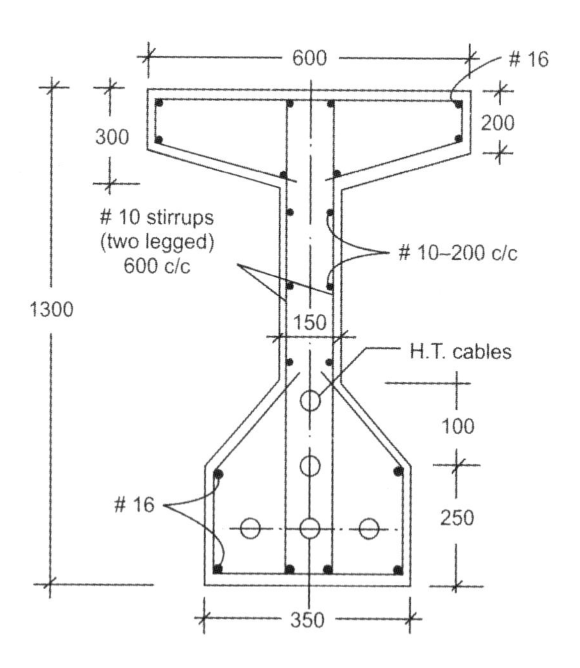

Fig. 15.12: Reinforcement in P.S.C. girder

Providing 10 mm diameter two legged stirrups,

$$\text{Spacing} \quad S_v = \left[\frac{(1000 \times 2 \times 78.5)}{225}\right] = 697 \text{ mm}$$

But spacing $\not>$ clear depth of web or 4 b_w, whichever is smaller
$\not>$ 650 mm or (4 × 150) = 600 mm

Hence, provide 10 mm diameter two legged stirrups at a spacing of 600 mm throughout the span.

Problem 15.3

Design a two span continuous post-tensioned prestressed concrete beam and slab bridge deck for a national highway crossing using the following data:

1. *Data*

   Two continuous spans each of 40 m between the bearings.

   Width of roadway = 7.5 m

   Krebs: 600 mm wide on each side

   Thickness of wearing coat = 80 mm

   Live load: IRC Class AA tracked vehicle

   For cast *in situ* deck slab adopt M-25 grade concrete

   For prestressed concrete girders adopt M-60 grade concrete with cube strength of concrete at transfer as 40 N/mm^2

   Loss ratio = 0.8

   High tensile steel strands of 15.2 mm diameter conforming to IS: 1786–1979 are available for use.

   Design the salient structural components of the bridge deck as Class 1 type structure conforming to the codes IRC: 6-2000, IRC: 18-2000 and IRC: 21-2000. Sketch the typical details of reinforcements and cables in the beam and slab deck.

2. *Permissible stresses*

   The permissible stresses for M-25 grade concrete and Fe-415 HYSD bars according to IRC: 21-2000 are compiled below.

   For M-20 Grade concrete and Fe-415 HYSD bars, we have the design coefficients as

$$\sigma_{cb} = 8.3 \text{ N/mm}^2$$
$$Q = 1.10$$
$$\sigma_{st} = 200 \text{ N/mm}^2$$
$$j = 0.90$$
$$m = 10$$

For M-60 grade concrete used in P.S.C. girder (IRC: 21-2000)

$$f_{ck} = 60\ \text{N/mm}^2$$
$$f_{ci} = 45\ \text{N/mm}^2$$
$$f_{ct} = 0.5$$
$$f_{ci} = (0.5 \times 45) = 22.5\ \text{N/mm}^2 \text{ limited to } 20\ \text{N/mm}^2$$
$$f_{cw} = 0.33$$
$$f_{ck} = (0.33 \times 60) \simeq 20\ \text{N/mm}^2$$
$$f_{tt} = f_{tw} = 0 \text{ (Class 1 type member–No tensile stress)}$$
$$E_c = 5000\sqrt{f_{ck}} = 5000\sqrt{60} = 38730\ \text{N/mm}^2 = 38.73\ \text{kN/mm}^2$$

3. *Cross-section of bridge deck*

Spacing of main girders = 2.5 m
Spacing of cross girders = 5.0 m
Thickness of deck slab = 250 mm
Thickness of wearing coat = 80 mm

Kerbs 600 mm wide by 300 mm deep are provided at each end.
Overall depth of main girder is assumed at 50 mm per metre of span

$\therefore$ Overall depth of girder $h = 50 \times 40 = 2000$ mm

Width of top and bottom flange $= (0.4h) = 0.4 \times 2000 = 800$ mm

Thickness of web to house a cable of 100 mm diameter with side covers of 75 mm on either side 100 + 75 + 75 = 250 mm

The section properties of the main girder are as follows:

Cross-sectional area $A = 0.94\ \text{m}^2$

Second moment of area $I = 0.45\ \text{m}^4$

$$y_b = y_t = 1\ \text{m}$$

Section modulus $Z_t = Z_b = Z = 0.45\ \text{m}^3$

The cross-sectional details of the bridge deck and main girders are shown in Figs 15.13 and 15.14 respectively.

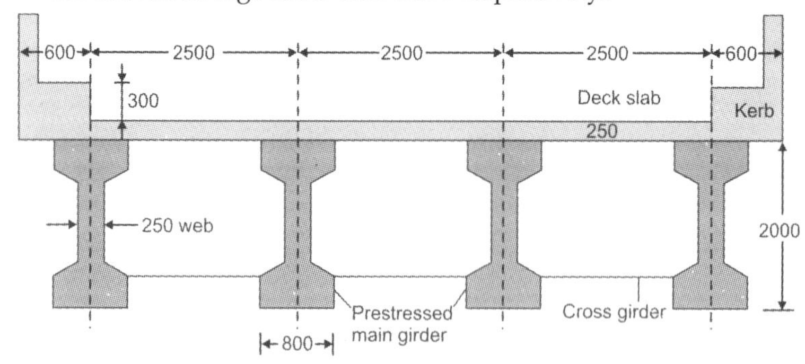

Fig. 15.13: Cross-section of prestressed concrete continuous bridge deck

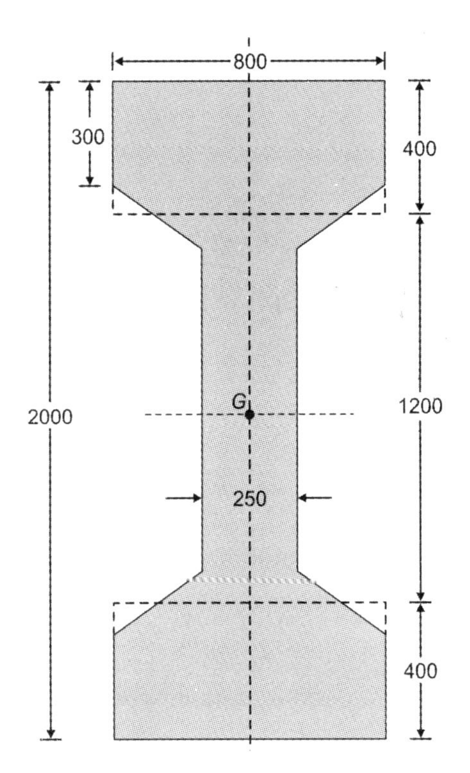

Fig. 15.14: Cross-section of continuous prestressed concrete girder

4. *Design of interior slab panel*

a. **Bending moments**

Size of slab panel = 5 m by 2.5 m.

Hence, L = 5 m and B = 2.5 m

Thickness of slab = 250 mm

Dead weight of slab = $(1 \times 1 \times 0.25 \times 25)$

= 6.25 kN/m^2

Dead weight of wearing coat = (0.08×22) = 1.76 kN/m^2

Total dead load g = 8.00 kN/m^2

Live load is IRC Class AA tracked vehicle. One wheel is placed at the centre of slab panel as shown in Fig. 15.15.

$$u = (0.85 + 2 \times 0.08) = 1.01 \text{ m}$$

$$v = (3.60 + 2 \times 0.08) = 3.75 \text{ m}$$

$$(u/B) = (1.01/2.5) = 0.404$$

$$(v/L) = (3.76/5.0) = 0.752$$

$$K = (B/L) = (2.5/5.0) = 0.5$$

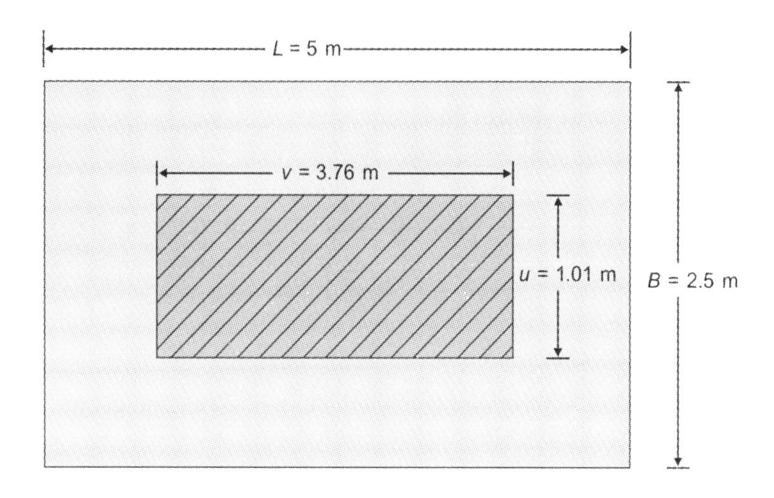

Fig. 15.15: Position of IRC Class AA wheel load for maximum moment

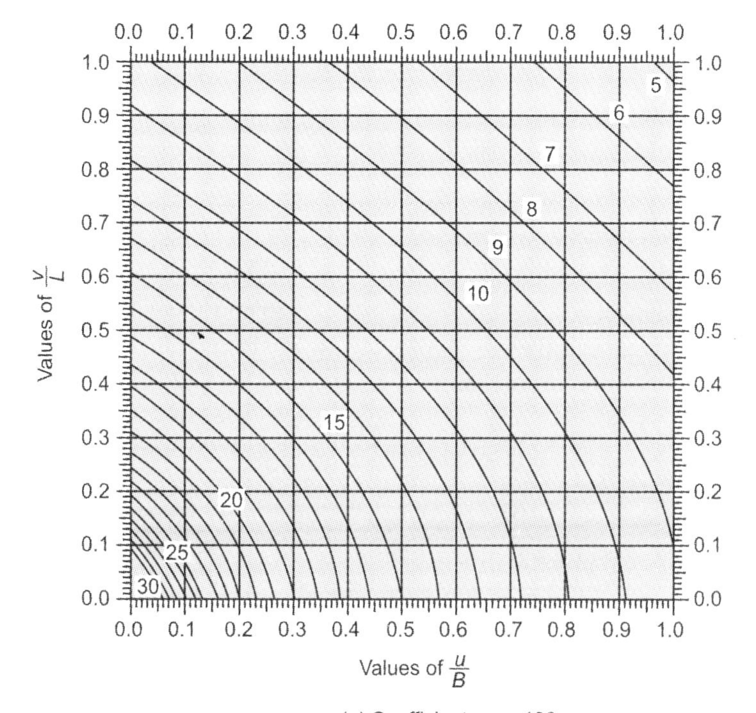

(a) Coefficient $m_1 \times 100$

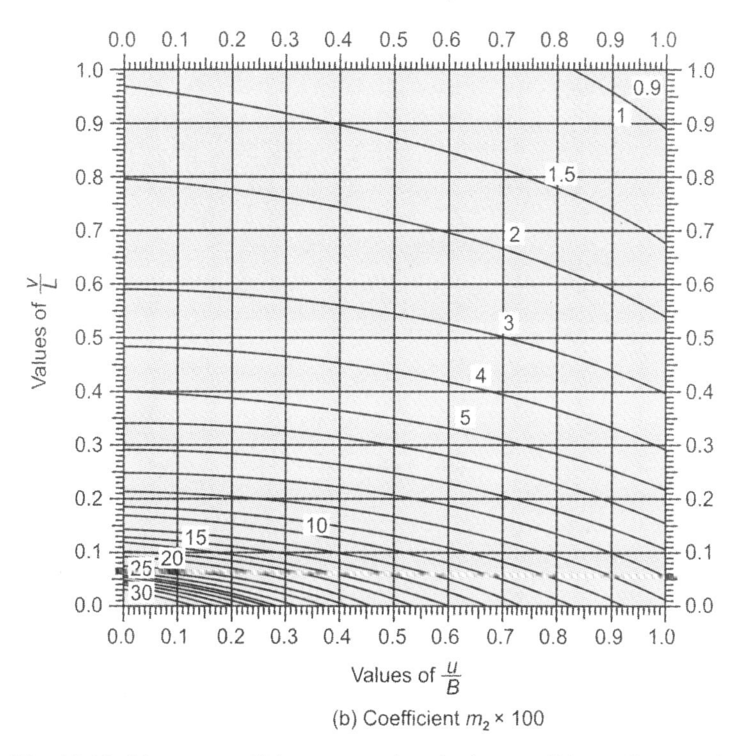

(b) Coefficient $m_2 \times 100$

Fig. 15.16: Moment coefficients m_1 and m_2 for $k = 0.5$ (Pigeaud's curves)

Referring to Pigeaud's curves (Fig. 15.16), read out the values of coefficients as

$$m_1 = 0.098 \quad \text{and} \quad m_2 = 0.02$$

$$\therefore \quad M_B = W[m_1 + \mu m_2]$$

$$= 350[0.098 + (0.15 \times 0.02)] = 35.35 \text{ kN.m}$$

As the slab is continuous, design B.M. $= 0.8\, M_B$

Design bending moment including the impact and continuity factors and is given by

M_B (short span) $= (1.25 \times 0.8 \times 35.35) = 35.35$ kN.m

Similarly,

M_L (long span) $= W[m_2 + \mu\, m_1]$

$= 350[0.02 + 0.15 \times 0.098] = 12.14$ kN.m

Design bending moment,

$$M_L = 1.25 \times 0.8 \times 12.14 = 12.14 \text{ kN.m}$$

b. **Shear forces**

Dispersion in the direction of span

$$= [0.85 + 2(0.08 + 0.25)] = 1.51 \text{ m}$$

For maximum shear force, the load is kept such that the whole dispersion is in the span.

The load is kept at (1.51/2) = 0.755 m from the edge of the beam as shown in Fig. 15.17.

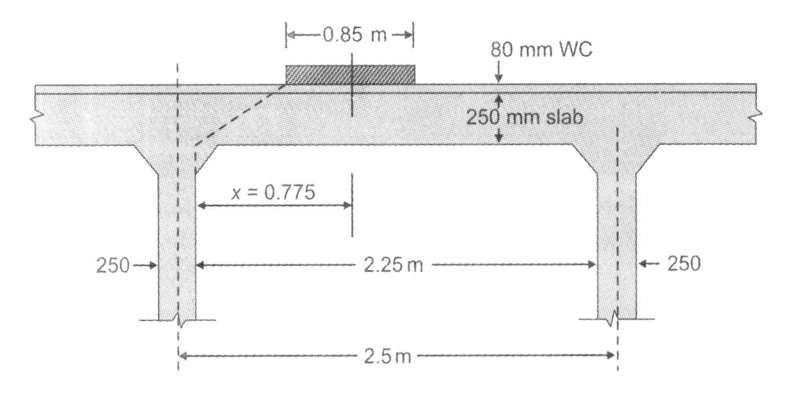

Fig. 15.17: Position of wheel load for maximum shear

Effective width of the slab = $kx\,[1 - (x/L)] + b_w$

Breadth of the cross girder = 200 mm

Clear length of the panel = (5 – 0.2) = 4.8 m

∴ (B/L) = (4.8/2.3) = 2.08

The value of 'k' for continuous slab = 2.60 (refer Table 15.1)

Effective width of the slab = (2.6 × 0.755) [1 – (0.755/2.3) + 3.6 + (2 × 0.08)] = 5.079 m

Load per metre width = (350/5.079) = 70 kN

Shear force/metre width = [70 (2.3 – 0.755)/2.3] = 47 kN

Shear force with impact = (1.25 × 47) = 58.75 kN

c. **Dead load bending moment and shear forces**

Dead load = 8 kN/m^2

Total load on panel = (5 × 2.5 × 8) = 100 kN

$(u/B) = 1$ and $(v/L) = 1$

as the panel is loaded with a uniformly distributed load.

$k = (B/L) = (2.5/5.0) = 0.5$ and $(1/k) = 2.0$

From Pigeaud's curves (Fig. 15.18), read out the moment coefficients as

$m_1 = 0.047$ and $m_2 = 0.01$

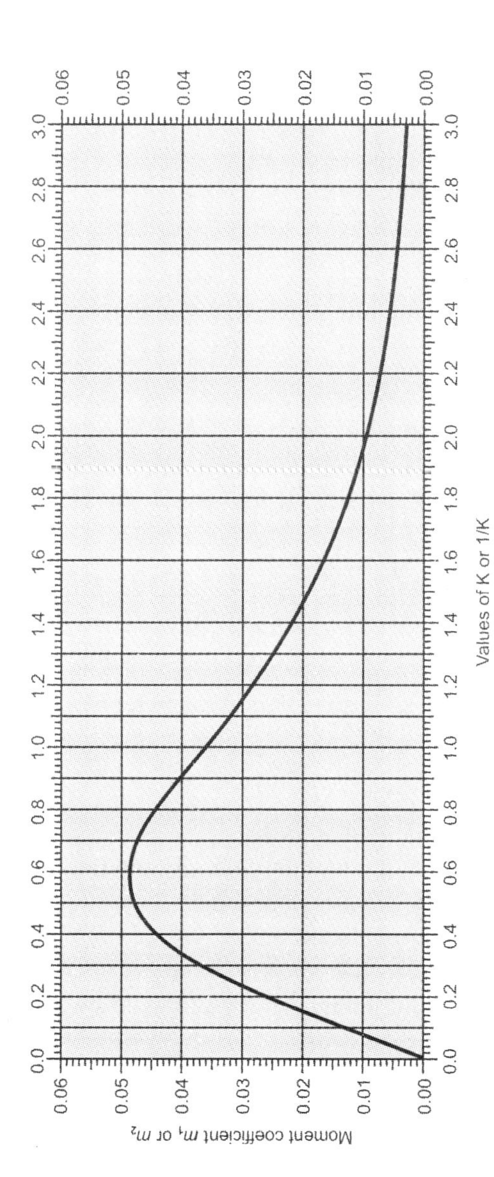

Fig. 15.18: Moment coefficients for slab completely loaded with uniformly distributed load (coefficient m_1 for K and m_2 for $1/K$)

$$M_B = 100\,(0.047 + 0.15 \times 0.01) = 4.85 \text{ kN.m}$$
$$M_L = 100\,(0.1 + 0.15 \times 0.047) = 1.70 \text{ kN.m}$$

Design B.M including the continuity factor

$$M_B = (0.8 \times 4.85) = 3.88 \text{ kN.m}$$
$$M_L = (0.8 \times 170) = 1.36 \text{ kN.m}$$

Dead load shear force $= (0.5 \times 8 \times 2.3) = 9.2$ kN

d. **Design moments and shear forces**

Total $M_B = (35.35 + 3.88) = 39.23$ kN.m

Total $M_L = (12.14 + 1.36) = 13.50$ kN.m

Total $V = (58.75 + 9.20) = 67.95$ kN

e. **Design of slab section and reinforcements**

$$\text{Effective depth } d = \sqrt{\frac{M}{Qb}} = \sqrt{\frac{39.23 \times 10^6}{1.10 \times 10^3}} = 189 \text{ mm}$$

According to IRC: 21-2000, the minimum clear cover to steel reinforcement should not be less than 40 mm.

Hence, adopt effective depth $= d = 200$ mm

$$A_{st} = \left[\frac{M}{\sigma_{st} jd}\right] = \left[\frac{39.23 \times 10^6}{200 \times 0.90 \times 200}\right] = 1090 \text{ mm}^2$$

Use 16 mm diameter bars at 180 mm centres $A_{st} = 1117$ mm^2,

Effective depth for long span using 10 mm diameter bars

$$= (200 - 8 - 50) = 187 \text{ mm}$$

$$\therefore \qquad A_{st} = \left(\frac{13.50 \times 10^6}{200 \times 0.90 \times 187}\right) = 401 \text{ mm}^2$$

But minimum reinforcement according to IRC: 18-2000 is 0.15 percent of the cross-sectional area. Hence area of steel in the long span direction is computed as

A_{st} (long span) $= 0.0015 \times 1000 \times 250 = 375$ mm^2

Adopt 10 mm diameter bars at 180 mm centres $A_{st} = 436$ mm^2,

Check for shear strength (as per IRC: 21-2000)

$$\text{Nominal shear stress} \quad \tau_V = \left(\frac{V}{bd}\right) = \left(\frac{67.95 \times 1000}{1000 \times 200}\right)$$
$$= 0.337 \text{ N/mm}^2$$

$$\text{Ratio} \left(\frac{100\,A_{st}}{bd}\right) = \left(\frac{100 \times 1117}{1000 \times 200}\right) = 0.558$$

From Table 12B of IRC: 21-2000, read out the permissible shear stress in concrete for M-25 grade concrete as 0.32 N/mm^2.

For solid slabs, permissible shear stress is $k\tau_c$, where $k = 1.10$ for 250 mm thick slab from Table 12C of IRC: 21-2000.

$\therefore$ Permissible shear stress

$$= (1.1 \times 0.32) = 0.352 \text{ N/mm}^2 > \tau_V..$$

Hence safe.

5. *Design of continuous longitudinal girder*

a. **Reaction factors**

Using Courbon's theory, the IRC Class AA tracked loads are arranged for maximum eccentricity as shown in Fig. 15.19. The reaction factor maximum for any of the longitudinal girder is computed using the general relation,

$$R_X = \frac{\Sigma W}{n}\left[1 + \left\{\frac{\Sigma I}{\Sigma d_x^2\, I}\right\} d_x . e\right]$$

where, R_X = Reaction factor for the girder under consideration

I = Second moment of area of each longitudinal girder

d_X = Distance of the girder under consideration from the centroidal axis of the bridge

SW = Total concentrated live load

n = Number of longitudinal girders

e = Eccentricity of live load with respect to the axis of the bridge

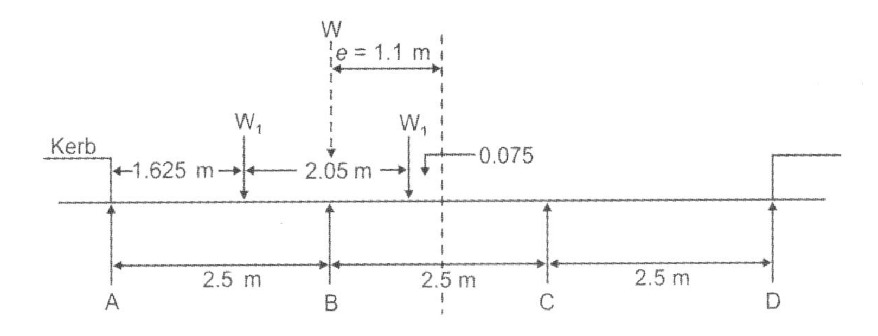

Fig. 15.19: Transverse disposition of IRC Class AA tracked vehicle loads

In the present case, the corresponding values for the exterior girder A are as follows:

$$\Sigma W = 2\,W_1$$

$$d_X = 3.75 \text{ m}$$

$$e = 1.1 \text{ m}$$

$$\therefore \quad R_A = \frac{2W_1}{4}\left[1+\frac{4I\times 3.75\times 1.1}{(2I\times 3.75^2)+(2I\times 1.25^2)}\right]$$

$$= 0.764\,W_1$$

If $\quad W = \text{axle load} = 700 \text{ kN}$

$$W_1 = 0.5\,W$$

$$\therefore \quad R_A = (0.764 \times 0.5\,W) = 0.382\,W$$

b. **Loads acting on the main girder**

Self weight of slab and wearing coat = 8 kN/m^2

Load from slab and W.C on girder (8 × 25) = 20 kN/m

Self weight of main girder = (0.94 × 25) = 23.5 kN/m

Weight of cross girders assumed to act as uniformly distributed load = 3.5 kN/m

∴ Total dead load of main girder

$$g = (20 + 23.5 + 3.5) = 47 \text{ kN/m}$$

c. **Dead load moments and shear forces**

Referring to bending and shear force coefficients given in Appendix-1 of the monograph by the author [*Advanced Reinforced Concrete*-2nd Edn. (IS: 456-2000), CBS Publishers, New Delhi, 2005, p.345], we have the dead load moment at mid support section computed as,

$$M_{gB} = 0.125\,gL^2 = 0.125 \times 47 \times 40^2 = 9400 \text{ kN.m}$$

Dead load moment at mid span section is calculated as

$$M_{gD} = 0.071\,gL^2 = 0.071 \times 47 \times 40^2 = 5340 \text{ kN.m}$$

Dead load shear is maximum near the mid support section and is computed as

$$V_g = 0.62\,gL = 0.62 \times 47 \times 40 = 1166 \text{ kN}$$

d. **Live load moments in girder**

Referring to the influence line diagram for bending moment shown in Fig. 15.20, the live load is placed near the maximum ordinate so as to yield the maximum positive live load moment which is computed as

$$M_D = \left(\frac{7.3 + 8.12}{2}\right) 700 = 5397 \text{ kN.m}$$

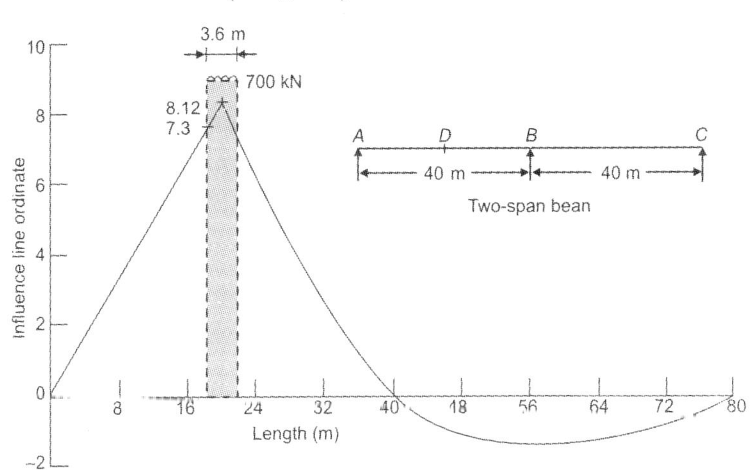

Fig. 15.20: Influence line for bending moment at mid span section

The maximum negative moment due to live loads developing at mid support B is obtained from the influence line diagram shown in Fig. 15.21. The maximum negative bending moment at B is computed as

$$M_B = (3.76 \times 700) = 2632 \text{ kN.m}$$

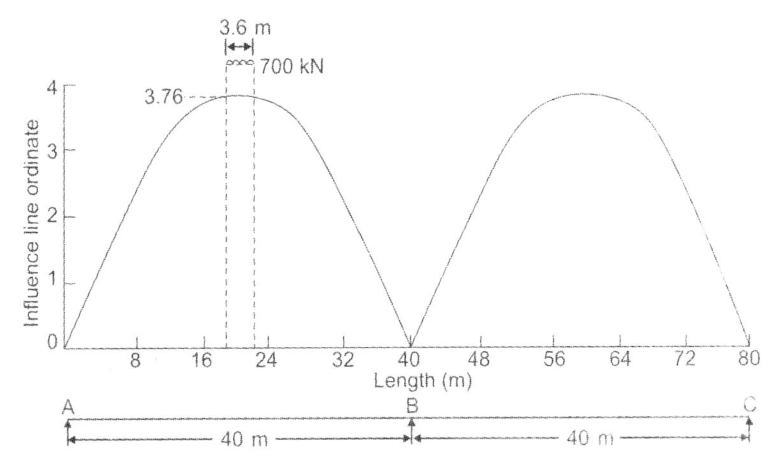

Fig. 15.21: Influence line for bending moment at mid support B

The live load bending moments including the reaction and impact factors for the exterior girder are

$$M_{qD} = (5397 \times 1.1 \times 0.382) = 2268 \text{ kN.m}$$
$$M_{qB} = (2632 \times 1.1 \times 0.382) = 1106 \text{ kN.m}$$

e. **Live load shear forces in girder**

Maximum live load shear develops in the interior girder when the IRC class AA loads are placed near the mid support *B* as shown in Fig. 15.22.

$$\text{Reaction of } W_2 \text{ on girder } B = \left(\frac{350 \times 0.45}{2.5}\right) = 63 \text{ kN}$$

$$\text{Reaction of } W_2 \text{ on girder } A = \left(\frac{350 \times 2.05}{2.5}\right) = 287 \text{ kN}$$

∴ Total load on girder $B = (350 + 63) = 413$ kN

Maximum reaction in the inner girder

$$B = \left(\frac{413 \times 38.2}{40}\right) = 394 \text{ kN}$$

Maximum live load shear with impact factor in girder *B*

$$= 394 \times 1.1 = 434 \text{ kN}$$

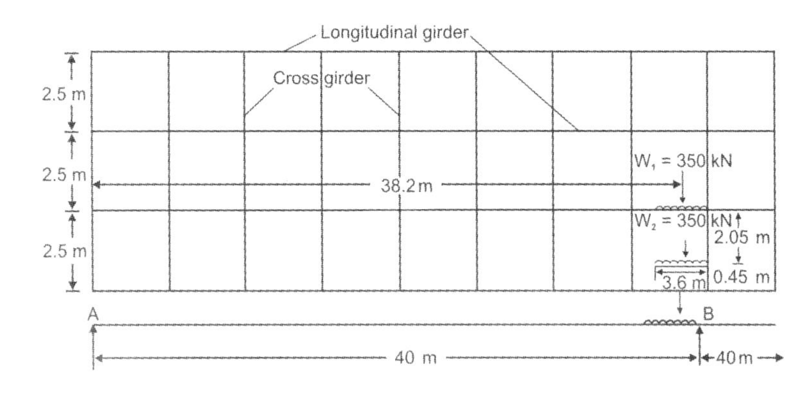

Fig. 15.22: Position of IRC Class AA loads for maximum shear in inner girder

f. **Design bending moments and shear forces**

The design bending moments and shear forces at service loads and at the limit state of collapse as stipulated in IRC: 18-2000 are compiled in Table 15.2.

Table 15.2: Working and ultimate load, moments and shear forces in longitudinal girders

a. Bending moments (outer girder)					
	Dead load B.M	Live load B.M	Total working Load B.M	Required M_u	Units
Mid span	(M_g)	(M_q)	$(M_g + M_q)$	$(1.5\,M_g + 2.5\,M_q)$	
section (D)	5340	2268	7608	13680	kN.m
Mid support	(M_g)	(M_q)	$(M_g + M_q)$	$(1.5\,M_g + 2.5\,M_q)$	
section (B)	9400	1106	10506	16865	kN.m

b. Shear forces (inner girder)					
	Dead load S.F	Live load S.F	Total working load S.F	Required V_u	Units
Near mid	(V_g)	(V_q)	$(V_g + V_q)$	$(1.5\,V_g + 2.5\,V_q)$	
support	1166	434	1600	2834	kN
section					

g. **Check for minimum section modulus**

Maximum design moments developed at mid support section B

$$M_g = 9400 \text{ kN.m}$$

$$M_q = 1106 \text{ kN.m}$$

$$\therefore \quad M_d = (M_g + M_q) = 10506 \text{ kN.m}$$

$$f_{br} = (\eta f_{ct} - f_{tw}) = [(0.8 \times 20) - 0] = 16 \text{ N/mm}^2$$

$$f_b = \left[\frac{f_{tw}}{\eta} + \frac{M_d}{\eta Z_b}\right] = \left[0 + \frac{(10506 \times 10^6)}{(0.8 \times 0.45 \times 10^9)}\right]$$

$$= 29.2 \text{ N/mm}^2$$

$$Z_b \geq \left[\frac{M_q + (1-\eta) M_g}{f_{br}}\right]$$

$$\geq \left[\frac{(1106 \times 10^6) + (1 - 0.8) 9400 \times 10^6}{16}\right]$$

$$\geq (0.186 \times 10^9) \text{ mm}^3,$$

which is less than (0.45×10^9) mm^3. Hence, the section provided is adequate.

h. **Prestressing force**

In the present case, $M_g > M_q$ and consequently the minimum prestressing force computed will lie outside the section. Hence, the maximum possible eccentricity with due respect to cover is provided and the enhanced prestressing force is calculated using the relation,

$$P = \left[\frac{Af_b Z_b}{Z_b + Ae}\right]$$

Maximum possible eccentricity at support section B is determined by assuming three Freyssinet cables of type 19K–15 (19 strands of 15.2 mm diameter) in 95 mm diameter cable ducts with a clear spacing not less than the diameter of the cable and with a clear cover of not less than 75 mm according to section 16 of IRC: 18-2000.

Hence, maximum possible eccentricity at support B is computed as,

$$e = [1000 - 75 - 95 - 100 - 50] = 680 \text{ mm}$$

Modified prestressing force is evaluated from the relation

$$P = \left[\frac{Af_b Z_b}{Z_b + Ae}\right] = \left[\frac{(0.94\times10^6\times29.20\times0.45\times10^9)}{(0.45\times10^9)+(0.94\times10^6\times680)}\right]$$

$$= 11338597 \text{ N} = 11338 \text{ kN}$$

Force in each cable = $19 \times 0.8 \times 265 = 4028$ kN

Provide 3 cables carrying an initial prestressing force

$$P = 3 \times 4000 = 12000 \text{ kN}$$

Area of each strand of 15.2 mm diameter = 140 mm^2

Area of 19 strands in each cable = $(19 \times 140) = 2660$ mm^2

Total area of high tensile steel in 3 cables

$$A_p = 3 \times 2660 = 7980 \text{ mm}^2$$

i. **Concordant cable profile**

For the two span continuous beam, a concordant parabolic cable profile is adopted with a maximum eccentricity e at mid support section B and $(e/2)$ at mid span section. Hence, the cables are arranged in parabolic concordant profile with an effective eccentricity of 680 mm towards the top fibre at mid support section B and an eccentricity of 340 mm towards the soffit at mid span section D as shown in Fig. 15.23.

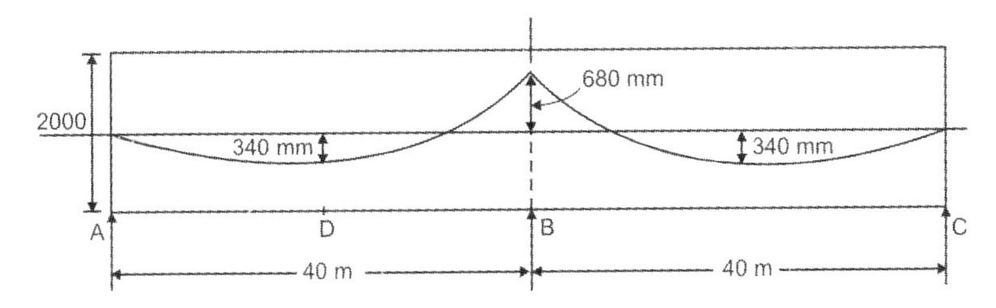

Fig. 15.23: Concordant cable profile

The centroid of the cables is concentric at the end supports A and C. The selected parabolic profiles of the three cables along the span with cross-sections at critical sections are shown in Fig. 15.24.

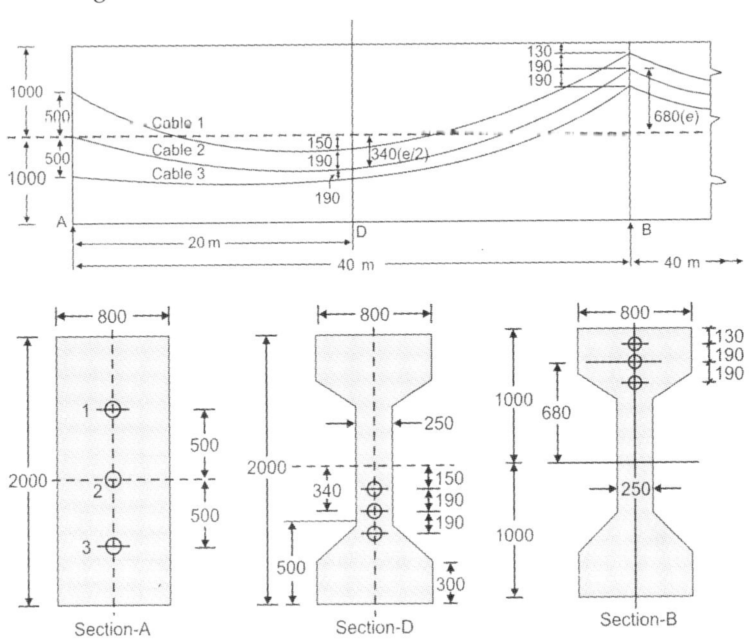

Fig. 15.24 Cable layout in two span prestressed concrete girder

6. *Check for stresses at service loads*

a. **Centre of span section**

$P = 12000$ kN $\qquad M_g = 5340$ kN.m

$e = 340$ mm $\qquad M_q = 2268$ kN.m

$A = (0.94 \times 10^6)$ mm^2 $\qquad \eta = 0.80$

$Z = (0.45 \times 10^9)$ mm^3

$$\left(\frac{P}{A}\right) = \left(\frac{12000\times10^3}{0.94\times10^6}\right) = 12.76\ \text{N/mm}^2$$

$$\left(\frac{Pe}{Z}\right) = \left(\frac{12000\times10^3\times340}{0.45\times10^9}\right) = 9.06\ \text{N/mm}^2$$

$$\left[\frac{M_g}{Z}\right] = \left[\frac{5340\times10^6}{0.45\times10^9}\right] = 11.86\ \text{N/mm}^2$$

$$\left[\frac{M_q}{Z}\right] = \left[\frac{2268\times10^6}{0.45\times10^9}\right] = 5.04\ \text{N/mm}^2$$

At the transfer stage, the extreme fibre stresses are

$$\sigma_t = \left[\left(\frac{P}{A}\right)-\left(\frac{Pe}{Z}\right)+\left(\frac{M_g}{Z}\right)\right]$$

$$= [12.76 - 9.06 + 11.86] = 15.56\ \text{N/mm}^2$$

$$\sigma_b = \left[\left(\frac{P}{A}\right)+\left(\frac{Pe}{Z}\right)-\left(\frac{M_g}{Z}\right)\right]$$

$$= [12.76 + 9.06 - 11.86] = 9.96\ \text{N/mm}^2$$

At the limit state of service loads, the stresses are

$$\sigma_t = \left[\eta\left(\frac{P}{A}\right)-\eta\left(\frac{Pe}{Z}\right)+\left(\frac{M_g}{Z}\right)+\left(\frac{M_q}{Z}\right)\right]$$

$$= [10.21 - 7.24 + 11.86 + 5.04]$$

$$= 19.87\ \text{N/mm}^2 < 20\ \text{N/mm}^2$$

$$\sigma_b = \left[\eta\left(\frac{P}{A}\right)+\eta\left(\frac{Pe}{Z}\right)-\left(\frac{M_g}{Z}\right)-\left(\frac{M_q}{Z}\right)\right]$$

$$= [10.21 + 7.24 - 11.86 - 5.04] = 0.55\ \text{N/mm}^2$$

b. **Mid support section**

$P = 12000$ kN $\qquad M_g = 9400$ kN.m

$e = 680$ mm $\qquad M_q = 1106$ kN.m

$A = (0.94 \times 10^6)$ mm $\qquad \eta = 0.80$

$Z = (0.45 \times 10^9)$ mm^3

$$\left(\frac{P}{A}\right) = 12.76\ \text{N/mm}^2$$

$$\left(\frac{Pe}{Z}\right) = \left(\frac{12000\times10^3\times680}{0.45\times10^9}\right) = 18.12\ \text{N/mm}^2$$

$$\left(\frac{M_g}{Z}\right) = \left(\frac{9400 \times 10^6}{0.45 \times 10^9}\right) = 20.88 \text{ N/mm}^2$$

$$\left(\frac{M_q}{Z}\right) = \left(\frac{1106 \times 10^6}{0.45 \times 10^9}\right) = 2.45 \text{ N/mm}^2$$

At the stage of transfer:

$$\sigma_t = [12.76 + 18.12 - 20.88] = 10.00 \text{ N/mm}^2$$

$$\sigma_b = [12.76 - 18.12 + 20.88] = 15.52 \text{ N/mm}^2$$

At the limit state of service loads:

$$\sigma_t = [0.8\,(12.76 + 18.12) - 20.88 - 2.45]$$
$$= 1.37 \text{ N/mm}^2$$

$$\sigma_b = [0.8\,(12.76 - 18.12) + 20.88 + 2.45]$$
$$= 19.04 \text{ N/mm}^2$$

The stresses at various critical stages are within the maximum permissible limits of 20 N/mm.

7. *Check for ultimate flexural strength*

a. **Centre of span section**

$A_p = 7980 \text{ mm}^2$ $\quad f_{ck} = 60 \text{ N/mm}^2$

$b = 800 \text{ mm}$ $\quad f_p = 1862 \text{ N/mm}^2$

$b_w = 250 \text{ mm}$ $\quad d = 1340 \text{ mm}$

$D_f = 400 \text{ mm}$, M_u (required) = 13680 kN.m

According to IRC: 18-2000 specification,

i. ***Failure by yielding of steel***

$$M_u = (0.9\, d.A_p f_p) = 0.9 \times 1340 \times 7980 \times 1862$$
$$= (17920 \times 10^6) \text{ N.mm} = 17920 \text{ kN.m} > 13680 \text{ kN.m}$$

ii. ***Failure by crushing of concrete***

$$M_u = [0.176\, b_w d^2 f_{ck}] + [(0.67 \times 0.8(b - b_w)\,(d - 0.5\,D_f)\,D_f f_{ck}]$$
$$= [(0.176 \times 250 \times 1340^2 \times 60)] + [(0.67 \times 0.8)\,(800 - 250)\,(1340 - 0.5 \times 400)\,(400 \times 60)]$$
$$= 12806 \times 10^6 \text{ N.mm}$$
$$= 12806 \text{ kN.m} < 13680 \text{ kN.m}$$

Since, the actual $M_u < M_u$ (required), additional untensioned reinforcements are designed to resist the balance moment.

Balance moment = M_{bal} = [13680 – 12806] = 874 kN.m

Using Fe-415 grade HYSD bars at an effective depth of 1900 mm,

$$A_{us} = \left[\frac{M_{bal}}{0.87 f_y (d - 0.5 D_f)}\right]$$

$$= \left[\frac{874 \times 10^6}{0.87 \times 415 (1900 - 0.5 \times 400)}\right] = 1425 \text{ mm}^2$$

Provide 6 bars of 20 mm diameter (A_{st} = 1885 mm²)

b. **Mid support section**

$A_p = 7980$ mm² $\qquad f_{ck} = 60$ N/mm²
$b = 800$ mm $\qquad f_p = 1862$ N/mm²
$b_w = 250$ mm $\qquad d = 1680$ mm
$D_f = 400$ mm, M_u (required) = 16865 kN.m

$$M_u = [0.176\, b_w\, d^2 f_{ck}] + [(0.67 \times 0.8(b - b_w)(d - 0.5\, D_f)\, D_f f_{ck}]$$
$$= [(0.176 \times 250 \times 1680^2 \times 60)] + [(0.67 \times 0.8)(800 - 250)(1680 - 0.5 \times 400)(400 \times 60)]$$
$$= (17922 \times 10^6) \text{ N.mm}$$
$$= 17922 \text{ kN.m} > 16865 \text{ kN.m}$$

The ultimate flexural strength of mid support section is greater than the required design ultimate strength.

8. *Check for ultimate shear strength*

Design shear force $V_u = 2834$ kN

According to IRC: 18-2000 code, the ultimate shear resistance of the support section uncracked in flexure is given by

$$V_{cw} = 0.67\, b_w\, h \sqrt{f_t^2 + 0.8 f_{cp} f_t} + \eta P \sin\theta$$

where $b_w = 250$ mm
$h = 2000$ mm
$f_t = 0.24\sqrt{f_{ck}} = 0.24\sqrt{60} = 1.85$ N/mm²

$$f_{cp} = \left(\frac{\eta P}{A}\right) = \left(\frac{0.8 \times 12000 \times 1000}{0.94 \times 10^6}\right) = 10.2 \text{ N/mm}^2$$

$$\theta = \left(\frac{4e}{L}\right) = \left(\frac{4 \times 680}{40 \times 1000}\right) = 0.068$$

$$V_{cw} = [(0.67 \times 250 \times 2000 \times \sqrt{1.85^2 + 0.8 \times 10.2 \times 1.85} + (0.8 \times 12000 \times 10^3 \times 0.068)]$$
$$= (2093 \times 10^3) \text{ N}$$
$$= 2093 \text{ kN} < 2834 \text{ kN}$$

Balance shear force = (2834 – 2093) = 741 kN

Using 12 mm diameter two legged stirrups

$$\text{Spacing} \quad S_V = \left[\frac{0.87 \times 415 \times 2 \times 113 \times 1900}{741 \times 1000}\right] = 209 \text{ mm}$$

Provide 12 mm diameter two legged stirrups at a spacing of 200 mm centres near the supports which is gradually increased to 300 mm centres towards the centre of span.

9. *Supplementary reinforcements*

According to IRC: 18-2000, longitudinal reinforcements of not less than 0.18% (for f_{ck} > M-45) of the gross cross-sectional area are to be provided to safeguard against shrinkage cracking.

$$A_{st} = \left[\frac{0.18 \times 0.94 \times 10^6}{100}\right] = 1692 \text{ mm}^2$$

Provide 15 bars of 12 mm diameter distributed in the top and bottom flanges and web. The reinforcements provided at mid span and mid support sections are shown in Fig. 15.25.

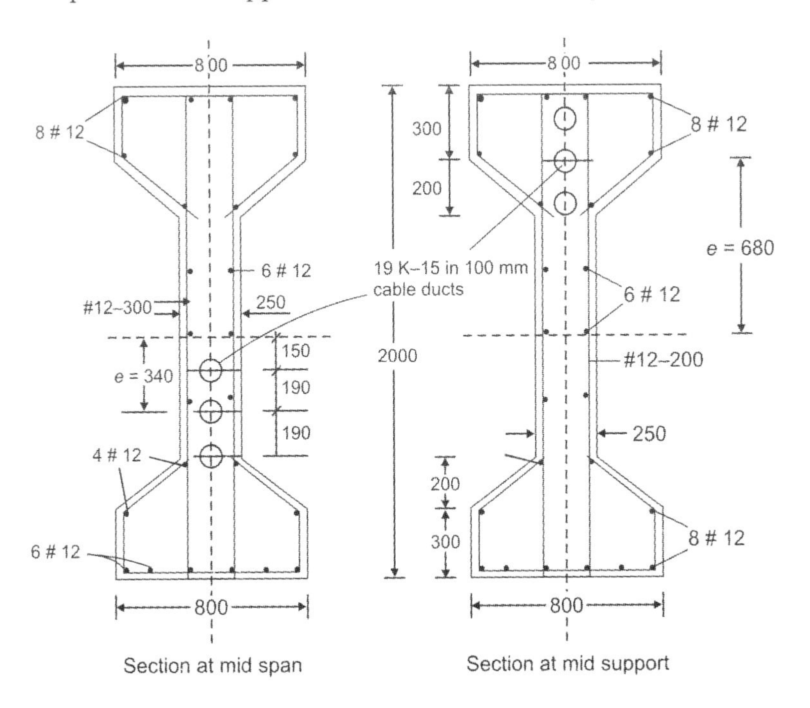

Fig. 15.25: Reinforcement and cable details at mid span and mid support sections

10. *Design of end bocks*

At the end supports *A* and *C*, solid end blocks, 800 mm wide by 2000 mm deep are provided for a length of 2 m from each face of the girder. The equivalent prisms on which the anchorage forces are considered to be effective are shown in Fig. 15.26(a). The bursting tension in the end block is evaluated using the equation recommended in IS: 1343 code.

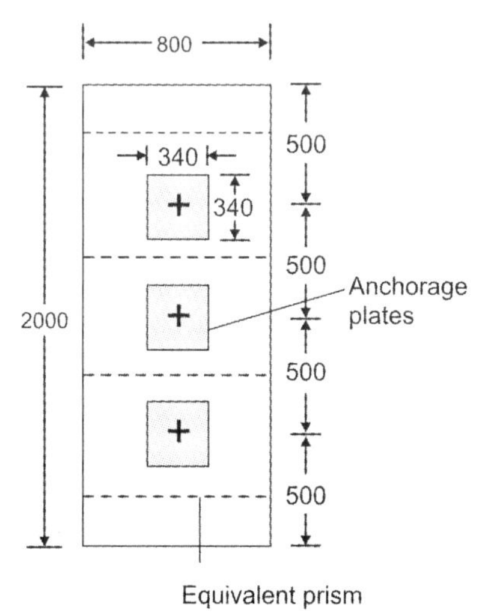

Fig. 15.26 (a): Anchorage zone reinforcement in end block

In the horizontal plane, we have

$$2y_{po} = 340 \text{ mm}$$
$$2y_o = 800 \text{ mm}$$
$$P_k = 4000 \text{ kN}$$

$$\text{Ratio} \left(\frac{y_{po}}{y_o}\right) = \left(\frac{340}{800}\right) = 0.425$$

Bursting tension,

$$F_{bst} = P_k \left[0.35 - 0.3\left(\frac{y_{po}}{y_o}\right)\right]$$
$$= 4000\,[0.32 - 0.3\,(0.425)] = 770 \text{ kN}$$

Using Fe-415 HYSD bars

$$A_{st} = \left[\frac{770 \times 1000}{0.87 \times 415}\right] = 2132 \text{ mm}^2$$

Provide 16 mm diameter bars at 150 mm centres in the horizontal plane distributed in the region from $0.2y_o$ to $2y_o$ (80 mm to 800 mm) as shown in Fig. 15.26(b). In the vertical plane, the ratio (y_{po}/y_o) being larger, the magnitude of bursting tension is less. However, the same reinforcements are provided in the vertical plane in the form of a mesh to resist the bursting tension.

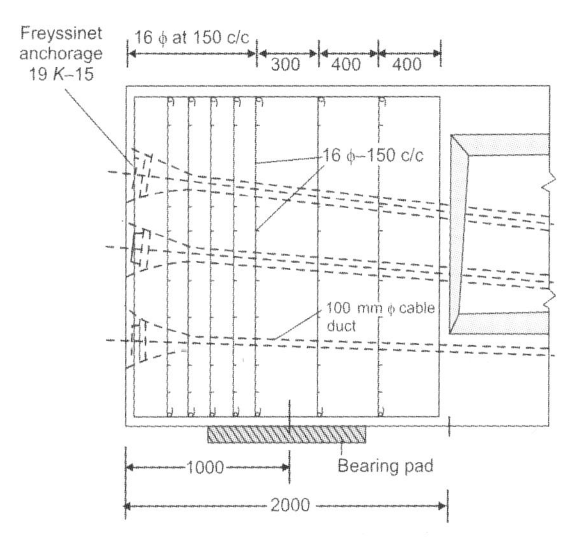

Fig. 15.26 (b): Anchorage zone reinforcement in end block

11. *Cross girders*

At intervals of 5 m along the span, cross girders of width 200 mm and depth 1600 mm are provided with nominal reinforcements of 0.18% of the cross-section comprising 12 mm diameter bars spaced at 300 mm centres, distributed near both vertical faces of cross-girder. 10 mm diameter ties are also provided. Two cables housing 12 numbers of 7 mm diameter high tensile wires are positioned at mid third points along the depth with a nominal prestress to provide lateral stiffness to the bridge deck system.

Problem 15.4

A cellular multicelled prestressed concrete box girder deck is to be designed for a national highway crossing. The proposed bridge deck is made up of two continuous spans each of 50 m. The road width is 7.5 m with footpaths 1.25 m on each side. The box girder is proposed to have 4 cells 2 m wide by 2 m deep and should support

IRC Class AA tracked vehicle loading. Design the cellular bridge deck adopting M-60 grade concrete, Fe-415 HYSD bars and high tensile steel strands of 15.2 mm diameter conforming to the relevant Indian standards.

Solution

1. *Data*

   Span = 50 m
   Cross-section: Multicelled box girder
   Cell dimensions = 2 m wide by 2 m deep
   Road width = 7.5 m
   Footpaths: 1.25 m wide on either side of roadway
   Wearing coat = 80 mm
   Thickness of web = 300 mm to house 27 K-15 Freyssinet type anchorages (27 strands of 15.2 mm diameter in 110 mm diameter cables)
   Thickness of top and bottom slabs = 300 mm
   Concrete grade: M-60
   Loss ratio = 0.80
   Type of tendons: High tensile strands of 15.2 mm diameter conforming to IRC: 6006 – 2000
   Type of supplementary reinforcements: Fe-415 HYSD bars
   Design the bridge deck as Class-1 type structure conforming to the codes– IRC: 6-1966, IRC: 18-2000 and IRC: 21-2000.

2. *Permissible stresses*

   For M-60 grade concrete and Fe-415 HYSD bars, the permissible stresses according to IRC: 21-2000 are:

$$\sigma_{cb} = 11.5 \text{ N/mm}^2 \qquad j = 0.879$$
$$\sigma_{st} = 200 \text{ N/mm}^2 \qquad Q = 1.844$$
$$m = 10$$

   For M-60 grade concrete, the permissible stresses according to IRC: 18-2000 are:

$$f_{ck} = 60 \text{ N/mm}^2$$
$$f_{ci} = 45 \text{ N/mm}^2$$
$$f_{ct} = 0.45 f_{ci} = (0.45 \times 45) = 20 \text{ N/mm}^2$$
$$f_{cw} = 0.33 f_{ck} = (0.33 \times 60) = 20 \text{ N/mm}^2$$
$$f_{tt} = f_{tw} = 0 \text{ (Class 1 type member)}$$
$$E_c = 5000\sqrt{f_{ck}} = 5000\sqrt{60} = 38730 \text{ N/mm}^2$$
$$= 38.7 \text{ kN/mm}^2$$

3. *Cross-section of box girder*

Overall depth of the box girder

$$= \left(\frac{\text{Span}}{25}\right) = \left(\frac{50}{25}\right) = 2 \text{ m}$$

Width of roadway = 7.5 m

Width of footpaths = (2 × 1.25) = 2.5 m

Total width of box girder at road level = (7.5 + 2.50) = 10 m

Spacing between webs = 2 m

4 celled box girder is adopted

Thickness of web as per Clause 9.3.2.1 of IRC: 18-2000 is computed as

t_w = [200 + diameter of cable duct for housing 27 K-15 strands]

= [200 + 100] = 300 mm

At end supports where anchorages are located, web thickness is increased to 400 mm

Thickness of top and bottom slabs = 300 mm

The multicelled box girder section selected is shown in Fig. 15.27.

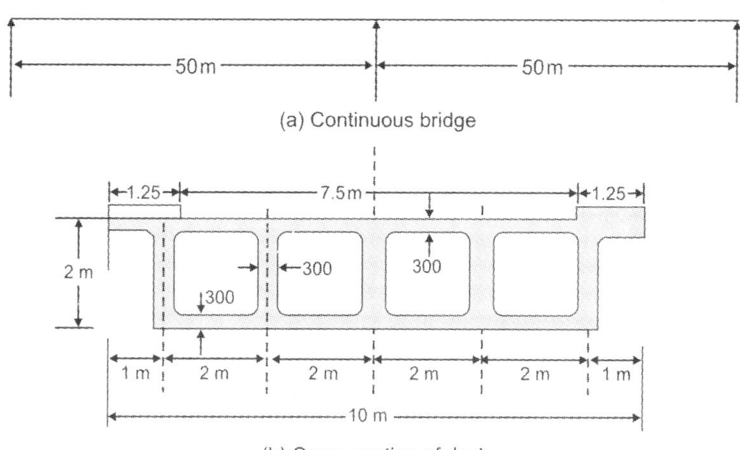

Fig. 15.27: Two span box girder bridge deck

Section properties of the symmetrical I-girder shown in Fig. 15.28 are as follows:

Cross-sectional area $A = 1.62 \text{ m}^2$

Second moment of area $I = 0.94 \text{ m}^4$

Distance of the extreme fibre from centroid

$$y = y_t = y_b = 1 \text{ m}$$

Section modulus $Z = Z_t = Z_b = (I/y) = 0.94 \text{ m}^3$

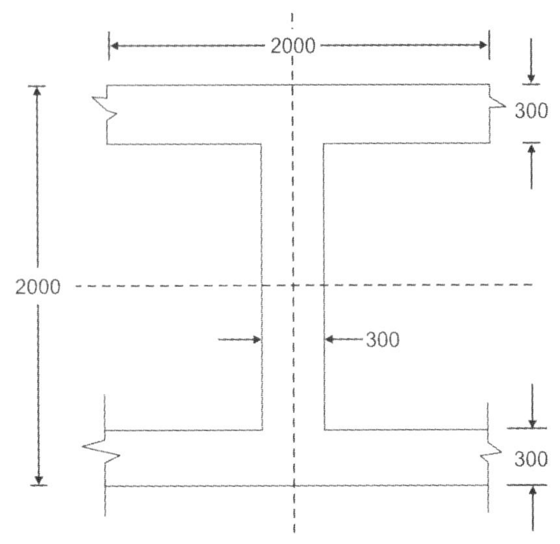

Fig. 15.28: Cross-section of web girder

4. *Design of slab panel*

   a. **Dead load bending moments**

   Dead weight of slab = $(1 \times 1 \times 0.3 \times 24) = 7.20 \text{ kN/m}^2$

   Dead weight of W.C = $(0.08 \times 22) = 1.76$

   Total dead load $g = 8.96 = 9.00 \text{ kN/m}^2$

   Referring to the bending moment coefficients compiled in a separate monograph by the author (*Advanced Reinforced Concrete*-2nd Edn. (IS: 456-2000), CBS Publishers, New Delhi, 2005, p.345) and the Fig. 15.29,

   Maximum negative bending moment due to dead load at supports

   $$= (0.107 \times gL) = (0.107 \times 9 \times 2)$$
   $$= 1.93 \text{ kN.m/m}$$

   Maximum positive bending moment at the centre of span

   $$= (0.077 \times gL) = (0.77 \times 9 \times 2)$$
   $$= 1.38 \text{ kN.m/m}$$

   Maximum shear force

   $$= (0.60 \times gL) = (0.60 \times 9 \times 2) = 10.8 \text{ kN}$$

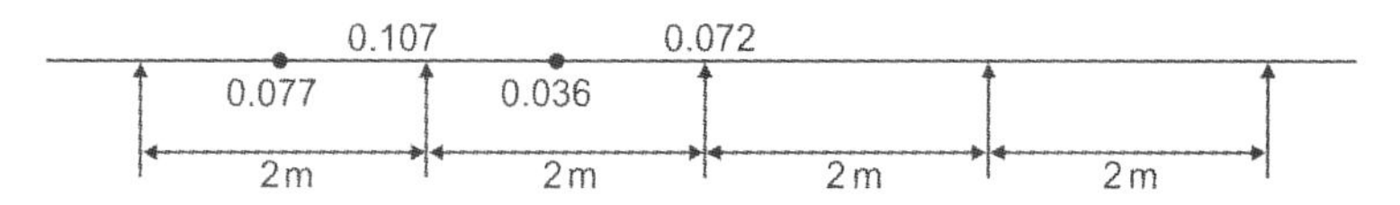

Fig. 15.29: Dead load bending moment coefficients in four continuous slab

b. **Live load bending moments**

The slab pane is continuous over webs in the transverse direction and free in the longitudinal direction. The slab spanning in the transverse direction is designed for IRC Class AA tracked loading using the procedure specified in IRC: 21-2000. When IRC Class AA tracked vehicle traverse on the deck, maximum bending moment in the transverse direction of the slab will develop when one tracked wheel occupies the centre of slab as shown in Fig. 15.30.

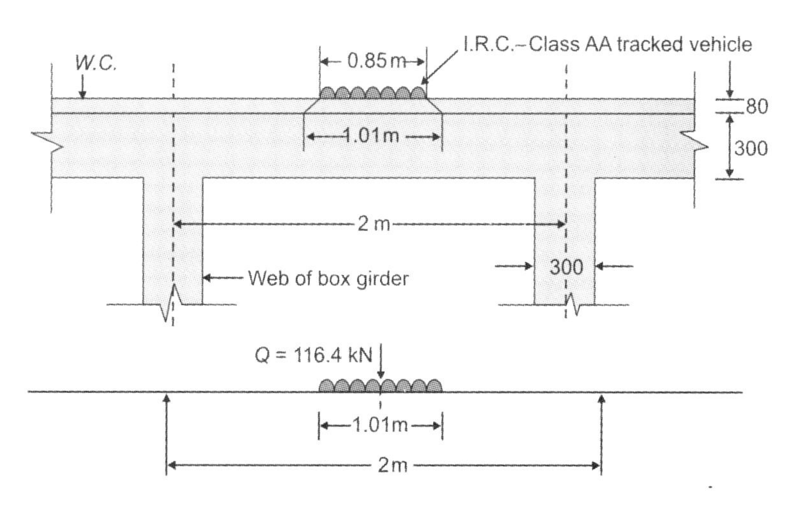

Fig. 15.30 Position of IRC Class AA load for maximum B.M. in slab

The effective width of dispersion of the wheel through the wearing coat is computed as

$$u = [0.85 + (2 \times 0.08)] = 1.01 \text{ m}$$

$$v = [3.60 + (2 \times 0.080] = 3.76 \text{ m}$$

Average intensity of wheel load with impact factor

$$= \left[\frac{1.25 \times 350}{3.76 \times 1.01}\right] = 115.20 \text{ kN/m}^2$$

Concentrated load acting at the centre of span in the transverse direction is computed as

$$Q = (115.20 \times 1.01) = 116.4 \text{ kN}$$

Referring to the bending moment and shear force coefficients compiled in Fig. 15.31,

Maximum positive B.M at middle of end span

$$= [0.210\, QL] = [0.210 \times 116.4 \times 2] = 48.9 \text{ kN.m}$$

Maximum negative B.M at penultimate support

$$= [0.181\, QL] = [0.181 \times 116.4 \times 2] = 42.13 \text{ kN.m}$$

Maximum shear force

$$= [0.60\, Q] = [0.60 \times 116.4] = 69.8 \text{ kN}$$

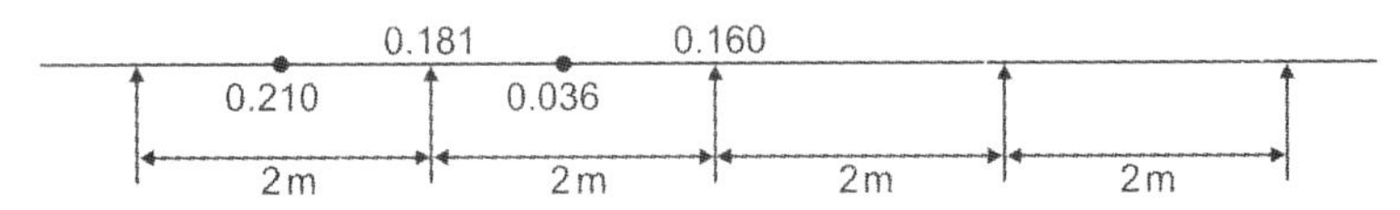

Fig. 15.31: Live load bending moment and shear force coefficients in four span continuous slab

c. **Design bending moments and shear forces**

Total positive bending moment

$$= [\text{DLBM} + \text{LLBM}] = [1.38 + 48.9] = 50.28 \text{ kN.m}$$

Total negative bending moment

$$= [\text{DLBM} + \text{LLBM}] = [1.93 + 42.13] = 44.06 \text{ kN.m}$$

Maximum shear force

$$= [\text{DLSF} + \text{LLSF}] = [10.8 + 69.8] = 80.6 \text{ kN}$$

Design of top slab section and reinforcements

$$\text{Effective depth } d = \sqrt{\frac{M}{Qb}} = \sqrt{\frac{50.28 \times 10^6}{1.844 \times 10^3}} = 165.12 \text{ m}$$

Effective depth provided

$$d = 250 \text{ mm with 50 mm cover}$$

Area of reinforcement

$$A_{st} = \left(\frac{M}{\sigma_{st} jd}\right) = \left(\frac{50.28 \times 10^6}{200 \times 0.879 \times 250}\right) = 1144 \text{ mm}^2$$

Provide 16 mm diameter bars at 160 mm centres ($A_{st} = 1257$ mm^2)

Referring to Clause 15.4 of IRC: 18-2000, for top slab of box girder decks, minimum supplementary reinforcement in the longitudinal direction is 0.18% of the gross cross-sectional area.

Distribution reinforcement

$$= (0.0018 \times 1000 \times 300) = 540 \text{ mm}^2$$

Provide 12 mm diameter bars at 160 mm centres ($A_{st} = 707 \text{ mm}^2$)

d. **Check for shear stress in slab**

Design shear force $V = 80.6$ kN

$$\text{Design shear stress} \quad \tau = \left(\frac{V}{bd}\right) = \left(\frac{80.6 \times 1000}{1000 \times 250}\right)$$

$$= 0.32 \text{ N/mm}^2$$

$$\left(\frac{1000\,A_{st}}{bd}\right) = \left(\frac{100 \times 1257}{1000 \times 250}\right) = 0.50$$

According to IRC: 21-2000, the permissible shear stress (τ_c) in concrete is obtained from Table 12B of IRC: 21-2000 as $\tau_c = 0.32 \text{ N/mm}^2$.

For solid slabs, permissible shear stress = $K\tau_c$, where K is a constant depending upon the thickness of the slab compiled in Table 12C of IRC: 21-2000.

In this case, the thickness of the slab

$$= 300 \text{ mm and hence } K = 1.00$$

Permissible shear stress in slab

$$K\tau_c = (1.00 \times 0.32) = 0.32 \text{ N/mm}^2$$

Since, τ does not exceed $K\tau_c$, shear reinforcements are not required.

5. *Design of web girder*

a. **Dead load bending moment and shear forces**

The continuous box girder is treated as an assemblage of I-sections with web serving the function of a main girder and flanges of symmetrical size as shown in Fig. 15.28.

Self weight of flanges $= (2 \times 0.3 \times 24) = 14.40 \text{ kN/m}^2$

Self weight of W.C $= (1 \times 1 \times 22) = 1.76$

Total load $= 16.16$

Self weight of web $= (1.4 \times 0.3 \times 24) = 10.08$

Total load on each I-girder

$$g = [(2 \times 16.16) + 10.08] = 43 \text{ kN/m}$$

The dead load bending moment coefficients for a two span continuous beam is shown in Fig. 15.32. The dead load bending moments at mid support and mid span sections are computed as:

$$M_{gB} = 0.125\,gL^2 = (0.125 \times 43 \times 50^2) = 13438 \text{ kN.m}$$

$$M_{gD} = 0.071\, gL^2 = (0.071 \times 43 \times 50^2) = 7633 \text{ kN.m}$$

Dead load shear is maximum near the mid support section and is computed as

$$V_g = 0.62\, gL = (0.62 \times 43 \times 50) = 1333 \text{ kN}$$

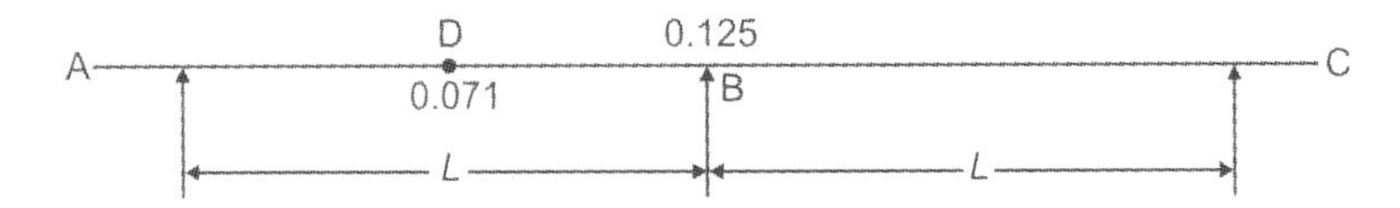

Fig. 15.32: Dead load bending moment coefficients

b. **Live load bending moments in continuous web girder**
Maximum live load reaction occurs in web girder when the transverse disposition of the IRC Class AA tracked vehicle load is arranged to have the maximum eccentricity with respect to the centre of the bridge deck as shown in Fig. 15.33. Maximum reaction due to live loads in girder B is computed as

$$R_B = \left[\frac{W \times 1.1}{2}\right] = 0.55\, W = (0.55 \times 700) = 385 \text{ kN}$$

Hence, the concentrated load $Q = 385$ kN.

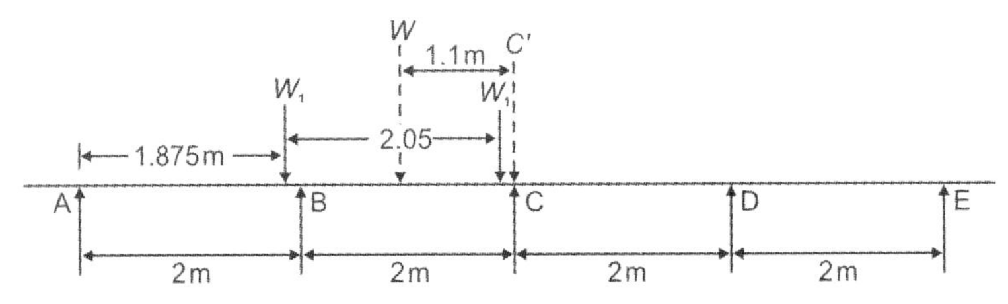

Fig. 15.33: Position of IRC class AA live loads for maximum reaction in girder

This load acting over a length of 3.6 m in the longitudinal direction is positioned at the centre of span of the two span continuous beam is shown in Fig. 15.34 to compute the maximum positive and negative moments.

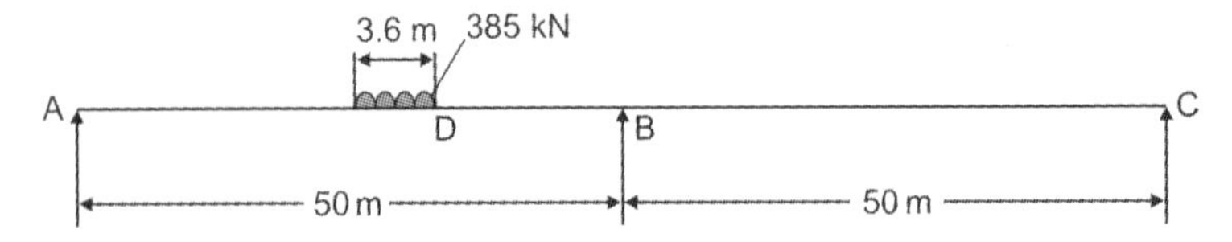

Fig. 15.34: Position of live load for maximum moments in two span continuous beam

The live load bending moment coefficients for maximum positive and negative moments in a two span continuous beam are shown in Fig. 15.35.

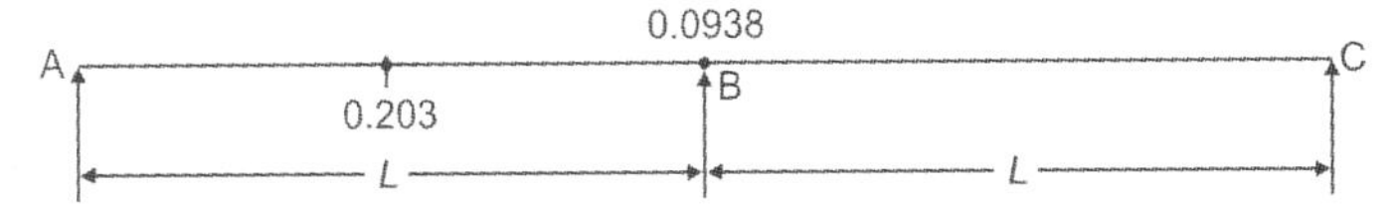

Fig. 15.35: Live load bending moment coefficients for a two span continuous girder

Maximum positive live load bending moment with impact factor at the centre of span is computed as

M_{max}(positive) $= (\text{I.F})\ (0.203\ QL)$

$= (1.10)\ (0.203 \times 385 \times 50) = 4298$ kN.m

Maximum negative live load bending moment with impact factor at mid support is computed as

M_{max}(negative) $= (\text{I.F})\ (0.0938\ QL)$

$= (1.1)\ (0.0938 \times 385 \times 50) = 1986$ kN.m

c. **Live load shear force in girder**

The maximum live load shear force develops in the interior webs when the IRC Class AA loads are placed near the mid support as shown in Fig. 15.36.

Reaction of load W on interior girder

$$= \left(\frac{350 \times 48.2}{50}\right) = 338 \text{ kN}$$

Maximum live load shear force with impact

$= (338 \times 1.1) = 372$ kN

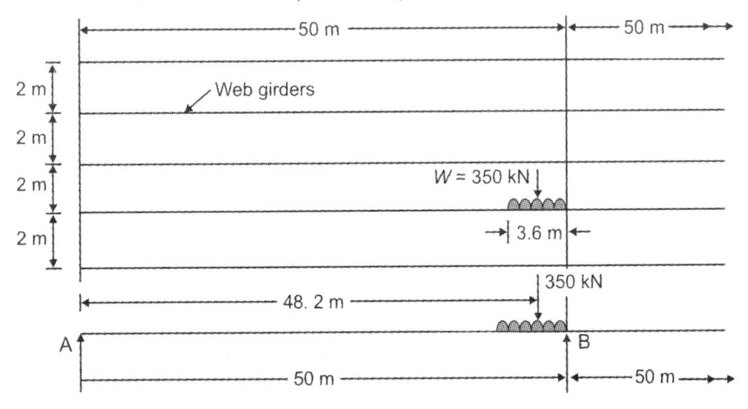

Fig. 15.36: Position of IRC class AA loads for maximum shear force in web girder

d. **Design bending moments and shear forces**

The design bending moments and shear forces at service and ultimate loads as stipulated in IRC: 18-2000 are compiled in Table 15.3.

Table 15.3: Service and ultimate load moments and shear forces in web girder

a. Bending moments (outer web girder)					
	D.L.B.M	**L.L.B.M**	**Working B.M**	**Ultimate B.M**	**Units**
Mid span section (D)	M_g 7633	M_q 4298	$(M_g + M_q)$ 11931	$M_u = (1.5\,M_g + 2.5\,M_q)$ 22195	kN.m
Mid support section (B)	13438	1986	15429	25122	kN.m

b. Shear forces (inner web girder)					
	D.L.S.F	**L.L.S.F**	**Working S.F**	**Ultimate S.F**	**Units**
Near mid support section (B)	V_g 1333	V_q 372	$(V_g + V_q)$ 1705	$V_u = (1.5\,V_g + 2.5\,V_q)$ 2930	kN

e. **Check for minimum section modulus**

At the mid support section B, the dead and live load moments are listed as:

$$M_{gB} = 13438 \text{ kN.m}$$

$$M_{QB} = 1986 \text{ kN.m}$$

$$M_{dB} = [M_{gB} + M_{QB}] = [13438 + 1986] = 15424 \text{ kN.m}$$

$$f_{br} = [\eta f_{ct} - f_{tw}] = [(0.8 \times 20) - 0] = 16 \text{ N/mm}^2$$

$$f_{inf} = \left[\frac{f_{tw}}{\eta} + \frac{M_g}{\eta Z_b}\right] = \left[0 + \frac{13438 \times 10^6}{0.8 \times 0.94 \times 10^9}\right] = 17.86 \text{ N/mm}^2$$

$$Z_b \geq \left[\frac{M_q + (1-\eta) M_g}{f_{br}}\right]$$

$$\geq \left[\frac{(1986 \times 10^6) + (1-0.8)13438 \times 10^6}{16}\right]$$

$$\geq (0.292 \times 10^9) \text{ mm}^3 < (0.94 \times 10^9) \text{ mm}^3$$

(section provided)

Hence, section provided is adequate.

f. **Prestressing force**

For the two continuous spans *AB* and *BC*, a concordant cable profile is selected such that the secondary moments are zero. The cable profile selected with eccentricity at mid support twice that at mid spans is shown in Fig. 15.37.

Providing an effective cover = 300 mm,

Maximum possible eccentricity at support

$$e = (1000 - 300) = 700 \text{ mm}$$

Prestressing force is computed from the relation,

$$P = \left[\frac{Af_bZ_b}{Z_b + Ae}\right] = \left[\frac{(1.62\times10^6)(17.86)(0.94\times10^9)}{(0.94\times10^9)+(1.62\times10^6\times700)}\right]$$

$$= 13109932 \text{ N} = 13110 \text{ kN}$$

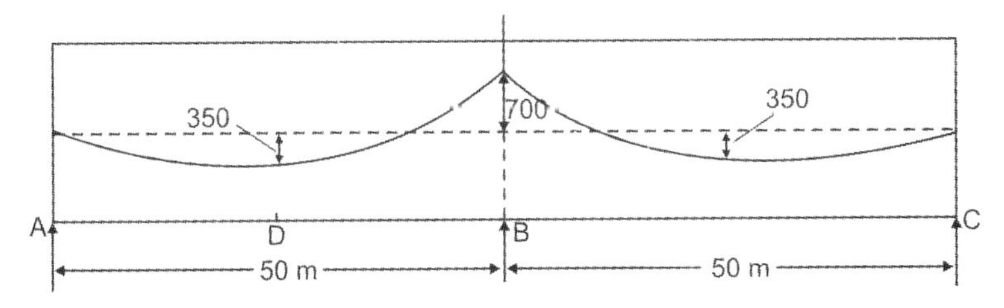

Fig. 15.37: Concordant cable profile

Using Freyssinet system with anchorage type 27K-15 (27 strands of 15.2 mm diameter) in 110 mm diameter cable ducts,

Force in each cable

$$= (27 \times 0.8 \times 265) = 5724 \text{ kN}$$

Provide three cables carrying an initial prestressing force of

$$P = (3 \times 5000) = 15000 \text{ kN}$$

Area of each strand of 15.2 mm diameter tendon = 140 mm^2

Area of 27 strands in each cable = (27 × 140) = 3780 mm^2

Total area in 3 cables A_p = (3 × 3780) = 11340 mm^2

The cables are arranged in a parabolic concordant profile so that the centroid of the group of cables has an eccentricity of 700 mm towards the top fibre at mid support section *B* and an eccentricity of 350 mm towards the soffit at mid span section *D*. The centroid of the cables is concentric at the end supports *A* and *C*. The selected cable profile is shown in Fig. 15.38.

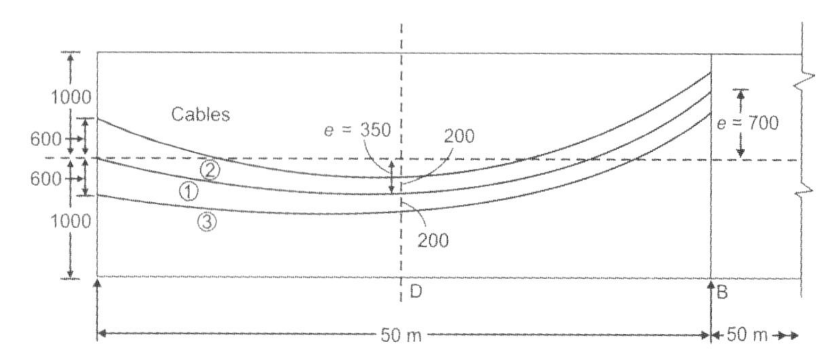

Fig. 15.38: Profiles of individual cables in span

g. **Check for stresses at service loads**

Centre of span section

$P = 15000$ kN $\qquad \eta = 0.80$

$e = 350$ mm $\qquad M_g = 7633$ kN.m

$A = (1.62 \times 10^6)$ mm^2 $\qquad M_q = 4298$ kN.m

$Z = (0.94 \times 10^9)$ mm^3

$$\left(\frac{P}{A}\right) = \left[\frac{15000 \times 10^3}{1.62 \times 10^6}\right] = 9.25 \text{ N/mm}^2$$

$$\left(\frac{Pe}{Z}\right) = \left[\frac{15000 \times 10^3 \times 350}{0.94 \times 10^9}\right] = 5.58 \text{ N/mm}^2$$

$$\left(\frac{M_g}{Z}\right) = \left[\frac{7633 \times 10^6}{0.94 \times 10^9}\right] = 8.12 \text{ N/mm}^2$$

$$\left(\frac{M_q}{Z}\right) = \left[\frac{4298 \times 10^6}{0.94 \times 10^9}\right] = 4.57 \text{ N/mm}^2$$

At transfer stage, the stresses at extreme fibres are computed as

$$\sigma_t = \left[\frac{P}{A} - \frac{Pe}{Z} + \frac{M_g}{Z}\right] = [9.25 - 5.58 + 8.12]$$

$$= 11.79 \text{ N/mm}^2$$

$$\sigma_b = \left[\frac{P}{A} + \frac{Pe}{Z} - \frac{M_g}{Z}\right] = [9.25 + 5.58 - 8.12]$$

$$= 6.71 \text{ N/mm}^2$$

At service load stage, the stresses at extreme fibres are computed as

$$\sigma_t = \left[\frac{\eta P}{A} - \frac{\eta Pe}{Z} + \frac{M_g}{Z} + \frac{M_q}{Z}\right]$$

$$= [0.8(9.25 - 5.58) + 8.12 + 4.57]$$

$$= 15.62 \text{ N/mm}^2 < 20 \text{ N/mm}^2$$

$$\sigma_b = \left[\frac{\eta P}{A} + \frac{\eta Pe}{Z} - \frac{M_g}{Z} - \frac{M_q}{Z}\right]$$

$$= [0.8(9.25 + 5.58) - 8.12 - 4.57]$$

$$= -0.8 \text{ N/mm}^2 \text{ (negligible tension)}$$

h. **Mid support section**

$P = 15000$ kN $\qquad \eta = 0.80$

$e = 700$ mm $\qquad M_g = 13438$ kN.m

$A = (1.62 \times 10^6)$ mm^2 $\qquad M_q = 1986$ kN.m

$Z = (0.94 \times 10^9)$ mm^3

$$\left(\frac{P}{A}\right) = \left[\frac{15000 \times 10^3}{1.62 \times 10^6}\right] = 9.25 \text{ N/mm}^2$$

$$\left(\frac{Pe}{Z}\right) = \left[\frac{15000 \times 10^3 \times 700}{0.94 \times 10^9}\right] = 11.16 \text{ N/mm}^2$$

$$\left(\frac{M_g}{Z}\right) = \left[\frac{13438 \times 10^6}{0.94 \times 10^9}\right] = 14.29 \text{ N/mm}^2$$

$$\left(\frac{M_q}{Z}\right) = \left[\frac{1986 \times 10^6}{0.94 \times 10^9}\right] = 2.11 \text{ N/mm}^2$$

At transfer stage, the stresses at extreme fibres are computed as

$$\sigma_t = [9.25 + 11.16 - 14.29] = 6.12 \text{ N/mm}^2$$

$$\sigma_b = [9.25 - 11.16 + 14.29] = 12.38 \text{ N/mm}^2$$

At service load stage, the stresses at extreme fibres are computed as

$$\sigma_t = [0.8\,(9.25 + 11.16) - 14.29 - 2.11]$$

$$= -0.072 \text{ N/mm}^2$$

$$\sigma_b = [0.8\,(9.25 - 11.16) + 14.29 + 2.11]$$

$$= 14.87 \text{ N/mm}^2$$

All the stresses are well within the maximum permissible limits of 20 N/mm^2 and no tensile stresses develop at transfer and service load stages.

i. **Check for ultimate flexural strength**

1. Centre of span section

$A_p = 11340$ mm^2 $\quad D_f = 300$ mm

$b = 2000$ mm $\quad f_{ck} = 60$ N/mm^2

$b_w = 300$ mm $\quad f_p = 1862$ N/mm^2

$d = 1350$ mm, $\quad M_u$ (required) $= 17292$ kN.m

According to IRC: 18-2000

i. ***Failure by yielding of steel***

$M_u = [0.9\, d\, A_p f_p]$

$= [0.9 \times 1350 \times 11340 \times 1862]$

$= [25654 \times 10^6]$ N.mm

$= 25654$ kN.m > 22195 kN.m (hence, safe)

ii. ***Failure by crushing of concrete***

$M_u = [0.176\, b_w\, d^2 f_{ck} + 0.67 \times 0.8\, (b - b_w)\, (d - 0.5\, D_f)\, D_f f_{ck}]$

$= [(0.176 \times 300 \times 1350^2 \times 60]$

$+ [(0.67 \times 0.8)\, (2000 - 300)\, (1350 - 0.5 \times 300)\, (300 \times 60)]$

$= (25454 \times 10^6)$ N.mm

$= 25454$ kN.m > 22195 kN.m (hence, safe)

2. Mid support section

$A_p = 11340$ mm^2 $\quad D_f = 300$ mm

$b = 2000$ mm $\quad f_{ck} = 60$ N/mm^2

$b_w = 300$ mm $\quad f_p = 1862$ N/mm^2

$d = 1700$ mm $\quad M_u$ (required) $= 22957$ kN.m

i. ***Failure by yielding of steel***

$M_u = [0.9\, d\, A_p f_p]$

$= [0.9 \times 1700 \times 11340 \times 1862]$

$= [32306 \times 10^6]$ N.mm

$= 32306$ kN.m > 25122 kN.m (hence, safe)

ii. ***Failure by crushing of concrete***

$M_u = [0.176\, b_w\, d^2 f_{ck} + 0.67 \times 0.8\, (b - b_w)\, (d - 0.5\, D_f)\, D_f f_{ck}]$

$= [(0.176 \times 300 \times 1700^2 \times 60] + [(0.67 \times 0.8)$

$(2000 - 300)\, (1700 - 0.5 \times 300)\, (300 \times 60)]$

$= (34577 \times 10^6)$ N.mm

$= 34577$ kN.m > 25122 kN.m (hence, safe)

The ultimate flexural strength of the centre of span and mid support sections are greater than the required design ultimate moment. Hence, the design satisfies limit state of collapse.

j. **Check for ultimate shear strength**

Design shear force $V_u = 2930$ kN

According to IRC: 18-2000, the ultimate shear resistance of the support section uncracked in flexure is given by the relation,

$$V_{cw} = 0.67\, b_w h \sqrt{f_t^2 + 0.8\, f_{cp} f_t} + \eta P \sin\theta$$

where,

$$b_w = 300 \text{ mm}$$

$$h = 2000 \text{ mm}$$

$$f_t = 0.24\sqrt{f_{ck}} = 0.24\sqrt{60} = 1.85 \text{ N/mm}^2$$

$$f_{cp} = \left(\frac{\eta P}{A}\right) = \left(\frac{0.8 \times 15000 \times 10^3}{1.62 \times 10^6}\right) = 7.4 \text{ N/mm}^2$$

$$\theta = \left(\frac{4e}{L}\right) = \left(\frac{4 \times 700}{50 \times 1000}\right) = 0.056$$

$$V_{cw} = [0.67 \times 300 \times 2000\sqrt{1.85^2 + (0.8 \times 7.4 \times 1.85)}\,] + [0.8 \times 15000 \times 10^3 \times 0.056]$$

$$= (2195 \times 10^3) \text{ N} = 2195 \text{ kN}$$

Balance shear force = (2930 – 2195) = 735 kN

Using 12 mm diameter two legged stirrups, the spacing is calculated as

$$S_V = \left\{\frac{0.87 \times 415 \times 2 \times 113 \times 1900}{735 \times 10^3}\right\} = 211 \text{ mm}$$

Adopt 12 mm diameter two legged stirrups at a spacing of 200 mm centres at supports gradually increased to 300 mm towards the centre of span.

k. **Supplementary reinforcement**

Longitudinal reinforcements of not less than 0.18 percent of the gross-sectional area are to be provided according to the specifications of IRC: 18-2000 to safeguard against shrinkage cracking in webs.

$$A_{st} = (0.0018 \times 1000 \times 300) = 540 \text{ mm}^2$$

Provide 12 mm diameter bars at 160 mm centres (A_{st} = 707 mm^2).

The reinforcements provided at mid support and centre of span sections are shown in Fig. 15.39.

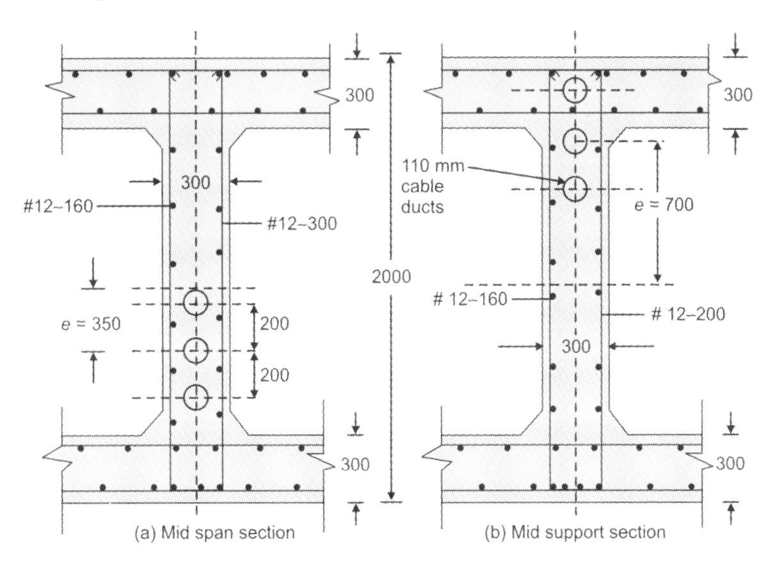

Fig. 15.39: Details of reinforcements and cables at mid span and support sections

1. **Design of end blocks**

   Near the end supports where anchorages are located, the web thickness is increased to 600 mm to accommodate the anchorage bearing plates of size 400 mm by 400 mm in conjunction with Freyssinet 27K-15 type prestressing strands as shown in Fig. 15.40.

   The bursting tension is computed using the data in Table 8 of IRC: 18-2000.

   The bursting tension in the horizontal plane is computed using the parameters,

$$2y_{po} = 400 \text{ mm}$$

$$2y_o = 600 \text{ mm}$$

$$\text{Ratio}\left(\frac{y_{po}}{y_o}\right) = \left(\frac{400}{600}\right) = 0.66$$

$$\left(\frac{F_{bst}}{P_k}\right) = 0.12$$

$$\text{Bursting tension} = F_{bst} = (0.12 \times 5000) = 600 \text{ kN}$$

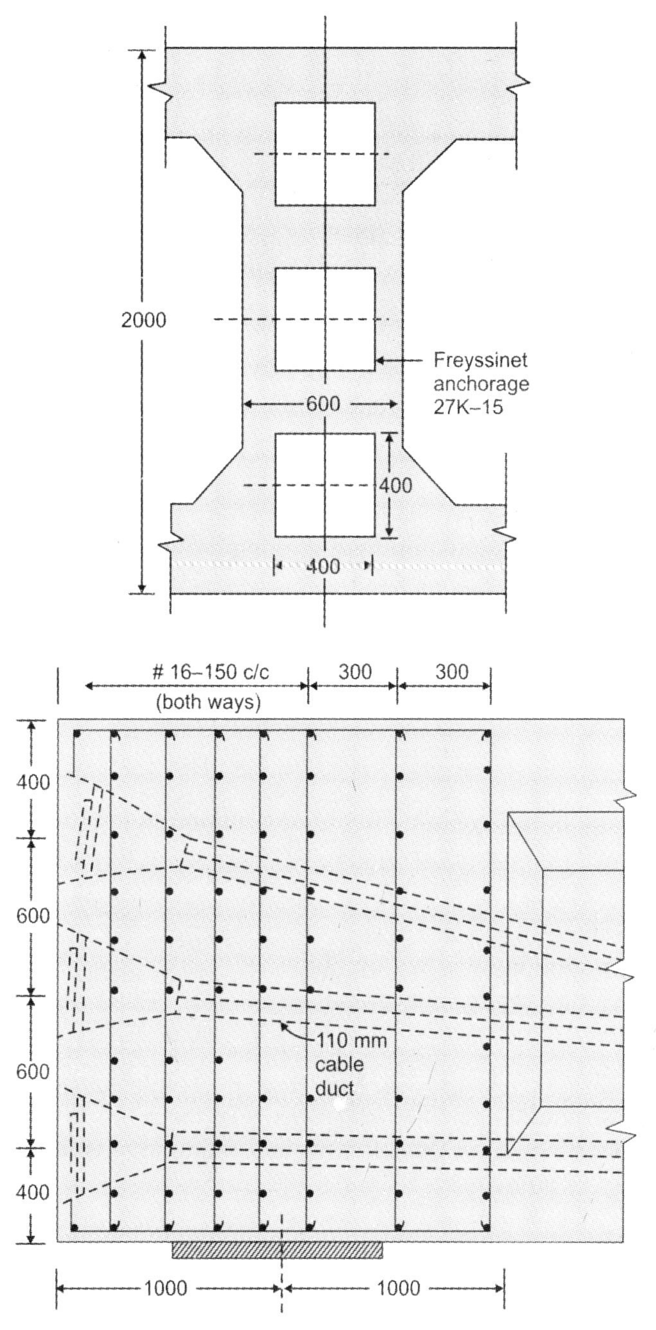

Fig. 15.40: Anchorage zone reinforcement in end block

Using Fe-415 HYSD bars, anchorage zone reinforcement required is

$$A_{st} = \left(\frac{600 \times 10^3}{0.87 \times 415}\right) = 1662 \text{ mm}^2$$

Provide 16 mm diameter bars at 150 mm centres in the horizontal and vertical planes distributed in the region from $0.2y_o$ to $2y_o$ (60 mm to 600 mm) as shown in Fig. 15.40.

Problem 15.5

Design a suitable prestressed concrete cable stayed bridge deck for crossing a deep valley of span length 200 m to suit the following data:

1. *Data*

   Effective span of bridge deck = 200 m
   Width of the roadway = 7.5 m
   Width of footpaths = 2.5 m on either side
   Wearing coat thickness = 80 mm

   Type of loading: IRC Class AA tracked vehicle

   The deck is proposed to be made up of a prestressed concrete solid slab with longitudinal stiffening girders supported by stay cables.

   Spacing of stay cables = 6 m intervals

   Type of concrete: M-60 grade concrete

   Type of steel: Fe-415 HYSD bars for use as supplementary reinforcement

   Standard Freyssinet prestressing cables are available for use as stay cables and for prestressing the deck slab and stiffening girders.

   Design the deck slab, longitudinal girders and stay cables.

   The structural concrete components of the bridge should be designed as Class 1 type without developing any tensile stresses under service loads.

   Sketch the details of reinforcements and cables in the deck slab and girders and the elevation of the bridge showing the stay cable configuration.

   The design should conform to the national codes: IRC: 6, IRC: 18 and IRC: 21.

2. *Permissible stresses*

The permissible compressive stress in concrete at transfer and service loads as recommended in IRC: 18 are as follows:

For M-60 grade concrete,

$$f_{ck} = 60 \text{ N/mm}^2$$
$$f_{ci} = 45 \text{ N/mm}^2$$
$$f_{ct} = 0.45 f_{ci} = (0.45 \times 45) = 20 \text{ N/mm}^2$$
$$f_{cw} = 0.33 f_{ck} = (0.33 \times 60) = 20 \text{ N/mm}^2$$
$$f_{tt} = 0 \text{ (Class 1 type member)}$$
$$E_c = 5000 \sqrt{f_{ck}} = 5000\sqrt{6} = 38730 \text{ N/mm}^2$$
$$= 38.73 \text{ kN/mm}^2$$

3. *Selection of dimensions of bridge deck*

The overall span of the bridge being 200 m, it is proposed to have a single tower at the centre of span with the deck comprising a prestressed concrete solid slab supported on two longitudinal stiffening girders which are suspended by stay cables in two planes at intervals of 6 m along the girder.

Width of two lane carriageway = 7.5 m

Footpath with kerb on either side = 2.5 m

Effective span of the deck slab

$$B = (7.5 + 2.5) = 10 \text{ m}$$

Thickness of slab assumed as 40 mm/m of span

$$= (40 \times 10) = 400 \text{ mm}$$

The depth of longitudinal girders supporting the slab is selected, based on the wind stability criteria proposed by Leonhardt.

The stiffening girder dimensions should be such that

$$B \geq 10H$$

where, B = width of the girder

H = depth of girder

L = span of girder = 200 m

In this example,

$$B = 10 \text{ m}$$

$$\therefore \quad H = \left(\frac{B}{10}\right) = \left(\frac{10}{10}\right) = 1 \text{ m} = 1000 \text{ m}$$

$$\text{Also,} \quad B \geq \left(\frac{L}{30}\right) \geq \left[\frac{200}{30}\right] \geq 6.66 \text{ m}$$

Both the criteria are satisfied.

Hence, adopt the depth of stiffening girder $H = 1000$ mm

Width of girder $= 0.5\,H$ to $0.6\,H$

$= (0.5 \times 1000)$ to (0.6×1000)

$= 500$ mm to 600 mm

Adopt a stiffening girder of size 600 mm wide by 1000 mm deep. The cross-section of the bridge deck is shown in Fig. 15.41.

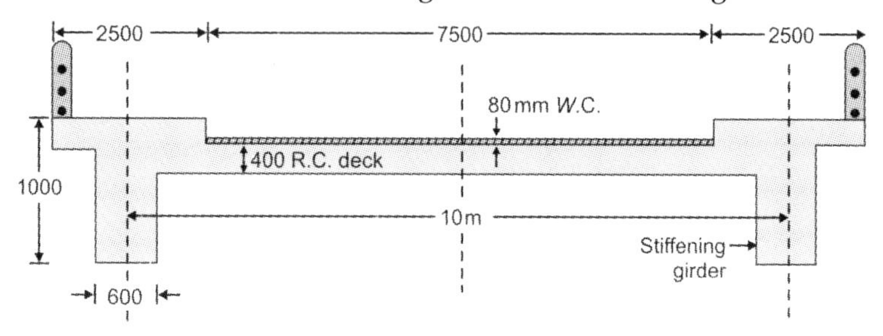

Fig. 15.41: Cross-section of bridge deck

4. *Sectional properties of slab and girder*

   a. **Deck slab**

   Thickness of slab $= 400$ mm

   Cross-sectional area per metre width

   $$A = (400 \times 10000) = (4 \times 10^5)\ \text{mm}^2$$

   $$\text{Sectional modulus } Z = Z_t = Z_b = \left(\frac{1000 \times 400^2}{6}\right)$$

   $$= (26.66 \times 10^6)\ \text{mm}^3$$

   b. **Stiffening girder**

   Width of stiffening girder $= 600$ mm

   Overall depth of girder $= 1000$ mm

   Cross-sectional area $A = (600 \times 1000) = (6 \times 10^5)\ \text{mm}^2$

   $$\text{Section modulus} \quad Z = \left(\frac{600 \times 1000^2}{6}\right) = 10^8\ \text{mm}^3$$

5. *Design of deck slab*

   a. **Dead load bending moments**

   The deck slab is rigidly connected to the edge stiffening girder.

   Effective span $L = 10$ m

   Thickness of slab $= 400$ mm

   Self weight of slab $= (0.4 \times 24) = 9.60\ \text{kN/m}^2$

   Self weight of W.C $= 90.08 \times 22 = 1.76$

   Footpaths, kerbs and finishes (L.S) $= 0.64$

Total dead load (g) = 12.00 kN/m^2

Dead load bending moment at support

$$M_g = \left(\frac{gL^2}{12}\right) = \left(\frac{12 \times 10^2}{12}\right) = 100 \text{ kN.m}$$

Dead load bending moment at centre

$$M_g = \left(\frac{gL^2}{24}\right) = \left(\frac{12 \times 10^2}{24}\right) = 50 \text{ kN.m}$$

b. **Live load bending moments**

The slab is monolithically cast with the edge girders. Hence, it is considered as fixed at supports over a span of 10 m. The IRC Class AA tracked load is positioned on the span as shown in Fig. 15.42 to yield maximum moments.

Live load per metre length $W = (350/3.6) = 97.22$ kN

Maximum negative bending moment at support is obtained as

$$M_A = \left(\frac{W\,ab}{L}\right) = \left(\frac{97.22 \times 3.975 \times 6.025}{10}\right)$$

$$= 233 \text{ kN.m}$$

Design negative bending moment at support with impact factor is computed as

$$M_q = (1.1 \times 233) = 256 \text{ kN.m}$$

Maximum positive bending moment at centre of span with impact factor is calculated as

$$M_q = 1.1\,(386.4 - 233) = 169 \text{ kN.m}$$

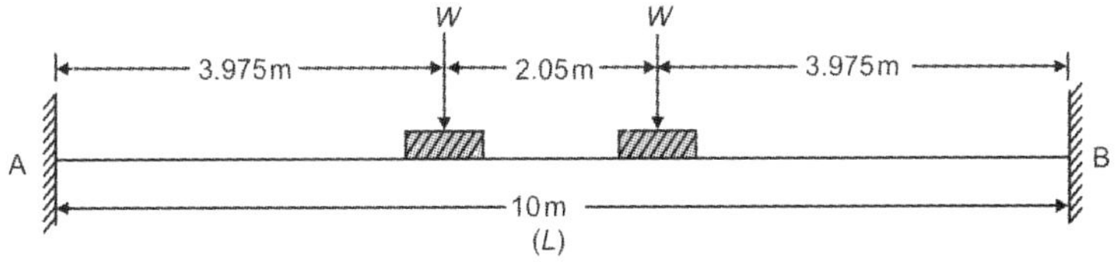

Fig. 15.42: Position of IRC Class AA loads for maximum moments in deck slab

c. **Dead and live load shear forces**

Dead load on slab g = 12 kN/m^2

Effective span = 10 m

Dead load shear force $V_g = \left(\frac{12 \times 10}{2}\right) = 60$ kN

Maximum live load shear force occurs when the IRC Class AA tracked vehicle loads are positioned, as shown in Fig. 15.43.

Maximum live load shear force with impact is computed as

$$V_q = 1.1\left[\frac{194.44 \times 6.1}{10}\right] = 130.5 \text{ kN}$$

Design ultimate shear force

$$V_u = (1.5\, V_g + 2.5\, V_q)$$
$$= [(1.5 \times 60) + (2.5 \times 130.5)] = 416.25 \text{ kN}$$

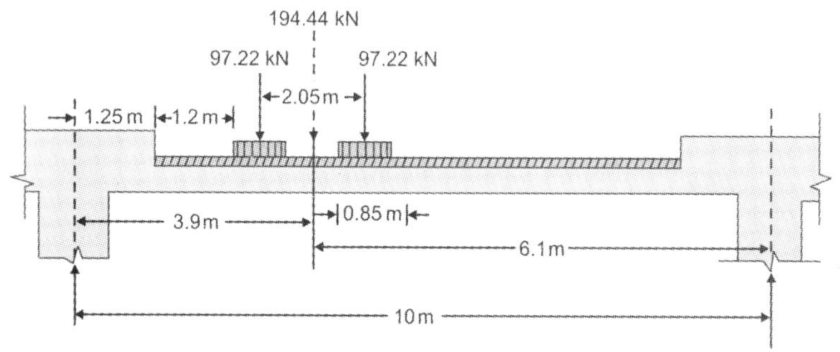

Fig. 15.43: Position of IRC Class AA loads for maximum shear force at support

d. **Check for minimum section modulus**

Dead load moment $M_g = 100$ kN.m

Live load moment $M_q = 256$ kN.m

$$Z = Z_t = Z_b = (26.66 \times 10^6) \text{ mm}^3$$

$$f_{ct} = f_{cw} = 20 \text{ N/mm}^2 \text{ and } f_{tw} = 0$$

Loss ratio $\eta = 0.8$

$$f_{br} = [\eta f_{ct} - f_{tw}] = [(0.8 \times 20) - 0] = 16 \text{ N/mm}^2$$

$$Z_b \geq \left[\frac{M_q + (1-\eta)M_g}{f_{br}}\right]$$

$$= \left[\frac{(256 \times 10^6) + (1-0.8)10^6}{16}\right]$$

$$\geq (17.25 \times 10^6) \text{ mm}^3 < (26.66 \times 10^6) \text{ mm}^3$$

(section modulus provided)

Hence, the section selected for the slab is adequate to resist the service loads safely without exceeding the permissible stresses.

e. **Minimum prestressing force**

The minimum prestressing force required is computed using the relation

$$P = \left[\frac{A(f_{\inf} Z_b + f_{\sup} Z_t)}{Z_t + Z_b}\right]$$

$$f_{\sup} = \left[f_{tt} - \frac{M_g}{Z_t}\right] = \left[0 - \frac{(100 \times 10^6)}{(26.66 \times 10^6)}\right]$$

$$= -3.75 \text{ N/mm}^2$$

$$f_{\inf} = \left[\frac{f_{tw}}{\eta} + \frac{M_g + M_q}{\eta Z_b}\right] = \left[0 + \frac{(100 + 256)10^6}{(0.8 \times 26.66 \times 10^6)}\right]$$

$$= 16.69 \text{ N/mm}^2$$

$$P = \left[\frac{(4 \times 10^5)(26.66 \times 10^6)(16.69 - 3.75)}{(2 \times 26.66 \times 10^6)}\right]$$

$$= (2588 \times 10^3) \text{ N} = 2588 \text{ kN}$$

Using Freyssinet cables containing 12 wires of 8 mm diameter stressed to 1100 N/mm^2,

$$\text{Force in each cable} = \left[\frac{12 \times 50 \times 1000}{1000}\right] = 660 \text{ kN}$$

$$\text{Spacing of cables} = \left[\frac{1000 \times 660}{2588}\right] = 255 \text{ mm}$$

Provide cables at a spacing of 250 mm c/c.

f. **Eccentricity of cables**

The eccentricity of cables at support sections is obtained from the relation

$$e = \left[\frac{Z_t Z_b (f_{\inf} - f_{\sup})}{A(f_{\sup} Z_t + f_{\inf} Z_b)}\right]$$

$$= \left[\frac{(26.66 \times 10^6)^2 (16.69 + 3.75)}{(4 \times 10^5)(26.66 \times 10^6)(16.69 - 3.75)}\right]$$

$$= 105 \text{ mm (towards the top of slab)}$$

The positive bending moment at centre of span is of smaller magnitude and hence, the eccentricity required at the centre of span is computed by limiting the tensile stress at top fibre to zero under the loading condition of prestress together with the self weight stress.

At the centre of span section, we have

$$P = 2588 \text{ kN and } M_g = 50 \text{ kN.m}$$

$$\left[\frac{P}{A} - \frac{Pe}{Z_t} + \frac{M_g}{Z_t}\right] = 0$$

$$\left[\frac{2588 \times 10^3}{4 \times 10^5} - \frac{2588 \times 10^3 \times e}{26.66 \times 10^6} + \frac{50 \times 10^6}{26.66 \times 10^6}\right] = 0$$

Solving, $e = 86$ mm

The cables are arranged in a parabolic profile with an eccentricity of 105 mm towards the top of slab at supports varying to an eccentricity of 86 mm towards the soffit of slab at the centre of span.

6. *Check for stresses in deck slab under service loads*

a. **Stresses at support section**

$P = (2588 \times 10^3)$ N $\quad M_g = 100$ kN.m

$e = -105$ mm (towards top of slab) $\quad M_q = 256$ kN.m

$A = (4 \times 10^5)$ mm² $\quad \eta = 0.8$

$Z = (26.66 \times 10^6)$ mm³

$$\left(\frac{P}{A}\right) = \left(\frac{2588 \times 10^3}{4 \times 10^5}\right) = 6.47 \text{ N/mm}^2$$

$$\left(\frac{Pe}{Z}\right) = \left(\frac{2588 \times 10^3 \times 105}{26.66 \times 10^6}\right) = 10.19 \text{ N/mm}^2$$

$$\left(\frac{M_g}{Z}\right) = \left(\frac{100 \times 10^6}{26.66 \times 10^6}\right) = 3.75 \text{ N/mm}^2$$

$$\left(\frac{M_q}{Z}\right) = \left(\frac{256 \times 10^6}{26.66 \times 10^6}\right) = 9.60 \text{ N/mm}^2$$

Stress at transfer

Stress at top $= (6.47 + 10.19 - 3.75) = 12.91$ N/mm²

Stress at bottom $= (6.47 - 10.19 + 3.75) = 0.03$ N/mm²

Stress at working loads

Stress at top $= [0.8 (6.47 + 10.19) - 3.75 - 9.60]$
$= -0.022$ N/mm²

Stress at bottom $= [0.8 (6.47 - 10.19) + 3.75 + 9.60]$
$= 10.37$ N/mm²

b. **Stresses at centre of span section**

$P = (2588 \times 10^3)$ N $\quad M_g = 50$ kN.m

$e = 86$ mm (towards soffit) $\quad M_q = 169$ kN.m

$A = (4 \times 10^5)$ mm² $\quad \eta = 0.8$

$Z = (26.66 \times 10^6)$ mm³

$$\left(\frac{P}{A}\right) = \left(\frac{2588 \times 10^3}{4 \times 10^5}\right) = 6.47\ \text{N/mm}^2$$

$$\left(\frac{Pe}{Z}\right) = \left(\frac{2588 \times 10^3 \times 86}{26.66 \times 10^6}\right) = 8.34\ \text{N/mm}^2$$

$$\left(\frac{M_g}{Z}\right) = \left(\frac{50 \times 10^6}{26.66 \times 10^6}\right) = 1.88\ \text{N/mm}^2$$

$$\left(\frac{M_q}{Z}\right) = \left(\frac{169 \times 10^6}{26.66 \times 10^6}\right) = 6.34\ \text{N/mm}^2$$

Stress at transfer

Stress at top $= (6.47 + 8.34 + 1.88) = 0.01\ \text{N/mm}^2$

Stress at bottom $= (6.47 + 8.34 - 1.88) = 12.93\ \text{N/mm}^2$

Stress at working loads

Stress at top $= [0.8\,(6.47 - 8.34) + 1.88 + 6.34]$

$= 6.72\ \text{N/mm}^2$

Stress at bottom $= [0.8\,(6.47 + 8.34) - 1.88 - 6.34]$

$= 3.63\ \text{N/mm}^2$

The stresses in the slab at various limit states are well within the safe permissible limits.

7. *Design of stay cables*

   a. **Dead loads**

   Total dead load of slab $= 12\ \text{kN/m}^2$

   Weight of longitudinal girder $= (1 \times 0.6 \times 24) = 32\ \text{kN/m}$

   Weight of cross girders $= (0.3 \times 0.6 \times 42) = 4.32\ \text{kN/m}$

   b. **Live loads**

   For maximum reaction, the critical loading position of IRC Class AA tracked vehicle loads are shown in Fig. 15.44.

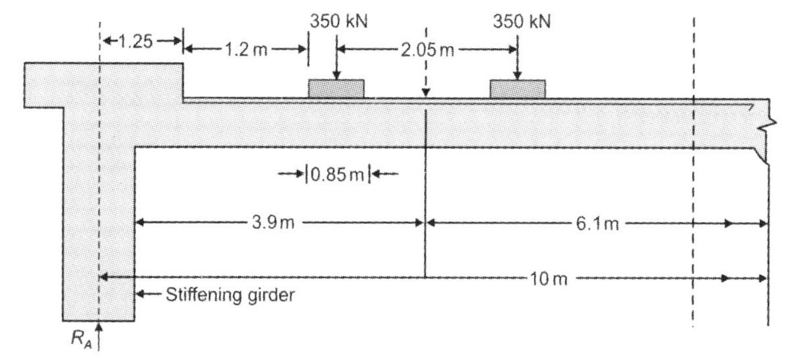

Fig. 15.44: Maximum live load reaction in stiffening girder

Allowing for an impact factor of 10 percent for live loads,

$$\text{Reaction} \qquad R_A = 1.1\left(\frac{700 \times 6.10}{10}\right) = 470 \text{ kN}$$

c. **Total weight of deck for 6 m length and 10 m wide**

Weight from deck slab = (6 × 10 × 12) = 720 kN

Weight of longitudinal girders = (2 × 14.4 × 6) = 172.8 kN

Weight of cross girder = (4.32 × 10) = 43.2 kN

Total load = 936.0 kN

Force transmitted on each cable due to deck loads

= (0.5 × 936) = 468 kN

d. **Total reaction on cable**

Due to live loads = 470 kN

Due to dead loads = 468 kN

Add for footpaths, railings, etc. = 62 kN

Total reaction (R) = 1000 kN

e. **Design of cable**

Referring to Fig. 15.45,

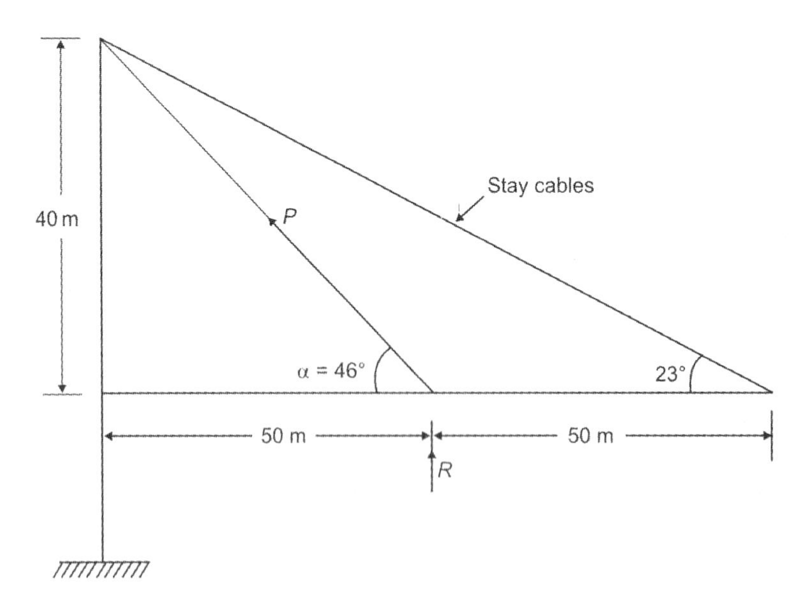

Fig. 15.45: Computation of force in cable stay

The angle subtended by the cable to the horizontal

$\alpha = 46°$ and $\sin \alpha = 0.72$

The tensile force developed in the cable is computed as

$$P = \left(\frac{R}{\sin \alpha}\right) = \left(\frac{1000}{0.72}\right) = 1389 \text{ kN}$$

Using 15.2 mm diameter high tensile strands initially stressed to 1500 N/mm^2,

$$\text{Force in each strand} = \left(\frac{140 \times 1500}{1000}\right) = 210 \text{ kN}$$

$$\text{Number of strands in the cable} = \left(\frac{1389}{210}\right) \cong 7$$

Adopt 7K-15 Freyssinet cable stays at intervals of 6 m.

f. **Design of stiffening girder**

The maximum bending moment developed in the stiffening girder including the effect of temperature is expressed as,

$$M = 0.05\,(\psi g + p + q)\,L^2$$

where, L = span length of stiffening girder (panel length)

g = weight of the stiffening girder per unit length

p = uniformly distributed weight of deck per unit length of span

q = uniformly distributed live load carried by a single girder per unit length

ψ = the construction coefficient of the stiffening girder = 1.4

In the present example, the corresponding values are

L = 6 m

p = 60 kN/m

q = 130 kN/m

$$M = 0.05\,(1.4 \times 14.4 + 60 + 130)\,6^2 = 378 \text{ kN.m}$$

The cross-sectional dimension of the stiffening girder assumed is 600 mm wide by 1000 mm deep.

Area of cross-section $A = (600 \times 1000) = (6 \times 10^5)$ mm^2

Breadth of section b = 600 mm

Overall depth h = 1000 mm

Effective depth d = 940 mm

(cover to main steel = 60 mm)

Using M-60 grade concrete and Fe-415 HYSD bars, the limiting moment of resistance of the section is computed as,

$$M_{u,\lim} = 0.138 f_{ck}\, b\, d^2$$

$$= (0.138 \times 60 \times 600 \times 940^2)$$

$$= (4389 \times 10^9) \text{ N.mm}$$

$$= 4389 \text{ kN.m} > 378 \text{ kN.m}$$

According to IRC: 21-2000, the minimum amount of reinforcement in the section should not be less than 0.2 percent of the gross cross-section. Providing 0.3 percent of reinforcement, we have

$$A_{st(\min)} = \left[\frac{0.3}{100} \times 600 \times 1000\right]$$

$$= 1800 \text{ mm}^2$$

Provide 4 bars of 25 mm diameter at top and bottom of the section (A_{st} = 1963 mm^2).

Provide 10 mm diameter 4-legged stirrups at 300 mm centres. Provide 10 mm diameter bars as side face reinforcement in web at 300 mm intervals.

Figures 15.46 to 15.49 show the longitudinal elevation and cross-section of the bridge deck.

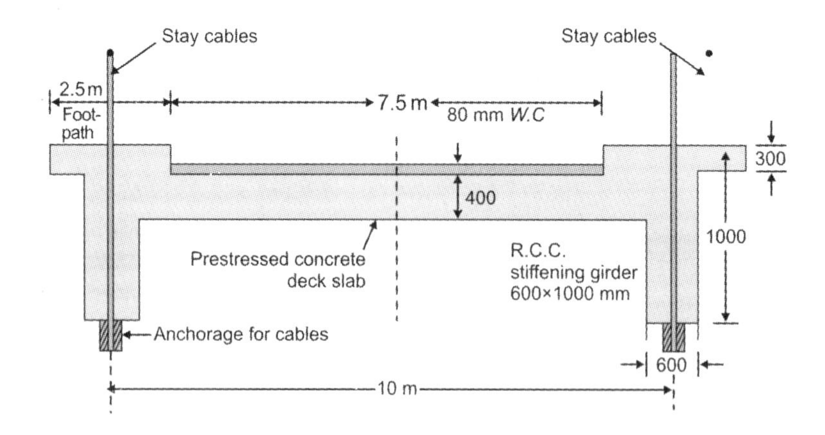

Fig. 15.46: Cross-section of cable stayed bridge deck

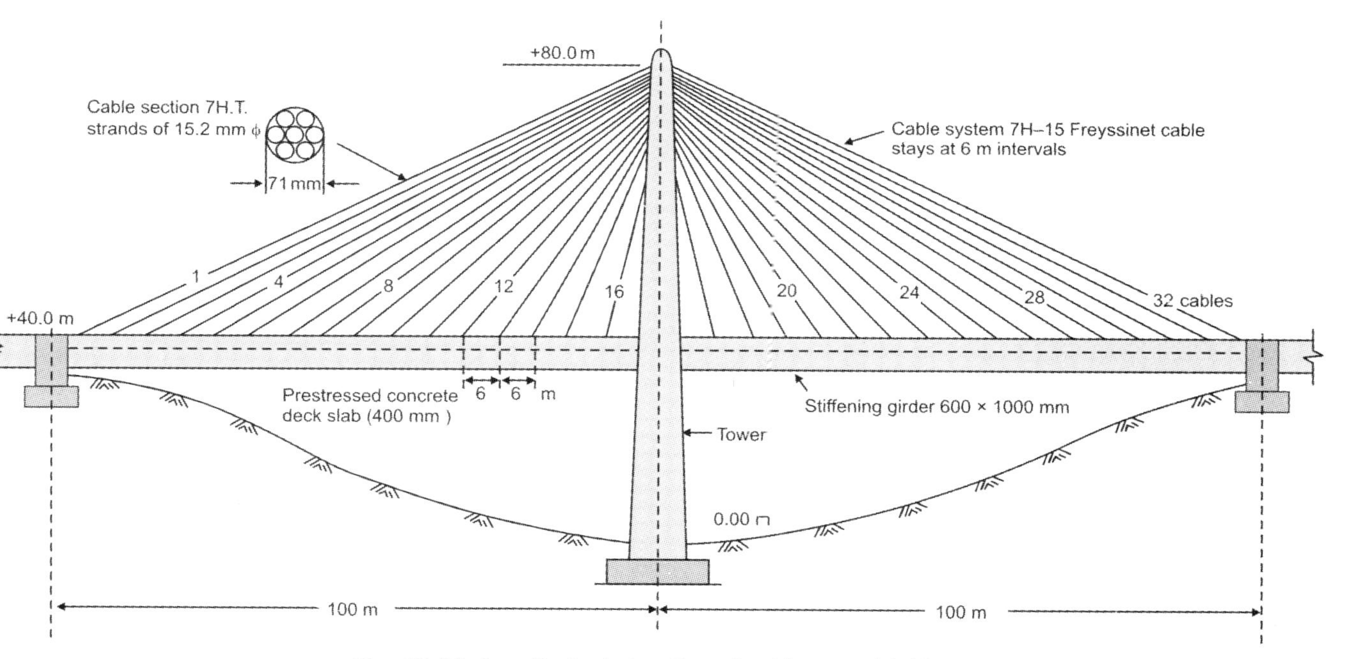

Fig. 15.47: Longitudinal elevation of cable stayed bridge

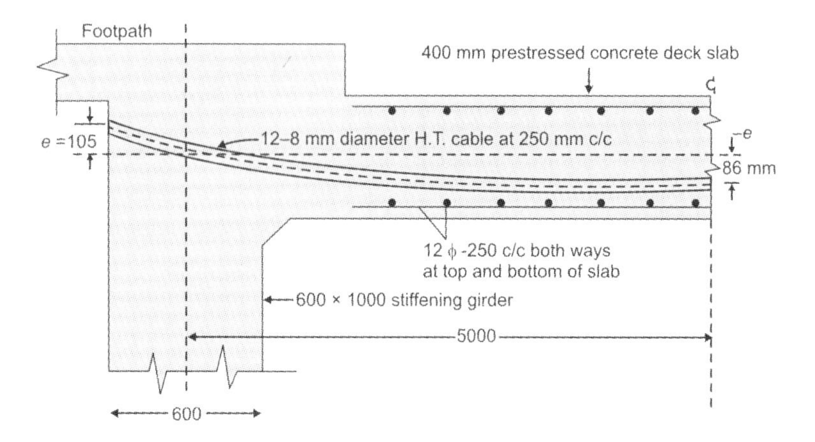

Fig. 15.48: Longitudinal section of prestressed concrete deck slab

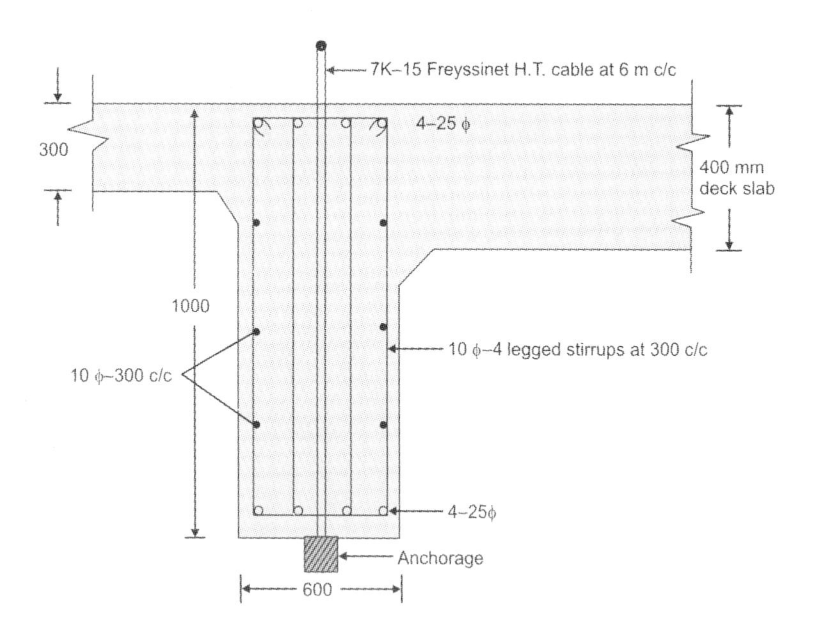

Fig. 15.49: Cross-section of stiffening girder

Suggested Reading and References

1. IS: 1343-2012 Indian Standard Code of Practice for Prestressed Concrete (Second revision), Indian Standards Institution, New Delhi, Nov. 2012.
2. Bennett, E.W., Structural Concrete Elements, Chapman & Hall, London, 1973.
3. Krishna Raju, N. Prestressed Concrete, Fifth Edition, Tata McGraw Hill Publishing, New Delhi, 2012.
4. Lin, T.Y. and Burns, Ned. H., Design of Prestressed Concrete Structures, Third Edition, John Wiley and Sons, New York, 1982.
5. Nilson Arthur. H., Design of Prestressed Concrete, Second Edition, John Wiley and Sons, New York, 1987.
6. Nawy Edward, G., Prestressed Concrete, A Fundamental Approach, Prentice Hall International, Englewood Cliffs, New Jersey, 1989.
7. Libby James R., Modern Prestressed Concrete, Second Edition, Van Nostrand Reinhold, New York, 1977.
8. Leonhardi, F., Prestressed Concrete, Design and Construction, Wilhelm Ernst and Sohn, Berlin, Second Edition, Berlin, 1964.
9. Dayaratnam, P., Prestressed Concrete, Oxford and I.B.H. New Delhi, Fourth Edition, 1985.
10. Bate, S.C.C. and Bennett, E.W., Design of Prestressed Concrete, Surrey University Press in Association with International Book Company Limited, London, 1976.
11. Ramaswamy G.S., Design and Construction of Concrete Shell Roofs, McGraw Hill Inc., New York, 1968.
12. Kong, F.K. and Evans, R.H., Reinforced and Prestressed Concrete, Third Edition, English Language Book Society, Van Nostrand Reinhold (International), London, 1987.
13. Rajagopalan, N., Prestressed Concrete, Narosa Publishing House, New Delhi, 2002.
14. Krishna Raju, N., Prestressed Concrete, Fifth Edition, Tata McGraw Hill Publishing, New Delhi, 2012.
15. IRC: 18-2000, Design Criteria for Prestressed Concrete Road Bridges (Post-tensioned Concrete), Third Edition (Revised), Indian Roads Congress, New Delhi, 2000.
16. Victor, J.D., Essentials of Bridge Engineering, Fifth Edition, Oxford and IBH Publishing, New Delhi, 2001.
17. Morice, P.B. and Cooley, E.H., Prestressed Concrete Theory and Practice, Sir Isaac Pitman & Sons, London, 1958.

18. IS: 3370-Part-III, Indian Standard Code of Practice for Concrete Structures for the Storage of Liquids, Part-III — Prestressed Concrete (Third Reprint), New Delhi, 1975.
19. IS: 3370-Part-IV, Indian Standard Code of Practice for Concrete Structures for the Storage of Liquids, Part-IV — Design Tables (Third Reprint), New Delhi, 1974.
20. Abeles, P.W., An Introduction to Prestressed Concrete, Vol. II, Concrete Publications, London, 1966.
21. Krishna Raju N., The Design of Prestressed Concrete Grid Floors, The Ind. Concrete J., Vol. 39, No. 1, January 1965.
22. Bennett, E.W., Prestressed Concrete Theory & Design, Chapman & Hall, London, 1964.
23. Krishna Raju, N. and Rama Murthy, L.N., Limit State Design of Partially Prestressed Pre-tensioned Concrete Poles, Proceedings of the International Symposium on Prestressed Concrete Pipes, Poles, Pressure Vessels and Sleepers, Madras, 1972.
24. Recommended Practice for Design, Manufacture and Installation of Prestressed Concrete Piling, Journal of the Prestressed Concrete Institute, 1977.
25. IS: 3201–1965, Indian Standard Specification: Criteria for Design & Construction of Precast Concrete Trusses, Bureau of Indian Standards, New Delhi, 1965.
26. Krishna Raju, N., Prestressed Concrete Bridges, First Edition, CBS Publishers and Distributors, New Delhi, 2009.
27. Raina, V.K., Concrete Bridge Practice, Analysis, Design and Economics, Tata McGraw Hill Publishing, New Delhi, 1991.
28. Rowe, R.E., Concrete Bridge Design, C.R. Books, First Edition, London, 1962.
29. Leonhardt, F. and Zellner, W., Cable Stayed Bridges, IABSE Surveys, S-13/80 IABSE Periodical 2/1989, May 1980.
30. Krishna Raju, N., Design of Bridges, Fourth Edition, Oxford & IBH Publishing, New Delhi, 2009.

Index

Reader's Note

Reader's Note

Reader's Note

Reader's Note